Materials Recycling

Materials Recycling

Edited by **Adele Cullen**

NY RESEARCH PRESS

New York

Published by NY Research Press,
23 West, 55th Street, Suite 816,
New York, NY 10019, USA
www.nyresearchpress.com

Materials Recycling
Edited by Adele Cullen

International Standard Book Number: 978-1-63238-511-6 (Hardback)

Printed in the United States of America.

Contents

Preface

In my initial years as a student, I used to run to the library at every possible instance to grab a book and learn something new. Books were my primary source of knowledge and I would not have come such a long way without all that I learnt from them. Thus, when I was approached to edit this book; I became understandably nostalgic. It was an absolute honor to be considered worthy of guiding the current generation as well as those to come. I put all my knowledge and hard work into making this book most beneficial for its readers.

The aim of this book is to present researches that have transformed the discipline of materials recycling and aided its advancement. It unravels the recent studies in this field and provides in-depth knowledge about the various techniques and concepts related to it. Materials recycling refers to the process of converting waste materials into re-useable goods, so as to reduce the rapid growth of waste and minimize the use of new raw materials and in turn control air, water, soil and other types of pollution. This process is based on the three R's namely re-use, re-cycle and reduce. This book consists of contributions made by international experts. It provides significant information of this discipline to help develop a good understanding of this area and related topics. It will help the readers in keeping pace with the rapid changes in this field. This book is appropriate for students seeking detailed information in this area as well as for engineers, researchers and professionals associated with materials recycling at various levels.

I wish to thank my publisher for supporting me at every step. I would also like to thank all the authors who have contributed their researches in this book. I hope this book will be a valuable contribution to the progress of the field.

Editor

1

Recycling Agave Bagasse of the Tequila Industry

C. G. Iñiguez[1], C. J. J. Bernal[2], M. W. Ramírez[2], N. J. Villalvazo[2]

[1]Departamento de Madera, Celulosa y Papel, Centro Universitario de Ciencias Exactas e Ingenierías, Universidad de Guadalajara, Guadalajara, México
[2]Departamento de Ingeniería de Proyectos, Centro Universitario de Ciencias Exactas e Ingenierías, Universidad de Guadalajara, Guadalajara, México
Email: giniguez@dmcyp.cucei.udg.mx

Abstract

This paper presents an overview of different handling systems and use of the agave bagasse. These systems have appeared from different research works always taking in account the environmental sustainability. It is mentioned that the agave bagasse can be used for animal feeding, for the elaboration of compound materials, as an element for agricultural and hydroponic vegetables cultivation purposes, and also as a means to treat biosolids, vinasses, and bagasses of slaughterhouses and tanneries.

Keywords

Agave Bagasse, Tequila Vinasses, Tequila

1. Introduction

Tequila production has contributed to the agricultural and industrial development in Mexico, especially in the State of Jalisco. In the last nine years, the agave consumption and tequila production have increased considerably. In 2003, the industrialization of the agave heads was of 412,900 tons, and by 2011 it was of 998,400 tons, with a decrease of 11.8% in 2012 (880,600 tons) and a maximum of 1,125,100 in 2008 [1]. Fortunately, this sudden growth of the tequila industry has not ignored the handling and disposal of two of the main by-products: agave bagasse and vinasses. With regard to vinasses, many actions are still to be taken due to its high pollution power: only few companies have successfully resolved their handling because of their financial capacity. The methods most commonly used to treat vinasses are the separation of settleable solids and the anaerobic digestion followed by aerated lagoons or activated sludge. In reference to the agave bagasse handling, the majority of the big and medium companies have chosen the composting process, as a resource for the disposal of vinasses, settleable solids or biosolids from the treatment plants. The objective of this paper is to highlight or provide with alternatives of handling and using the agave bagasse to give environmental sustainability to the tequila industry.

2. Agave Bagasse

The agave bagasse is the residual fibrous material remaining after the *Agave tequilana* Weber var. *azul* heads are shredded, cooked, milled and the sugars are water-extracted to produce tequila. The bagasse is primarily the rind and fibrovascular bundles dispersed throughout the interior of the agave head. Nowadays some tequila factories have upgraded their processes to obtain tequila, to smash only the agave heads so then with hot water extract the fermentable sugars, resulting in an agave bagasse with different characteristics than the first one. The bagasse is compounded of fiber and pith. The fiber is thick walled and long (10 - 12 cm). It represents about 40% of the total weight of the milled agave on a wet weight basis [2]. Bagasse is available all year in only two main regions of the tequila producing areas in Mexico: the Tequila region and the Jalisco Highlands. **Tables 1-4** show some of the physical and chemical characteristics of agave bagasse obtained from cooked agave heads.

2.1. Agave Bagasse for Animal Feeding

Because of its characteristics, the agave bagasse as it comes out of the tequila factory is difficult to be used for animal feed. The level of usefulness depends on the attack that it has in the rumination micro flora. The best mi-

Table 1. Chemical composition of dry agave bagasse (5% water content).

Item	(%)
Cellulose	43
Lignin	15
Hemicellulose	19
Total nitrogen	3
Pectins	1
Fats	1
Reducing sugars	5
Ash	6
Others	2

Source: [16].

Table 2. Physical composition of the dry agave bagasse (5% water content).

Texture	Not very rigid
Color	Brown-yellow
Fiber length	5 - 10 cm
Diameter	0.3 - 0.4 mm
Water absorption	6 mL·g^{-1}

Source: [16].

Table 3. Extracted sugar analysis (wet agave bagasse).

Item	Reducing Sugars (%, dry basis)	Fructose (%, dry basis)	Glucose (%, dry basis)
Tequila factory 1	5	3	0.4
Tequila factory 2	8	4	0.4
Tequila factory 3	10	5	0.5
Tequila factory 4	12	7	0.5

Source: [16].

Table 4. Free extracted sugars (dry agave bagasse).

Item	Extraction 70°C 3 h (%, dry matter)	Extraction 90°C 3 h (%, dry matter)
Fructose	1.00	1.20
Arabinose	0.02	0.04
Glucose	0.02	0.02
Galactose	0.02	0.02

Source: [16].

crobial use of cellulose and lignocellulosic agricultural by-products is limited by the close physical and chemical association between structural carbohydrates and lignin and the crystalline arrangement of the cellulose polymer in plant cell walls [3]-[5]. Lignin is the most important factor limiting degradation of cellulose by microorganisms [6]. However, to facilitate the optimal use of the bagasse for feeding animals, it can start by the physic separation of what has a low digestibility as the fiber and what has a higher digestibility as the pith. **Figure 1** shows a piece of equipment used for that, and **Table 5** shows the results of the effect of the screen mesh size on separation of agave bagasse from different tequila factories. It is seen that the amount of recovered pith depends on the manner in which agave heads were processed for fermentable sugar extraction. For example, there was a significant difference ($P < 0.05$) between the yield of recovered pith from the agave heads processed in "La Rojeña", Sauza and Camichines tequila factories. In the La Rojeña tequila factory, the bagasse comes from cooked, shredded and pressed agave heads in mills to extract the sugars. In the Sauza factory, the bagasse comes from shredded agave heads before cooking and sugars extraction in pressing mills. In the Camichines tequila factory, the bagasse comes from shredded agave heads, subjected to a sugars extraction process using hot water and in the processing mills. The extracted juice is cooked in a later step. **Table 6** shows the behavior of lambs when they were fed with 3 comparatives diets. Diet 1, or control diet, was basically formulated with a base of ground corn (79.3%), ground alfalfa (15%) and cane molasses (5%). The remains were mineral elements. Diet 2 was mainly 63.3% of agave pith, and diet 3 of 63.3% of corn [7]. Note that in the table, the weight gain and the daily feed consumption were statistically the same ($P > 0.05$) among the diets based in the pith and the corn-based diet. **Table 7** shows the results of a study in which a balanced diet based in agave bagasse pith was formulated. In this case, the animals average daily gain was 186 g, a result which is very similar to the one reported in **Table 6** with a difference in dry matter intake (783 vs. 1077 g/day) [7]. A better use of the agave bagasse pith for animals' feed implies that this should be eaten immediately to avoid its decomposition because of opportunist microorganisms. The pith silage can be an alternative conservation while not used. Silage is a conservation processes for forage based in a lactic fermentation of the grass that produces lactic acid and a decrease of the pH below 5. It allows holding original grass nutritive qualities much better than the dry forage.

2.2. Agave Bagasse for Composite Materials

The agave bagasse can be used for the manufacturing of composite materials provided that the pith is removed, as this can be used for animal feeding. In 2001 Iñiguez *et al.* [7] prepared medium and high density boards samples with short and long agave bagasse fibers. These samples presented comparable properties to boards' prepared with fibers and wood particles. They were stronger in flexion tests than the ANSI standards (American National Standards Institute) for hard boards. **Figure 2** shows board samples prepared with short and long agave bagasse fibers.

2.3. Agave Bagasse as Substrate for Agricultural Purposes

It results interesting to use agave bagasse as a substrate for agricultural purposes as long as the fiber is not completely degraded in the composting process. Iñiguez *et al.* [8], in a greenhouse test with tomatoes, found no significant statistics differences ($p \leq 0.5$) when two substrates of agave bagasse compost and two commercial substrates were used (coconut coir and "cocopeat") to evaluate the production and quality of tomatoes from the first to fifth cut after 55 days of transplantation (**Figure 3**). Martínez *et al.* [9] used three agave mezcalerobagasse compost (piling time: 0, 90 and 180 days) as organic substrate in tomatoes cultivation (Solanumlycopersicum L.). They reported that the fruit quality was not affected and the output was surpassed, when compared with coco powder substrate. With the 180 days pill-up bagasse the best output was obtained (3.5 kg per plant) and number

Table 5. Effects of screen mesh size on separation of agave bagasse from different sources.

Tequila Factory	Agave bagasse processed (kg)	Screen opening size (cm) screen[a]			Recovered pith (%)
		1	2	3	
La Rojeña	100	2.54	2.54	1.90	36.0[a]
Sauza	100	2.54	2.54	1.90	56.0[a]
Orendain	100	2.54	2.54	1.90	45.3[a]
Viuda de Romero	100	2.54	2.54	1.90	43.5[a]
Camichines	100	2.54	2.54	2.54	38.3[a]

[a]Mean of five runs; [b]Mean of eight runs; Source: [7].

Table 6. Comparison of sheep performance fed diets based on corn (1), agave bagasse pith (2) and corn stubble (3).

Item	Diet			SEM[a]
	1	2	3	
Average daily gain (g)	179.0[b]	96.4[c]	72.1[c]	14.826
Dry matter intake (g/day)	783.0[b]	774.0[b]	772.0[b]	12.339
Feed/gain	4.4[b]	8.32[c]	11.12[b]	0.926

[a] Standard error of the mean; [b,c] Means in the same row with different superscripts differ ($P<0.05$); Source: [7].

Table 7. Sheep performance fed diet based on agave bagasse pith.

Item	Value[a]	SEM[b]
Average daily gain (g)	186.0	12.617
Dry matter intake (g/day)	1.077	0.056
Feed/gain	5.87	0.417

[a]Represents the mean of five pens with three animals per pen; [b]Standard error of the mean; Source: [7].

Figure 1. Equipment for the physical separation fiber/pith of the agave bagasse.

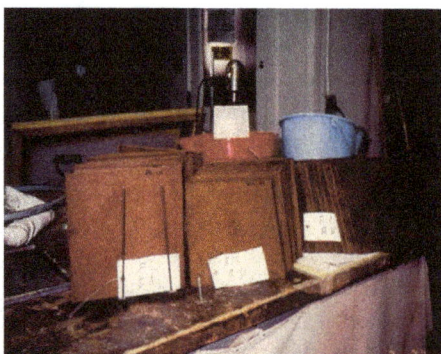

Figure 2. Board samples made with short and long agave bagasse fibers.

Figure 3. Tomato production using an agave bagasse substrate.

of commercial fruit (26.4 fruits per plant). Before planting in the field, Crespo *et al*. [10] used agave bagasse compost to adapt during 9 months (**Figure 4**), propagated agave seedling *in vitro* of approximately 16 months of age. The treatments were based in coco powder (PC) (Cocosnucifera), peat (T) (Sphagnum spp.) and agave bagasse compost (C). Those treatments consisted of: 1) 80% PC + 10% T + 10% C; 2) 100% PC; 3) 100% T; 4) 70% PC + 30% C; 5) 50% PC + 50% C; 6) 30% PC + 70% C; y 7) 100% C. The response of the agave plants to the different treatments was evaluated through the following morphologic parameters: Pine's diameter, (DPñ) and steam diameter (DTll); ratio DPñ/DTll; the longest leaf's length (LHj); longest leaf's width (AHj) (the widest section) and number of leaves (N Hj). Three treatments that were prepared based on agave bagasse compost produced meaningful differences and highly meaningful in the morphologic parameters evaluated and widely surpassed the effects of the coco powder and the peat. Treatments with 30 and 50% compost produced a greater effect over the pine's and steam diameter, as well as over the number of leaves and their longitude; however, all the mixtures with compost increased the leaf's width. The obtained results reflect the possibility of substitute the commercial substrate with the compost substrates, particularly with the mixtures 30, 50 y 70%. When this substitution was completed, abundant agave bagasse can be used, avoiding environmental pollution, and reducing production costs, as the bagasse compost can be locally elaborated. Rodríguez *et al*. [11] evaluated agave bagasse compost in comparison with commercial peat (Sunshine Mix$_3$ y Berger BM$_2$) for the production of tomatoes seedling(var. Hermosa) obtaining, with the agave bagasse compost, better height values, dry and fresh weight in the areal part and dry weight in the root.

2.4. Agave Bagasse as a Tool for the Biosolids Treatment and Tequila Vinasses

Rodríguez *et al*. [12] conducted a field study for the composting of the raw agave bagasse using biosolids from a tequila vinasses treatment plant to maintain the moisture of the process. Eight piles of 30 tons of bagasse each were put in composting. The piles 1 - 4 were moved each week to be ventilated and water or biosolids were added.1560 L of water were added to pile 1; and to piles 2 - 4, 1560, 3120 and 4680 L of biosolids were added, respectively. Piles 5-8 were gradually added with the same amount of water or biosolids, with the exception that they were moved every two weeks. At the end of the 19 weeks composting period, the color, smell and texture of the material of the eight piles became similar to the garden soil. The ratio of processed bagasse kg/L of water or biosolids added to the piles for the treatments 1 - 8 was of 1/0.728, 1/0.676, 1/1.04, 1/1.56, 1/0.52, 1/0.364, 1/0.572 y 1/0.858, respectively. If the last ratio is considered to add vinasses (as currently carried out) instead of biosolids to the agave bagasse compost piles, and if the 2011 CRT statistics [1] is considered regarding the bagasse and vinasses production, it is therefore concluded thatthe bagasse composting process could only treat 59.6 % of the vinasse generated during the tequila production; so, 40.4% would require of another handling system. The obtained compost is used to enhance the physical and chemical soil structure in the agave plantations. **Figure 5** shows a composting plant where the bagasse piles are used for the final disposal of the tequilavinasse. Iñiguez *et al*. [13] used the agave bagasse as a final disposal of the tequila vinasse. To do that, they put four

Figure 4. Use of an agavebagasse compost for the adaptation during 9 months of agave seedlingbefore planting in the field.

Figure 5. Composting plant where the bagasse piles are used for final disposal of the tequilavinasses.

bagasse piles in composting processand, during the process, two of them were irrigated with vinasses and the other two with water. The latter, at the beginning of the composting were added with urea to adjust the C:N ratio to 25:1. The test period for the piles with urea was of 228 days and during this time, 0.912 L of vinasses were added per kg of wet bagasse. The test period for the piles without urea was of 242 days and during this time, 0.55 L of vinasses were added per kg of wet bagasse. This showed that when the C:N ratio was initially adjusted to the recommended for good composting, the microbial activity was accelerated and a major water loss was noted, which enables the addition of more vinasses and resolves in a better way the final disposal. The obtained compost had similar characteristics to the garden soil without problems of phytotoxicity at the moment to be evaluated with cucumber seeds.

2.5. Agave Bagasse as a Disposal Tool of Slaughterhouses and Tanneries Waste

Agave bagasse was used as an alternative to stabilize the tannery residuals (hair and flesh material) through composting [14]. Two wood cells of 2.5 m wide by 2.5 m long and 1.5 m high were used alternating with tannery residuals and agave bagasse layers to a height of approximately 0.7 m with the result of achieving one ton of thread material in each cell. The biodegradation process of the thread material was followed by the regular measurement of the temperature changes. At the end of 154 days of thermophile degradation, the thread material was totally decomposed, having as a result a blackish product, with odor and texture similar to garden soil. There was a total loss of dry material of 67.3% in average for the two piles considering the initial and final weight of the ingredients. From the research, it was concluded that the biodegradation process by layers might convey a technicaland economically viable alternative to help the tannery industry in handling and final disposal of thread materials, obtaining a product with agronomic potential. Iñiguez y Vaca [15] researched the effective

Figure 6. Italian lettuce cultivation with BAPEN hydroponic system.

ness of composting swine large intestines and wet agave bagasse using a layering method. It consisted of placing alternate layers of 150 kg of wet agave bagasse and 100 kg of pig intestines. The pile formed was moved every week to facilitate the ventilation and the water addition. After 102 composting days, the product turned dark brown with a smell of soil. After that, these trials were conducted in field conditions showing no problem.

2.6. Agave Bagasse as a Substrate in Vegetable Hydroponics Cultivations

There are 6 basic types of hydroponic systems: wick, water culture, ebb and flow (flood and Drain), drip (recovery and non-recovery), N.F.T. (Nutrient Film Technique) and aeroponics. There are hundreds of variations on these basic types of systems, but all hydroponic methods are a variation (or combination) of these six. One of these variations can be the so called system BAPEN, which consists of packing a substrate called BAGACOMCO PIT in a system of "bolis" (horizontal grow bags) to avoid the substrate dispersion and water evaporation. The "boli" is put in a PVC gutter provided with a thin bed of gravel to facilitate the recirculation of a nutritive area from a deposit (with a submerged pump) to a gutter and from here to the container by gravity (**Figure 1**). Italian lettuce seedling was grown in four different experiments, long lettuce (Latuca sativa), radish (Raphanussativus) and chard (Beta vulgaris var. Cicla). As an example, **Figure 6** shows the results of the cultivation of Italian lettuce after 45 days in the development after the transplant. The average weight for plant (from 18) was of 217.5 g in comparison with 112.5 g of lettuce in the market.

3. Conclusion

Up to now, the agave bagasse has been used in the treatment of vinasses and final disposal of biosolids in the vinasses treatment plants. However, there is enough bibliographical information so the agave bagasse is used commercially as a substrate for the germination of different seeds, as well as for the development of seedling and mainly for the tomatoes production. On the other hand, due to the constant increase of fuels, tequila producers have been seriously considering to use agave bagasse for the generation of energy to support their own necessities.

References

[1]   Consejo Regulador del Tequila (CRT) (2012) www.crt.org.mx

[2]   Cedeño, C.M. (1995) Tequila Production. *Critical Reviews in Biotechnology*, **15**, 1-11.
       http://dx.doi.org/10.3109/07388559509150529

[3]   Morris, E.J. and Bacon, J.S.D. (1977) The Fate of Acetyl Groups and Sugar Components during the Digestion of Grass Cell Walls in Sheep. *Journal of Agricultural Science*, **89**, 327-340. http://dx.doi.org/10.1017/S0021859600028252

[4]   Hartley, R.D. and Jones, E.C. (1978) Effects of Aqueous Ammonia and Other Alkalis on the *in Vitro* Digestibility Barley Straw. *Journal of the Science of Food and Agriculture*, **29**, 92-98. http://dx.doi.org/10.1002/jsfa.2740290204

[5]   Scalbert, A., Monties, B., Lallemand, J.Y., Guittet, E. and Rolando, C. (1985) Ether Linkages between Phenolic Acids and Lignin Fractions from Wheat Straw. *Phytochemistry*, **24**, 1359-1362.

http://dx.doi.org/10.1016/S0031-9422(00)81133-4

[6]   Gould, J.M. (1984) Studies on the Mechanism of Alkaline Peroxide Delignification of Agricultural Residues. *Biotechnology and Bioengineering*, **27**, 225-231. http://dx.doi.org/10.1002/bit.260270303

[7]   Iñiguez, C.G., Lange, E.S. and Rowell, M.R. (2001) Utilization of by Products from the Tequila Industry: Part 1: Agave Bagasse as a Raw Material for Animal Feeding and Fiberboard Production. *Bioresource Technology*, **77**, 25-32. http://dx.doi.org/10.1016/S0960-8524(00)00137-1

[8]   Iñiguez, C.G., Martínez, G.A., Flores, P.A. and Virgen, C.G. (2011) Utilización de subproductos de la industria tequilera. Parte 9. Monitoreo de la evolución del compostaje de dos fuentes distintas de bagazo de agave para la obtención de un substrato para jitomate. *Revista Internacional de Contaminación Ambiental*, **27**, 47-59.

[9]   Martínez, G.G.A., Iñiguez, C.G., Ortíz, H.Y.D., López, C.J.Y. and Bautista, C.A. (2013) Tiempos de apilado del bagazo del maguey mezcalero y su efecto en las propiedades del compost para sustrato de tomate. *Revista Internacional de Contaminación Ambiental*, **29**, 209-216.

[10]   Crespo, G.M.R., González, E.D.R., Rodríguez, M.R., Rendón, S.L.A., Del Real, L.J.I. and Torres, M.J.P. (2013) "Evaluación de la composta de bagazo de agave como componente de sustratos para producir plántulas de agave azul tequilero. *Revista Mexicana de Ciencias Agrícolas*, **4**, 1161-1173.

[11]   Rodríguez, G.L.B., Fernández, C.F.A., Iñiguez, C.G., Rodríguez, G.E., Rodríguez, D.E. and Arriaga, R.M.C. (2010) Utilización de bagazo de agave como sustrato para producción de plántulas de tomate. *Sciencia-CUCBA*, **12**, 11-15.

[12]   Rodríguez, M.R., Jiménez, J.F., Del Real, L.J.I., Salcedo, P.E., Zamora, J.F. and Iñiguez, C.G. (2013) Utilización de subproductos de la industria tequilera. Parte 11. Compostaje de bagazo de agave crudo y biosólidos provenientes de una planta de tratamiento de vinazas tequileras. *Revista Internacional de Contaminación Ambiental*, **29**, 303-313.

[13]   Iñiguez, C.G., Acosta, N., Martínez, L., Parra, J. and González, O. (2005) Utilización de subproductos de la industria tequilera. Parte 7. Compostaje de bagazo de agave y vinazas tequileras. *Revista Internacional de Contaminación Ambiental*, **21**, 37-50.

[14]   Iñiguez, C.G., Flores, S. and Martínez, L. (2003) Utilización de subproductos de la industria tequilera. Parte 5. Biodegradación de material de descarne de la industria de curtiduría. *Revista Internacional de Contaminación Ambiental*, **19**, 83-91.

[15]   Iñiguez, C.G.and Pilar, V. (2001) Utilización de subproductos de la industria tequilera. Parte 4. Biodegradación del intestino grueso de cerdos con bagazo de agave húmedo. *Revista Internacional de Contaminación Ambiental*, **17**, 109-116

[16]   Alonso, M.S. and Rigal, L. (1997) Caracterización y valoración del bagazo de *Agave tequilana* Weber de la industria del tequila. *Revista Chapingo, Serie Horticultura*, **3**, 31-39.

Recycling of Cobalt by Liquid Leaching from Waste 18650-Type Lithium-Ion Batteries

Liangmou Yu[1], Bo Shu[2], Shiwen Yao[2]

[1]Faculty of Environmental Engineering, Kunming Metallurgy College, Kunming, China
[2]Yunnan Copper Co., Ltd., Kunming, China
Email: xiashubiao401@163.com

Abstract

In this work, we recover cobalt from waste 18650-type lithium-ion batteries by acid leaching. The cathode material is completely dissolved, after leaching waste batteries by using 10 mol/L industrial sulfuric acid at 70°C for 1 h. The rate of cobalt leaching is nearly 100%. Removal of sodium carbonate, iron, aluminum and other impurities from the leaching solution was well performed by adjusting the pH to 2 - 3 with stirring vigorously. Finally, under the conditions of 55°C - 60°C of 240 A/m² current density, electrodeposition current efficiency was 90.01%, the quality of the electrical output achieved cobalt 1A standard electrolytic cobalt, cobalt until greater than 90% yield. The process is easy and suitable for large-scale lithium-ion batteries used in the recovery of valuable metals.

Keywords

Component, Cobalt, 18650-Type Lithium-Ion Battery, Leaching, Electrodeposition

1. Introduction

Rechargeable lithium-ion batteries as a new generation of green, non-pollution chemical energy storage device are widely used in portable electronic apparatus and vehicles. However, waste LIBs contain heavy metals, organic chemicals and plastics, which will bring environmental pollution. Therefore, the recycling of major components from spent LIBs is considered to be a beneficial way to prevent environmental pollution and as alternative resource of cobalt [1]. 18650-type power batteries are widely used in small portable devices, increasing along with the rise of electric vehicles in Tesla [2]. The spent lithium ion batteries contain 5 - 15 wt.% cobalt and 2 - 7 wt.% lithium as important constituents, which are an active cathodic material. The waste 18650 type lithium ion batteries contain precious metal elements cobalt with a certain toxicity as important constituents, which are an active cathodic material. Since cobalt is a relatively expensive material compared to the other battery constituents and can be widely used in new LiBs electrode materials, its recovery is one of the primary objectives in the recycling of spent batteries [3] [4]. **Figure 1** is a 18650 type battery price polyvalent metallic element [5]. It is not difficult to see from **Figure 1** that the price of the cobalt is more expensive than the other battery constituents. Therefore the recovery and utilization of cobalt have significant economic and social environmental benefits [6].

Separation and Recovery Technology 18650 cobalt lithium ion battery mainly includes three steps: 1) the waste batteries discharge, stripped shell, simple crushing, after screening to obtain an electrode material, or simply crushed after firing to remove organic matter to obtain an electrode material [7]; 2) the material obtained in the first step of the electrode is dissolved in the leaching of various metals into solution wherein both the cobalt and nickel represent as trivalent forms. Leaching sub-step dissolution method and two-dissolution method: direct step by acid leaching dissolution method, all the metal was dissolved in acid, and then isolated using a number of different purification and recovery; two-step method is to use an alkali leaching aluminum and recovered, and then use acid leaching surplus metal oxide, followed by the similar treatment as the first step [8]; 3) the solution (leachate) dissolved metal elements in the solution is separated for recycling or direct synthesis anode material. Separation and recovery methods are chemical precipitation, salting out, ion exchange, extraction, electrochemical method, respectively cobalt or lithium-containing compound [9]. E.M.S. Barbieri *et al*. [10] have investigated recycling cobalt from the cathodes of spent Li-ion batteries as β-Co(OH)$_2$, obtaining Co$_3$O$_4$. β-Co(OH)$_2$ with a hexagonal structure by using chemical precipitation or electrochemical precipitation. This method does not directly recovered cobalt. In this paper, we recovered directly cobalt by electrodeposition.

2. Experimental

Experimental material: Waste 18650 lithium-ion battery, industries concentrated sulfuric acid, industrial grade sodium borate. Hand split waste 18650 lithium-ion batteries, remove the interior material with pliers, and finally the positive and negative material separated and dried 24 h at 60°C. The obtained positive electrode material with 10 mol/L sulfuric acid leaching 1h, and filtered to give leachate. Spectrophotometric determination of Co^{2+} concentration in the solution (722S spectrophotometer, Shanghai Precision Scientific Instrument Co., Ltd.) with cobalt ions spectral qualities, using X-ray diffraction (Japan Rigaku company DM/ax-IIIA) salting crop products phase analysis. Process is shown in **Figure 2**, ICP analysis content of the test solution of various metal elements.

3. Results and Discussion

3.1. Effect of pH on the Leaching Process

When the solution of sulfuric acid concentration is too low, even at boiling temperature, leaching reaction is also extremely slow. Therefore, the leaching temperature was set at 70°C, under the conditions of the reaction time of 1 h, acidity leaching test, the results shown in **Figure 3** shows that, as the solution acidity increase, cobalt and

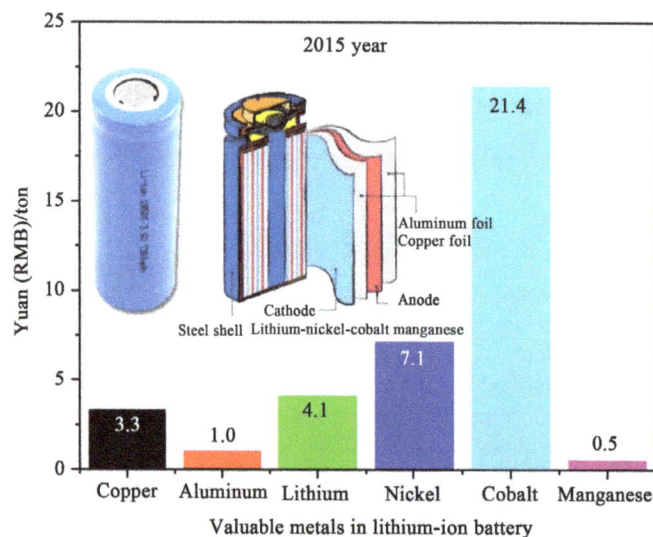

Figure 1. Price metal elements contained at 18650 lithium-ion batteries.

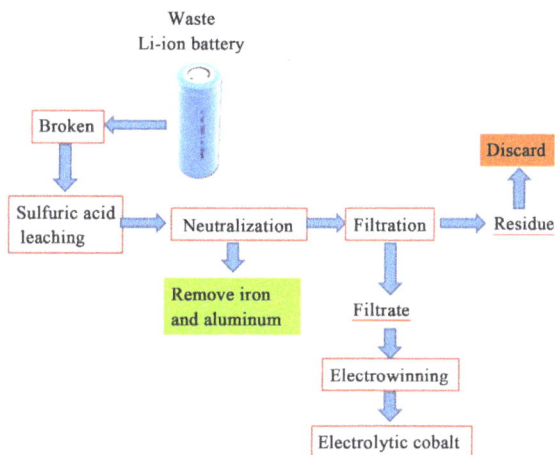

Figure 2. Process flow diagram of cobalt recovery by liquid leaching.

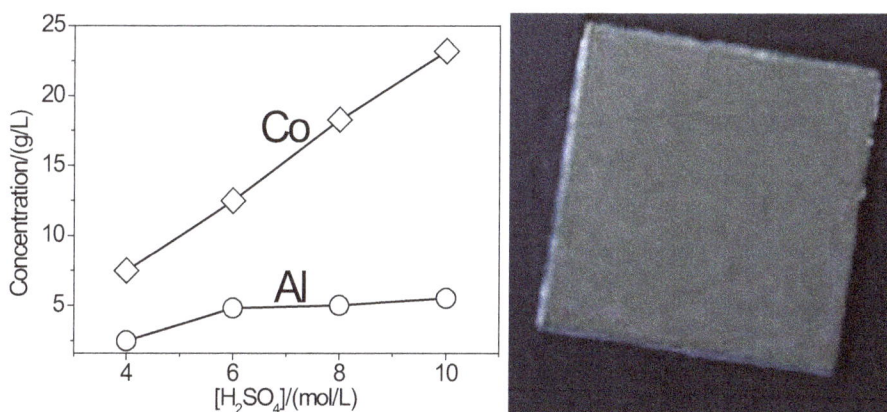

Figure 3. Effect of acid concentration on leaching rate and electrolytic cobalt photo.

aluminum solubility in the solution increases accordingly. With 10 mol/L sulfuric acid solution as an industrial leaching agent, the lithium-ion battery with the waste placed in 5 L beaker, heated to 70°C, leaching 1 h, the reaction was completely dissolved, ICP testing the resulting solution composition shown in **Table 1**.

3.2. Neutralization Reaction to Remove Iron, Aluminum

Leaching with sodium carbonate as a neutralizing agent resulting solution was adjusted to pH 2 - 3, and heated to 90°C, blast stirred solution of iron and aluminum precipitate. At the same time, also co-precipitation in the form of silicon is removed. After hydrolysis impurity solution composition shown in **Table 2**.

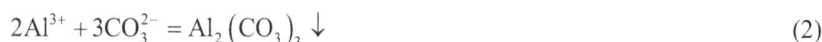

$$2Fe^{3+} + 3CO_3^{2-} = Fe_2(CO_3)_3 \downarrow \tag{1}$$

$$2Al^{3+} + 3CO_3^{2-} = Al_2(CO_3)_3 \downarrow \tag{2}$$

3.3. Electrodeposition

Starting with the production of cobalt cathode pole piece, a titanium plate as an anode, the impurity resulting solution directly electrowinning current density of 240 A/m^2, electrolyte temperature 55°C - 60°C, in return anolyte electrolysis and cleaning section the resulting electric cobalt surface roughness, the spectral analysis of its components, such as tables, **Table 3**, **Figure 3** also lists the quality standards 1 and 1A # # electric cobalt electric cobalt [11].

Table 3 shows full-immersion neutralized after hydrolysis direct electrowinning cobalt impurity, product quality can be made 1A electric cobalt standards. The electrolytic cobalt electricity obtained by the Formula (3)

Table 1. Content of each metal element in leaching solution.

Element	Co	Ni	Li	Al	Cu	Fe
Content (g/L)	23.67	0.022	1.72	6.10	0.001	0.004

Table 2. Content of each metal element in leaching solution after neutralization reaction to remove iron, aluminum.

Element	Co	Ni	Li	Al	Cu	Fe
Content (g/L)	47.12	0.033	2.84	0.01	0.001	0.001

Table 3. Cathode cobalt quality and criterion of GB6517-86.

Element	Content (Impurities ≤, Co ≥)/%		
	Electrolytic cobalt by experiment	Standard of Co1	Standard of Co1A
Co≥	99.88	99.98	99.65
Mn	0.0005	0.001	0.01
Fe	0.001	0.003	0.05
Ni	0.051	0.005	0.2
As	0.001	0.0005	0.001
Pb	0.0004	0.0003	0.001
Cu	0.001	0.001	0.02
Si	0.001	0.001	-
Al	0.01	0.001	-

obtain the current efficiency of the electrolysis process 90.01%.

$$\eta = G/(q \cdot A \cdot t \cdot n) = 460/(1.1 \times 4 \times 116 \times 1) \times 100\% = 90.01\% \tag{3}$$

where: η—current efficiency, %; G—precipitation Electric Co Quality, g; q—cobalt electrochemical equivalent, 1.1 g Ah; A—current intensity, A; t—electrolysis cycle, h; n—cell number. Direct yields greater than 90% cobalt.

4. Conclusion

The use of sulfuric acid leaching-EW treatment of waste lithium 18605 lithium-ion battery technologies is simple. Cobalt leaching rate basically reached 100%, absolute yield of more than 90%. Leach solution after hydrolysis impurity can be directly electrowinning cobalt, simple process. We get 1A # electric cobalt comply GB6517-86 by electrolyte. The current efficiency was 90.01%.

References

[1]   Debaraj, M., Kim, D.-J., Ralph, D.E., *et al.* (2008) Bioleaching of Metals from Spent Lithium Ion Secondary Batteries Using *Acidithiobacillus ferrooxidans*. *Waste Management*, **28**, 333-338.

[2]   Jin, Y.-J., Mei, G.-J. and Li, S.-Y. (2006) Study on Cobaltous Recovery from Cathode Leachate of Lithium-Ion Battery by Salting Out. *Acta Scientiae Circumstantiae*, **26**, 1122-1125. (In Chinese)
      Wang, X.-F., Kong, X.-H. and Zhao, Z.-Y. (2001) Rcovery of Metal in Lithium ion Battery. *Battery Bimonthly*, **31**, 14-15. (In Chinese)

[3]   Nan, J.-M., Han, D.-M., Cui, M., *et al.* (2004) Recycling of Valuable Metal from Spent Li-Ion Batteries by Solvent Extraction. *Battery Bimonthly*, **34**, 329-311. (In Chinese)

[4]   Lupi, C., Pasquali, M. and Dell'Era, A. (2005) Nickel and Cobalt Recycling from Lithium-Ion Batteries by Electrochemical Processes. *Waste Management*, **25**, 215-220. http://dx.doi.org/10.1016/j.wasman.2004.12.012

[5] Wang, C.-Y., Qiu, D.-F., Chen, Y.-Q., *et al.* (2004) Status of Spent Batteries Recovery and Recycle. *Nonferrous Metals*, **5**, 39-42. (in Chinese)

[6] Xie, G.Y., Ling, Y. and Zhong, S. (2009) Overview of Recovery Techniques of Spent Lithium-Ion Batteries. *Environmental Science & Technology*, **32**, 97-101.

[7] Li, J.H., Shi, P.X., Wang, Z.F., *et al.* (2009) A Combined Recovery Process of Metals in Spent Lithium-Ion Batteries. *Chemosphere*, **77**, 1132-1136. http://dx.doi.org/10.1016/j.chemosphere.2009.08.040

[8] Sun, L. and Qiu, K.Q. (2011) Vacuum Pyrolysis and Hydrometallurgical Process for the Recovery of Valuable Metals from Spent Lithium-Ion Batteries. *Journal of Hazardous Materials*, **194**, 378-384.

[9] Li, L., Ge, J., Wu, F., *et al.* (2010) Recovery of Cobalt and Lithium from Spent Lithium Ion Batteries Using Organic Citric Acid as Leachant. *Journal of Hazardous Materials*, **176**, 288-293.

[10] Barbieri, E.M.S., Lima, E.P.C., Lelis, M.F.F., *et al.* (2014) Recycling of Cobalt from Spent Li-Ion Batteries as β-$Co(OH)_2$ and the Application of Co_3O_4 as a Pseudocapacitor. *Journal of Power Sources*, **270**, 158-165. http://dx.doi.org/10.1016/j.jpowsour.2014.07.108

[11] Zhao, J.M., Shen, X.Y., Deng, F.L., *et al.* (2011) Synergistic Extraction and Separation of Valuable Metals from Waste Cathodic Material of Lithium Ion Batteries Using Cyanex272 and PC-88A. *Separation and Purification Technology*, **78**, 345-349. http://dx.doi.org/10.1016/j.seppur.2010.12.024

Using Microwave Heating to Completely Recycle Concrete

Heesup Choi[1,2], Myungkwan Lim[1,2]*, Hyeonggil Choi[3], Ryoma Kitagaki[3], Takafumi Noguchi[3]

[1]Department of Civil Engineering, Kitami Institute of Technology, Hokkaido, Japan
[2]Graduated School of Engineering, Hankyong National University, Ansung, Korea
[3]Department of Architecture, The University of Tokyo, Tokyo, Japan
Email: *limmk79@naver.com

Abstract

The aim of this study was to develop a technique for the complete recycling of concrete based on microwave heating of surface modification coarse aggregate (SMCA) with only inorganic materials such as cement and pozzolanic materials (silica fume, fly ash). The mechanical properties of SMCA, which was produced using original coarse aggregate (OCA) and inorganic admixtures, as well as its separation from the cement matrix and recovery performance were quantitatively assessed. The experimental results showed that micro structural reinforcement of the interfacial transition zone, which is a weak part of concrete, by coating the surface of the OCA with cement and admixtures such as pozzolanic materials can help suppress the occurrence of micro-cracks and improve the mechanical performance of the OCA. Microwave heating was observed to cause micro-cracking and hydrate decomposition. Increasing the void volume and weakening the hydrated cement paste led to the effective recovery of recycled coarse aggregate.

Keywords

Recycling, Surface Modification, Interfacial Transition Zone, Pozzolanic Reaction, Microwave, Recovery

1. Introduction

Concrete, which is used in large quantities in civil engineering and building construction, becomes weak with time; thus, old structures must be demolished and replaced [1] [2]. The handling of old concrete is a major problem

---

*Corresponding author.

for society to adhere to the 3R concept (reduce, reuse, and recycle). The accumulation and storage of concrete in huge piles cannot be a long-term solution because of the reduced natural resources and lack of space. Moreover, decreases in road construction work, which is the main use for recycled concrete, are expected to lead to less demand for sub-base coarse material for roads [3] [4]; in the long term, this calls for measures to expand and diversify the use of recycled concrete waste and to use recycled aggregates for concrete manufacture [5]. Thus, research on recycled aggregates is being conducted from various angles worldwide, and the Japanese Industrial Standards (JIS) has been revised for recycled aggregates [6]. However, there are still problems related to the production of high-quality recycled aggregates such as high energy consumption and the generation of large amounts of fine powder during crushing [7] [8]. On the other hand, using low-quality recycled aggregates can lower the concrete performance, which impedes the spread of recycled aggregate use [9] [10]. Because the aggregate resources that can be newly used are limited, an efficient and reliable mechanism for concrete recycling with low energy consumption is necessary. There are a variety of benefits to recycling concrete rather than dumping or burying it in a landfill [2] [11]:

- Keeping concrete debris out of landfills saves space.
- Using recycled material as gravel reduces the need for gravel mining.
- Recycling 1 ton of cement can save 1360 gallons of water and 900 kg of CO_2.
- Using recycled concrete as the base material for roadways reduces the pollution involved in trucking material.

2. Technical Overview

Concrete recycling via microwave heating is a completely new technique and is shown in **Figure 1** [21] [22]. Admixtures (e.g., pozzolanic materials) improve the chemical bonding and mechanical friction between aggregates in the coating layer of the original coarse aggregate (OCA) surface and cement matrices at the interfacial transition zone (ITZ), which is the weak part of concrete. Thus, recycled coarse aggregate (RCA) can be recovered for concrete structures because the mechanical performance of the concrete is improved, as shown in **Figure 2(a)** and **Figure 2(b)**. This technique involves coating the OCA with iron oxide (Fe_2O_3), which has a high dielectric constant, as a binder and then selectively heating and weakening the aggregate interface with microwaves to manufacture RCA following the dismantling of a structure, as shown in **Figure 3(a)** and **Figure 3(b)** [5]. This technique allows almost complete recycling of the aggregates by recovering high-quality RCA while using a small amount of energy [21] [22]. This technique allows for a trade off [9] between improvement in the concrete strength and aggregate recovery rate. Concrete fabricated with this technique comprises OCA, surface modification coarse paste (SMCP), surface modification coarse aggregate (SMCA), and Fe_2O_3, as shown in **Figure 1** [5].

3. Mechanical Performance of OCA-SMCP-Cement Matrix

3.1. Experiment Overview

In general, because the bond strength between the aggregate and paste is less than the individual tensile strengths of either the aggregate or paste, a crack tends to initiate from the aggregate-paste interface (*i.e.*, ITZ) owing to bleeding in the fresh concrete or a load-induced crack in the hardened concrete [12]. Therefore, the bond strength of aggregate-paste is somewhat directly related with the strength of the concrete, and this shear bond strength is generated by chemical and physical adhesion [13] [14]. In order to review the effectiveness of surface modification at improving the OCA-SMCP (interface)-cement matrix in detail, as shown in **Figure 2**, a shear bond strength experiment was conducted by coating the OCA surface with either cement or an admixture comprising cement and pozzolans. The changes in mechanical properties were assessed.

3.2. Experimental Method

In this experiment, specimens with OCA, SMCP, and a cement matrix structure were fabricated in order to characterize the chemical and physical bonds that form in the interface between the modified aggregate and cement matrices, as shown in **Figure 4**. The compressive and tensile shear bond strengths were measured at interface angles (α, β) of 30°, 45°, and 60°, and the failure load at each angle was compared and analyzed [15] [16]. The OCA specimens were cut from crushed hard sandstone (standard density: 2.66 g/cm^3, water absorption ratio: 0.70%); the compressive and tensile shear specimens had dimensions of 10 cm × 10 cm × 40 cm and 10 cm ×10

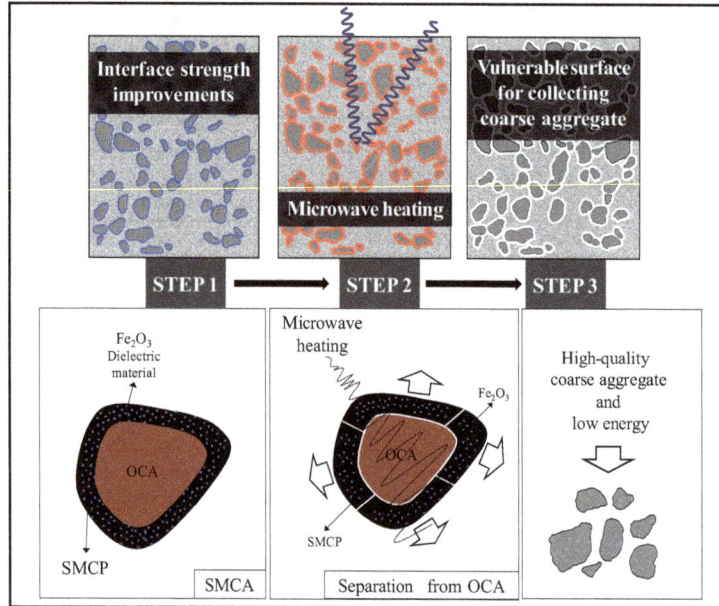

Figure 1. Improvement in concrete strength by modification and recovery by microwave heating [21] [22].

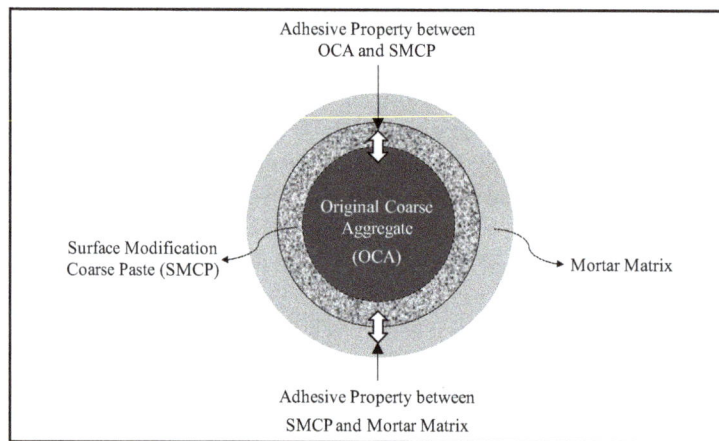

(a)

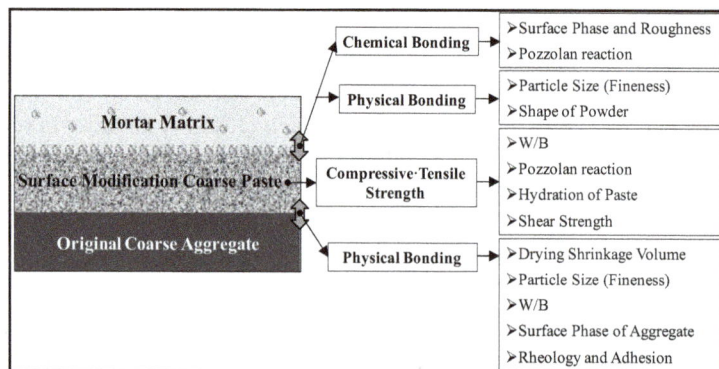

(b)

Figure 2. Mechanism of adhesion between modified aggregate and interface [5]. (a) Concept of surface adhesion of modified aggregate; (b) Control factor of surface adhesion of modified aggregate.

(a)

(b)

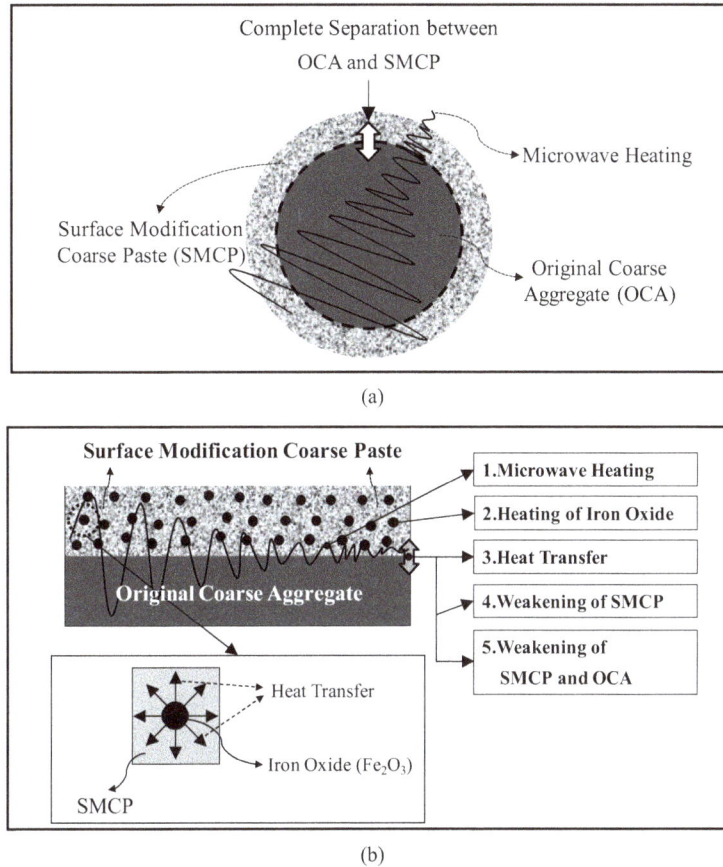

Figure 3. Mechanism of weakening between OCA and SMCP [5]. (a) Concept of surface separation between OCA and SMCP; (b) Control factor of surface separation of modification aggregate.

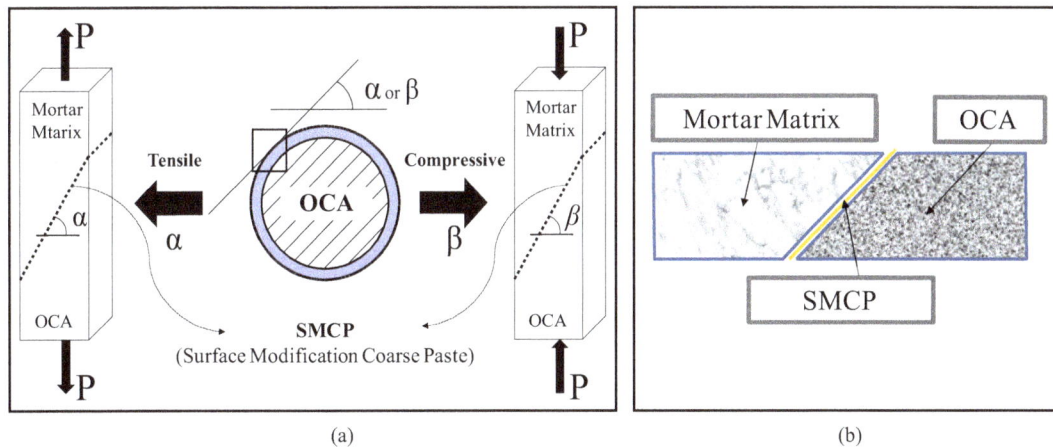

(a)   (b)

Figure 4. Outline of shear bond testing of SMCA concrete. (a) Concept of shear bond testing; (b) Manufacture of specimen.

cm × 20 cm, respectively (**Table 1**). The SMCP was mixed based on the mixing ratio presented in **Table 2**; for the materials, cement and cement substituted with pozzolanic materials (silica fume, fly ash) were used. The W/C ratio for the mortar was set to 55%; this ratio is generally applied in the field to satisfy the general strength condition given in **Table 3**. In the experiment, the modified aggregate specimens were cut in advance to ensure that the interface angle would be 30°, 45°, and 60°; these were installed in the form before SMCP was coated on

Table 1. Experimental factors and conditions.

Experimental factors		Conditions
SMCP	OCA	Cutting specimen of crushed hard sandstone (Standard density: 2.66 g/cm^3, Water absorption ratio: 0.70%)
	W/C	30%
	Modification materials	Cement, Fly ash (F/A), Blast furnace slag (BFS), Silica fume (S/F)
	Replace ratio of modification materials (%)	Cement = 100 Cement + Fly-ash (F/A) = 70:30 Cement + Blast Furnace Slag (BFS) = 70:30 Cement + Silica-fume (S/F) = 90:10
	Cement matrix	Normal strength W/C = 55%
	Interface angle	30°, 45°, 60°

Table 2. Mix proportions of SMCP.

Type	W/B (%)	W (ml)	Volume (ml/l)				Weight (g/l)				Fe$_2$O$_3$ (g/l)
			C	FA	BFS	SF	C	FA	BFS	SF	Fe$_2$O$_3$
C			513	0	0	0	1620	0	0	0	1620
FA30	30	487	359	154	0	0	1134	363	0	0	1497
BFS30			359	0	154	0	1134	0	446	0	1580
SF10			461	0	0	51	1458	0	0	113	1571

Note: Density of each material (unit: g/cm^3). C: Cement: 3.16; FA: Fly ash: 2.22; BFS: Blast furnace slag: 2.91; SF: Silica fume: 2.24.

Table 3. Mix proportion of mortar.

Type	W/C (%)	Air (%)	Unit weight (kg/m^3)			
			W	C	Fine aggregate	Admixture
OCA's concrete (O)	55	4.5 ± 1.5	175	318	833	1.59 (a[*])
SMCA's concrete (M)						

Note: O: OCA concrete; M: SMCA concrete; a[*]: Plasticizer.

the specimen surfaces. The specimens were then air-cured for 28 days in a room at constant temperature and humidity (20°C, 60%RH), and the mortar was placed. In order to prevent failure in the aggregate and cement matrix surface when the specimen was taken out of the form, the specimens were air- and water-cured for 5 and 28 days, respectively, in the room at the same constant temperature and humidity. Then, a load was applied at a rate of 1mm/min to measure the bond failure in the interface. In the tensile shear bond strength test, epoxy resin adhesive was used to attach the specimen to the tensile testing instrument, and the load was applied using a round loop to prevent eccentricity. **Table 4** presents the experimental levels of the SMCA concrete.

3.3. Compressive Shear Bond Strength

Figure 5 shows the results of the compressive shear bond strength test; the SMCA concrete demonstrated an approximately 50% increase in strength compared to the OCA concrete regardless of the interface angle. The compressive shear bond strength tended to decrease as the interface angle increased; this was deemed to be caused by the sliding effect resulting from the increase in the interface angle irrespective of the bonding surface. On the other hand, adding silica fume to the SMCP led to higher compressive shear bond strength because of the micro-filler effect and pozzolanic reaction. A 30% substitution with fly ash resulted in the second-highest compressive shear bond strength following the SMCP containing silica fume. Thus, the pozzolanic reaction caused by the

Table 4. Experimental levels.

Experiment		Modified paste	Compressive shear	Tensile shear
O		N/A		
	C			
	C + FA	A	P	P
M	C + BFS			
	C + SF			

Note: O: OCA; M: Modified coarse aggregate (SMCA concrete), C: Cement; C + FA: Cement + Fly ash; C + BFS: Cement + Blast furnace slag; C + SF: Cement + Silica fume, N/A: Not applicable; A: Application; P: Performed.

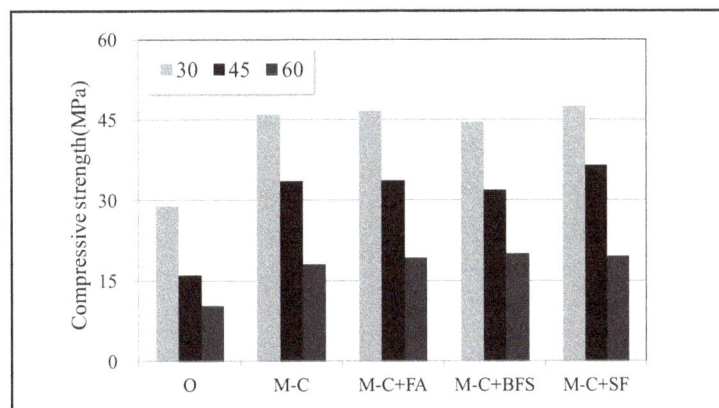

Figure 5. Compressive shear bond strength.

substituted fly ash may improve the strength.

3.4. Tensile Shear Bond Strength

Figure 6 shows the results of the tensile shear bond strength test; all of the modified aggregate specimens had improved shear bond strength compared to the OCA. In addition, the tensile shear bond strength was observed to increase with the interface angle. In other words, a larger bonding surface meant a larger area of the microstructure of ITZ was improved, which in turn improved the shear bond strength. In particular, when a pozzolanic material such as silica fume or fly ash was added to the SMCP, the tensile shear bond strength increased even further; this may be a result of the structural densification caused by the micro-filler effect and pozzolanic reaction. **Figure 7(a)** and **Figure 7(b)** show the fracture surfaces and scanning electron microscope images of the ITZ for the OCA and SMCA concretes. As shown in **Figure 7(a)**, failure occurred in the ITZ between the OCA and cement matrices, which were observed to contain calcium hydroxide (C-H) and ettringite. However, failure of the SMCA concrete occurred in the cement matrices and not in the interface, as shown in **Figure 7(b)**. This may indicate that a denser and stronger ITZ with a high level of calcium silicate hydrate (C-S-H) was strengthened by the surface modification treatment and pozzolanic reaction.

4. Weakening between OCA and SMCP by Microwave Heating

4.1. Experiment Overview and Method

An experiment was carried out to measure the weakening between OCA and SMCP caused by microwave heating (frequency of 2.45 GHz and high-frequency output of 1800 W) at heating times of 0, 60, 120, or 180 sin the SMCA concrete, as shown in **Figure 3**. The microwave heating characteristics and changes in the pores before and after heating were measured by mercury intrusion porosimetry (MIP). The temperature was measured using thermography before and after microwave heating was applied, and the temperature characteristics under each condition were assessed (**Figure 8**). The experimental specimens were manufactured by selecting O and M-C

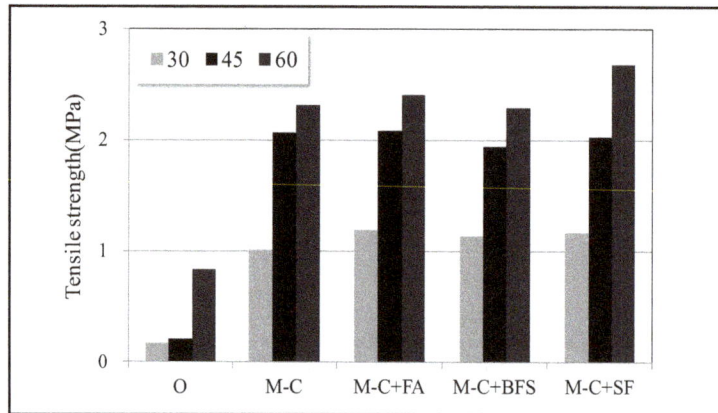

Figure 6. Tensile shear bond strength.

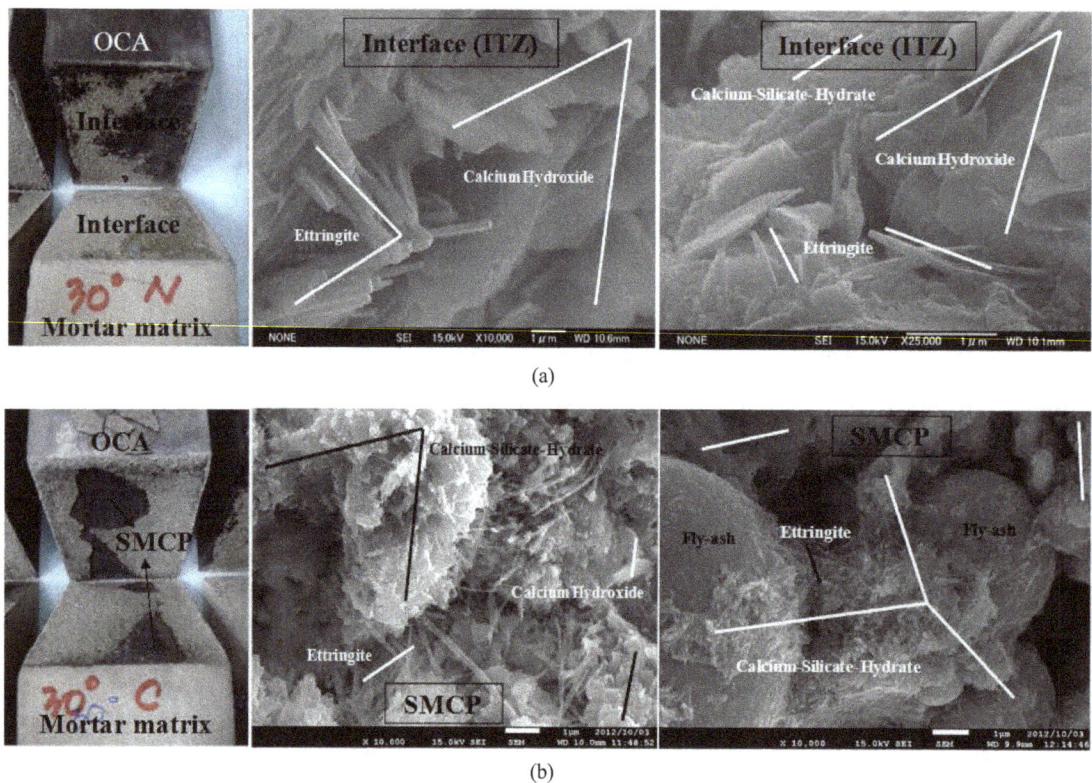

(a)

(b)

Figure 7. Fracture surfaces and scanning electron microscope images for ITZ of each concrete specimen. (a) OCA concrete; (b) SMCA concrete.

from the pozzolanic materials in Section 3. Also, aggregates with a significant amount of cement paste attached were collected from the concrete for use as test specimens in the MIP experiment, as shown in **Figure 9**.

4.2. Temperature Performance of SMCA by Microwave Heating

Figure 10 shows the experimental results for the OCA concrete. The temperature increased to 110°C after 60 s, 200°C, after 120 s, and 280°C after 180 s. In comparison, the temperature of the SMCA concrete increased to 190°C after 60 s, 280°C after 120 s, and 405°C after 180 s, as shown in **Figure 10**. The SMCA concrete showed a greater temperature increase of approximately 80°C - 130°C compared to the OCA concrete. Thus, microwave heating was determined to have a greater effect on the former. In particular, when the SMCA was heated for 180 s, the increase in temperature was approximately 130°C, higher than that of the OCA. This may be due to the

Figure 8. Microwave heating.

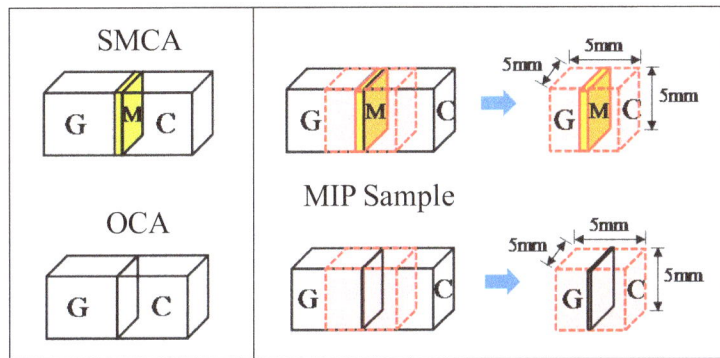

Figure 9. Microwave heating. Note: G: Gravel; M: SMCP; C: Cement matrix.

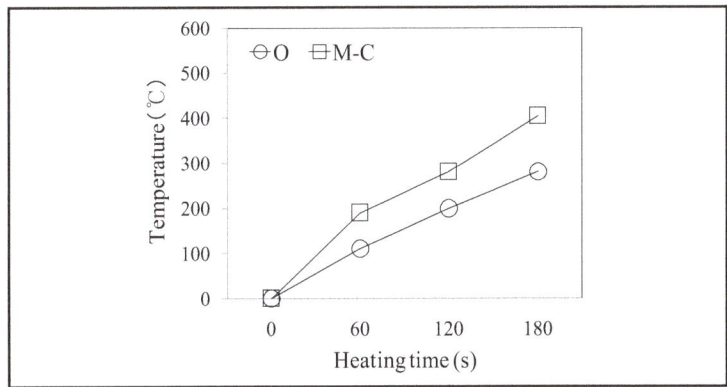

Figure 10. Temperature characteristics of SMCA and OCA with microwave heating.

effective heating of iron oxide [17], which was mixed in the SMCP as a dielectric material, as shown in **Figure 11**.

4.3. Weakening of SMCA Due to Changes in Void Volume

The experimental results showed that the temperatures of the specimens increased because of the thermal conductivity of the iron oxide being heated by the microwaves, as shown in **Figure 12** and **Figure 13**. The total increase in void volume and peak of the pore distribution moved to the part with the largest void volume. In particular, when the heating time was over 180 s, the number of pores with a diameter of 0.05 μm or less decreased, whereas pores with a diameter of 0.05 - 0.1 μm moved toward spores with a diameter of 0.1 μm as the void volume increased (**Figure 13**). As shown in **Figure 14**, when the heating time was over 180 s, the void volume

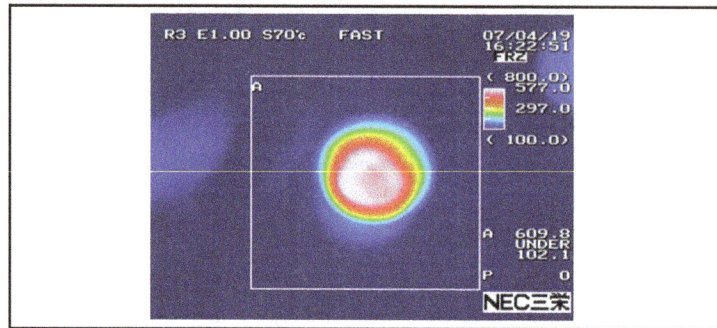

Figure 11. Heating of iron oxide by microwave heating (577°C after 30 s).

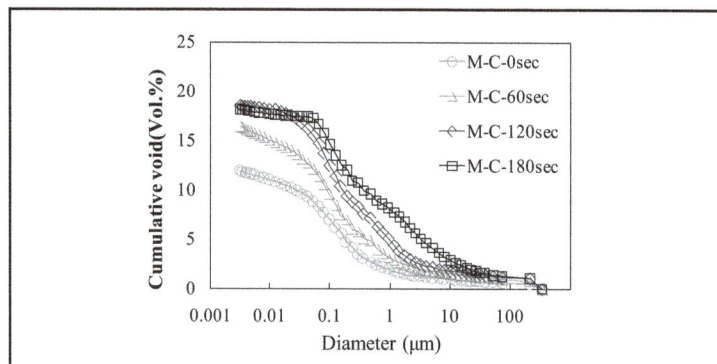

Figure 12. Cumulative void size distribution of SMCA.

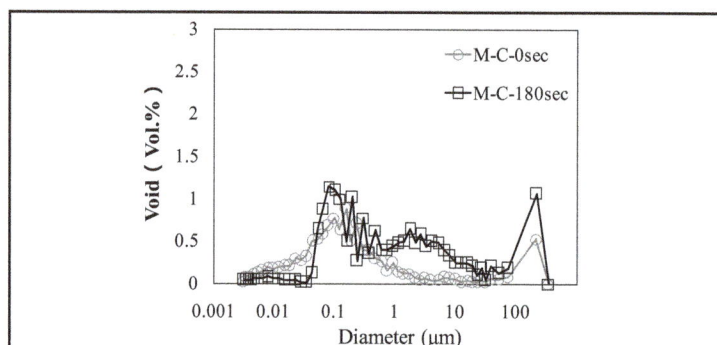

Figure 13. Void size distribution of SMCA.

of the SMCA concrete was slightly higher than that of the OCA concrete. The increase in the concrete porosity [18] [19] because of microwave heating may represent the weakening of the cement paste around the ITZ and aggregate. Thus, the increase in pores in the SMCA concrete because of microwave heating may result from the evaporation of bound water caused by the decomposition of calcium hydroxide and C-S-H hydrates and the occurrence of micro-cracks in the concrete [20].

5. Performance Review of SMCA Concrete

5.1. Experiment Overview and Method

Concrete is generally composed of cement, water, and fine and coarse aggregates. However, with respect to the SMCA concrete, the surface of the OCA was coated with cement paste. This required that the modified paste be reflected in the mixing design, as shown in **Figure 15**. In the mixing design, the amount of coarse aggregate coated with an admixture can be increased. Thus, the amount of SMCP coated on the coarse aggregate can be excluded from the amounts of cement and water in the mixing ratio, and the amounts of cement, water, and

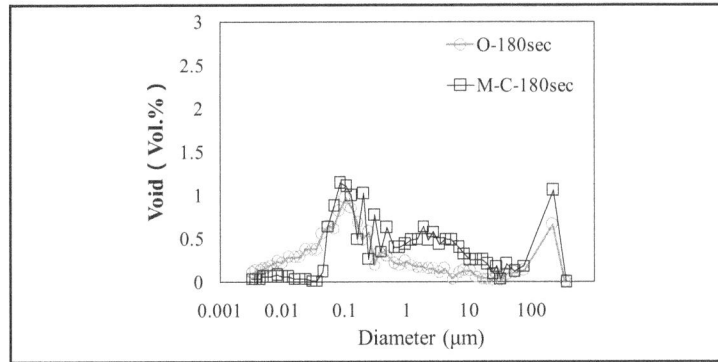

Figure 14. Void size distribution of each aggregate.

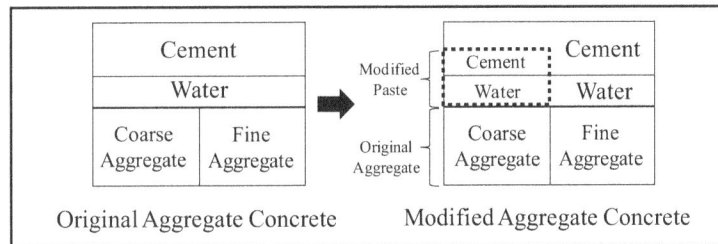

Figure 15. Ratios of constituents in concrete (Note: Materials/m^3) [5].

OCA for the OCA and SMCA concretes can be the same [5]. In this experiment, OCA and SMCA concrete specimens with a water-cement (W/C) ratio of 55% were compared. In order to achieve equal characteristics for the fresh specimens for the comparison, the W/C ratio of the SMCA concrete and amount of chemical admixture were adjusted as shown in **Table 5**. Using the above method, the same amount of cement was applied to the mixing design for the OCA and SMCA. As the binder for the surface admixture, only cement and cement with admixtures (e.g., pozzolanic material) were used with a W/C ratio of 30%. The mixing percentage of the surface admixture is shown in **Table 6**. Based on the mechanical performance and weakening of the SMCA concrete in Sections 3 and 4, the compressive and splitting tensile strengths of the SMCA concrete were examined in this experiment to assess the improvement in the mechanical properties. The recovery rates of RCA by microwave heating (frequency of 2.45 GHz and high-frequency output of 1800 W) for 0, 60, 120, and 180 s were measured to assess the aggregate recovery characteristics.

For the OCA used in this experiment, the parent materials were collected from a rocky mountain, crushed with a jaw crusher, and sorted by diameter from 5 mm to 20 mm. Then, the particle size distribution was adjusted to satisfy the standard diameter prescribed in JIS A 1102 (standard for aggregate sieving test), and the density and water absorption ratio of the OCA and SMCA specimens (only cement and cement with pozzolanic materials, respectively) were measured in accordance with JIS A 1110 (standard for coarse aggregate density and water absorption test). The results of the preliminary experiment showed that the density of OCA with a modified surface was higher than that with no surface modification by 1% - 1.5%, as shown in **Table 7**. This may be because the SMCP with a low W/C ratio contained high-density iron oxide, which resulted in the ITZ microstructure becoming more compact. On the other hand, the water absorption ratio also increased. This was due to the cement paste forming a thin film on the OCA surface, which resulted in the OCA and SMCP absorbing water. However, the OCA and SMCA in this experiment satisfied the criteria prescribed in JIS A 5021 (standard for concrete using recycled aggregate H) for H class of RCA, which can be used for structural aggregate. Thus, the water absorption ratio was not expected to cause any potential problems in this experiment.

5.2. Mechanical Properties of SMCA Concrete

The compressive and splitting tensile strength tests were performed based on JIS A 1108 and JIS A 1113. For the mechanical properties, the compressive and splitting tensile strengths of each SMCA concrete specimen were higher than those of the OCA concrete specimens by 5% - 12%, as shown in **Figure 16** and **Figure 17**. The

Table 5. Mix proportions of concrete.

Type	W/C (%)	Slump (mm)	Air (%)	G_{max} (mm)	Unit weight(kg/m³)				Admixture (%)
					W	C	S	G	
OCA's concrete (O)	55	180±25	4.5±1.5	20	175	318	833	961	C × 0.5 (a*)
SMCA's concrete (M)					175	311	833	982	C × 0.9 (a*)

Note: O: OCA concrete; M: SMCA concrete; a*: Plasticizer.

Table 6. Mix proportions of modification paste.

W/B (%)	Water (g)	Binder (g)	Fe_2O_3 (g)	Superplasticizer (g)	Table flow (mm)
30	21	70	B × 100%	B × (1.9% - 2.5%)	300

Note: Based on 1 kg of OCA.

Table 7. Type and properties of each coarse aggregate.

	Type	Density (g/cm³)	Water absorption ratio (%)
	O	2.66	0.70
	C	2.69	1.65
M	C + FA	2.70	1.63
	C + BFS	2.70	1.65
	C + SF	2.70	1.62

Note: O: OCA (OCA concrete); M: Modified coarse aggregate (SMCA concrete), C: Cement; C + FA: Cement + Fly ash; C + BFS: Cement + Blast furnace slag; C + SF: Cement + Silica fume.

improved strength of the modified aggregate concrete was due to the reinforced physical and chemical bonding between the modified paste and cement matrices, which was caused by the increased mechanical friction resulting from the particle size and shape of the iron oxide and the SMCP coating effect on the ITZ [5]. In addition, the SMCA concrete using pozzolanic materials (silica fume and fly ash) showed an increase in strength of approximately 8% after 7, 14, and 28 days of curing compared to the SMCA concrete using only cement (**Figure 16**). The splitting tensile strength test results showed an improvement in strength of approximately 5% after 7, 14, and 28 days of curing (**Figure 17**). As noted in Section 2, when pozzolanic materials such as silica fume and fly ash were added to the SMCP, the micro-filler effect and pozzolanic reaction caused the ITZ structure to become denser.

5.3. Recovery Properties of SMCA Concrete

As shown in **Figure 18(a)** and **Figure 18(b)**, when the microwave heating times were 0 and 60 s, the RCA recovery rate for the SMCA concrete was equivalent to or slightly lower than that of the OCA concrete. This may be due to the enhanced bonding caused by the SMCP coating between the OCA and cement matrices. On the other hand, when the microwave heating times were 120 and 180 s, the recovery rate for the SMCA concrete became substantially lower than that of the OCA concrete regardless of the admixture (cement only and cement with pozzolanic materials). When the heating time was120 s, the temperature reached 300°C, which is the weakening temperature of cement paste [5] [6]. This was the point at which the recovery rate began to differ. In particular, when the heating time was180 s, the recovery rate of RCA for the SMCA concrete reached nearly 100%; thus, the microwaves were concluded to effectively heat the dielectric material (iron oxide) present in the SMCP. This result could be predicted by the RCA recovery rates for both the OCA and SMCA concrete specimens after 180 s of microwave heating, as shown in **Figure 19**.

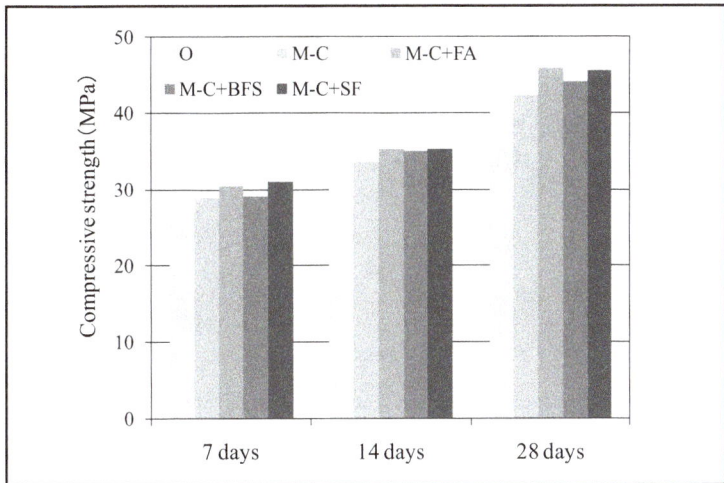

Figure 16. Comparison of strength.

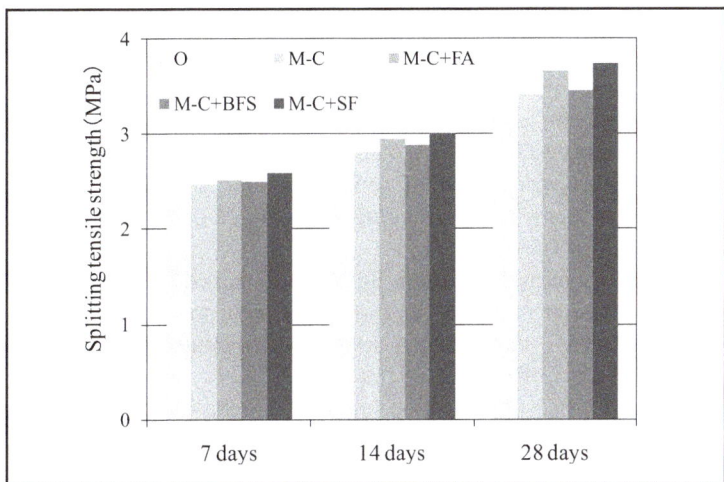

Figure 17. Splitting tensile strength.

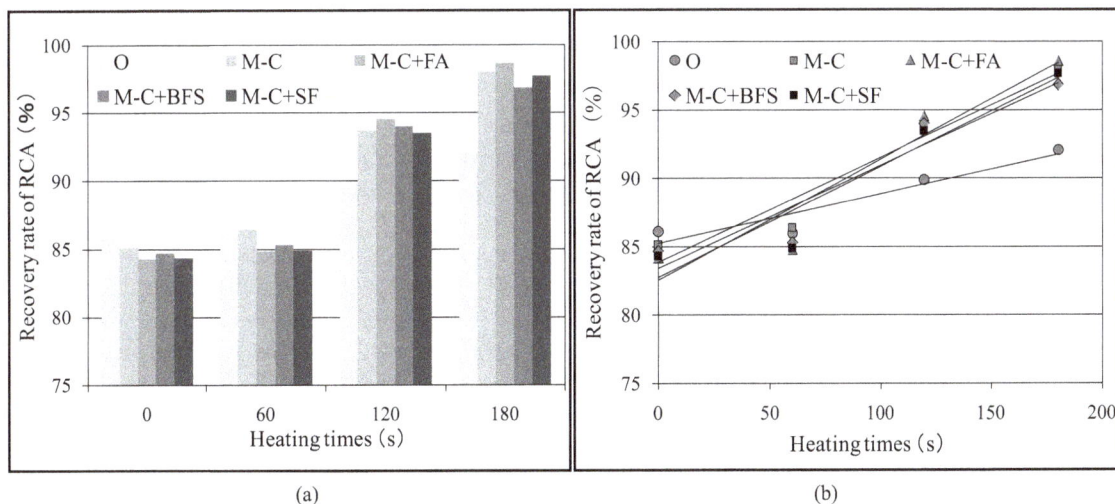

(a)

(b)

Figure 18. Recovery rate of RCA of each concrete specimen by microwave heating. (a) RCA recovery rates; (b) Comparison of RCA recovery rates.

Figure 19. Comparison of RCA from each type of concrete.

6. Conclusions

The following conclusions were made based on the results of the experiments conducted to examine the mechanical properties and recovery performance of the SMCA concrete with admixtures and microwave heating.

1) When the W/C ratio was 55%, the improvements in the compressive strength and split tensile strength, including the shear bond strength, were confirmed to be due to the SMCP coating; the ITZ structure was made denser by the admixtures (only cement and pozzolanic materials).

2) When the SMCA concrete containing iron oxide (Fe_2O_3), was heated with microwaves, the temperature increased more significantly compared to the OCA concrete. In particular, when the microwave heating time was 180 s, the maximum temperature was over 400°C, and micro-cracks occurred in the concrete along with an increased void volume caused by dehydration of the hydrates. Based on these results, the microwaves effectively heated the iron oxide contained in the SMCP.

3) The RCA recovered after microwave heating for 180 s contained less than 5% of paste and fine aggregate, regardless of the admixture. The recovered RCA was very similar to the OCA, which proves the feasibility of recovering high-quality RCA. Thus, microwave heating was determined to weaken binders containing a dielectric material for efficient recovery of RCA.

The mechanical performance of SMCA can be improved by the use of inorganic materials and microwave heating to effectively recover RCA.

References

[1]   Noguchi, T. and Tamura, M. (2001) Concrete Design towards Complete Recycling. *Structural Concrete Journal of the Fib*, **2**, 155-167.

[2]   Noguchi, T. (2008) Resource Recycling in Concrete: Present and Future. *Stock Management for Sustainable Urban Regeneration*, **4**, 255-274.

[3]   Hendriks, Ch.F. and Janssen, G.M.T. (2001) Construction and Demolition Waste: General Process. *HERON*, **46**, 79-88.

[4]   Shima, H., Tateyashiki, H., Matsuhashi, R. and Yoshida, Y. (2005) An Advanced Concrete Recycling Technology and its Applicability Assessment through Input-Output Analysis. *Journal of Advanced Concrete Technology*, **3**, 53-67. http://dx.doi.org/10.3151/jact.3.53

[5]   Choi, H.S., Kitagaki, R. and Noguchi, T. (2014) Effective Recycling of Surface Modification Aggregate using Microwave Heating. *Journal of Advanced Concrete Technology*, **12**, 34-45. http://dx.doi.org/10.3151/jact.12.34

[6]   Choi, H.S., Kitagaki, R. and Noguchi, T. (2012) A Study on the Completely Recovery of Surface Modification aggregate using Microwave and Effective Utilization. *Proceedings of the 5th ACF International Conference*, Pattaya, October 2012, Session 1-2, ACF2012-0093, 41-46.

[7]   Kunio, Y. (2003) A Study on the Manufacturing Technology of High-Quality Recycled Fine Aggregate. *Japan Concrete Institute*, **25**, 1217-1222.

[8]   Shima, H. and Tateyashiki, H. (1999) New Technology for Recovering High-Quality Aggregate from Demolished Concrete. *Proceedings of the 5th International Symposium on East Asian Recycling Technology*, The M.M.P.I. in Japan 1999, 106-109.

[9]   Tamura, M., Tomosawa, F. and Noguchi, T. (1997) Recycle-Oriented Concrete with Easy-to-Collect Aggregate. *Ce-*

ment Science and Concrete Technology, **51**, 494-499.

[10] Tsujino, M., Noguchi, T., Tamura, M., Kanematsu, M. and Maruyama, I. (2007) Application of Conventionally Recycled Coarse Aggregate to Concrete Structure by Surface Modification Treatment. *Journal of Advanced Concrete Technology*, **5**, 13-25. http://dx.doi.org/10.3151/jact.5.13

[11] Value Engineering Benefits (2010) Concrete Recycling.org. Retrieved 2010-04-05.

[12] Mehta, P.K. and Moneiro, P.J.M. (2006) Concrete: Microstructure, Properties and Materials. McGraw-Hill Companies, New York.

[13] Diamond, S. and Huang, J. (2001) The ITZ in Concrete. *Cement and Concrete Composite*, **23**, 59-64.

[14] Elsharief, A., Cohen, D. and Olek, J. (2003) Influence of Aggregate Size, Water Cement Ratio and Age on the Microstructure of the Interfacial Transition Zone. *Cement and Concrete Research*, **33**, 1837-1849. http://dx.doi.org/10.1016/S0008-8846(03)00205-9

[15] Robin, P.J. and Austin, S.A. (1995) A Unified Failure Envelope from the Evaluation of Concrete Repair Bond Tests. *Magazine of Concrete Research*, **47**, 57-68. http://dx.doi.org/10.1680/macr.1995.47.170.57

[16] Austin, S., Robins, P. and Pan, Y.G. (1999) Shear Bond Testing of Concrete Repair. *Cement and Concrete Research*, **29**, 1067-1076. http://dx.doi.org/10.1016/S0008-8846(99)00088-5

[17] McGill, S.L., *et al.* (1988) The Effects of Power Level on the Microwave Heating of Selected Chemicals and Minerals. *Proceedings of the MRS Symposium*, Nevada, April 1988, 124.

[18] Schneider, U. (1982) Behavior of Concrete at High Temperatures. Deutscher Ausschuss für Stahlbeton, Berlin, 28-33.

[19] Bazant, Z.P. and Kapaln, M.F. (1996) Concrete at High Temperatures: Material Properties and Mathematical Models. Prentice Hall, Upper Saddle River.

[20] Takeo, A., Fukujiro, F., Kuniyuki, T., Kenji, K. and Isao, K. (1999) Mechanical Properties of High-Strength Concrete at High Temperatures. *Architectural Institute of Japan*, **515**, 163-168.

[21] Tsujino, M., Noguchi, T., Kitagaki, R. and Nagai, H. (2010) Completely Recyclable Concrete of Aggregate-Recovery Type by a New Technique Using Aggregate Coating. *Architectural Institute of Japan*, **75**, 17-24.

[22] Tsujino, M., Noguchi, T., Kitagaki, R. and Nagai, H. (2011) Completely Recyclable Concrete of Aggregate-Recovery Type by Using Microwave Heating. *Architectural Institute of Japan*, **76**, 223-229.

4

Life Cycle Assessment and Life Cycle Cost of Waste Management—Plastic Cable Waste

Mats Zackrisson, Christina Jönsson, Elisabeth Olsson

Energy and Environment Group, Department of Materials, Swerea IVF AB, Mölndal, Sweden
Email: mats.zackrisson@swerea.se

Abstract

The main driver for recycling cable wastes is the high value of the conducting metal, while the plastic with its lower value is often neglected. New improved cable plastic recycling routes can provide both economic and environmental incentive to cable producers for moving up the "cable plastic waste ladder". Cradle-to-gate life cycle assessment, LCA, of the waste management of the cable scrap is suggested and explained as a method to analyze the pros and cons of different cable scrap recycling options at hand. Economic and environmental data about different recycling processes and other relevant processes and materials are given. Cable producers can use this data and method to assess the way they deal with the cable plastic waste today and compare it with available alternatives and thus illuminate the improvement potential of recycling cable plastic waste both in an environmental and in an economic sense. The methodology applied consists of: cradle-to-gate LCA for waste material to a recycled material (recyclate); quantifying the climate impact for each step on the waste ladder for the specific waste material; the use of economic and climate impact data in parallel; climate impact presented as a span to portray the insecurities related to which material the waste will replace; and possibilities for do-it-yourself calculations. Potentially, the methodology can be useful also for other waste materials in the future.

Keywords

Cable Recycling, Life Cycle Assessment, LCA, Life Cycle Cost, LCC, Economic Analysis, Climate Impact, Waste Recycling

1. Introduction

The main driver for recycling cables is the high value of the conducting metal (usually copper or aluminium), while the plastic with its lower value is often neglected. On the other hand, if it is not for the metal, the whole waste cable may be neglected, as it is often experienced with optical waste cables today [1]. This paper aims to provide primarily cable producers with a methodology to assess the way they deal with the cable plastic waste today and compare it with available alternatives and thus facilitate realizing the improvement potential of recycling cable plastic waste. The hypothesis is that it is possible to create a transparent methodology that provides additional insights and incentives of the value of recycling the plastic parts in addition to the metal core of the cable. Through using the methodology provided, the reader or user will be able to show the climate effects of improving the cable waste recycling (compared to how it is done today) and also to show the economic, technical and management implications of such improvements. The methodology as such can also be applicable to other waste materials.

The situation with small or negligible profit margins is similar for many other waste materials, for example textiles and construction waste [2]. It is therefore important to include economic data when analyzing waste recycling options. The simple knowledge that recycling a particular waste would lead to reduced environmental impact will not automatically lead to that it will be done; it will have to be economically beneficial (or enforced by law) otherwise it will not happen.

This paper and underlying report [3] have been compiled within the scope of the Wire and Cable project which is managed by the Swedish research institute Swerea IVF and financed by Vinnova, a Swedish governmental funding agency, and participating companies. The following cable manufacturers, polymer manufacturers, cable users and recycling companies are members of the project running from 2010 to 2013: Borealis AB, Draka Kabel Sverige AB, Ineos ChlorVinyls, Nexans Sweden AB, ABB AB, Stena Metall AB, Volvo Lastvagnar AB, Volvo Personvagnar AB and Ericsson AB. The main objective of the Cable project is to facilitate increased recycling of cable plastics. As from 2014, the Wire and Cable project will continue with many of the old members and some new ones.

2. Method

The methodology described and applied below consists of: cradle-to-gate LCA, life cycle assessment, for waste material to recyclate; quantifying the climate impact for each step on the waste ladder for the specific waste material; the reporting of economic and climate impact data in parallel; presentation of the climate impact as a span to portray the insecurities related to which material the waste will replace; and possibilities for do-it-yourself calculations. It has been developed in cooperation with the cable industry and used by them. It follows guidance about LCA of waste management issued by EUs Joint Research Centre [4], which in turn builds on the International Organization for Standardization (ISO) 14044 standard [5] for LCA and the International Reference Life Cycle Data System Handbook [6]. Similar methodology may be useful also for waste categories other than cable waste.

2.1. Life Cycle Assessment in General

LCA according to ISO 14044 [5] consist of four stages: scooping, inventory, environmental impact assessment and interpretation. All stages except the one for environmental impact assessment are considered obligatory. The stages are often repeated in an iterative way that gradually refines the assessment. None of the stages are unique to the LCA methodology. What makes LCA unique is that all (or as many as possible/relevant) life cycle phases of the analyzed object are included from raw material extraction to the product's end-of-life [7]. The life cycle phases are often referred to as raw material production, (own) manufacturing, use and end-of-life [8], see **Figure 1**.

When all life cycle phases are included in an LCA study, it is referred to as a cradle-to-grave study [4]. Studies that only include data about raw material production and own manufacturing are referred to as cradle-to-gate studies. Such cradle-to-gate LCA studies exist for most commodities like different steels, plastics etc.

2.2. Proposed LCA Application

Life cycle assessment of waste materials or waste management, though very common, has no special name in literature. A complete product life cycle as depicted in **Figure 1** is rarely involved [4]. Instead, focus lies on recycling processes after the use phase or directly after the manufacturing processes as shown in **Figure 2**. Also production of virgin materials is included in order to account for that material recycling avoids primary or virgin material production. As can be seen, LCA of waste materials span over two adjacent product life cycles.

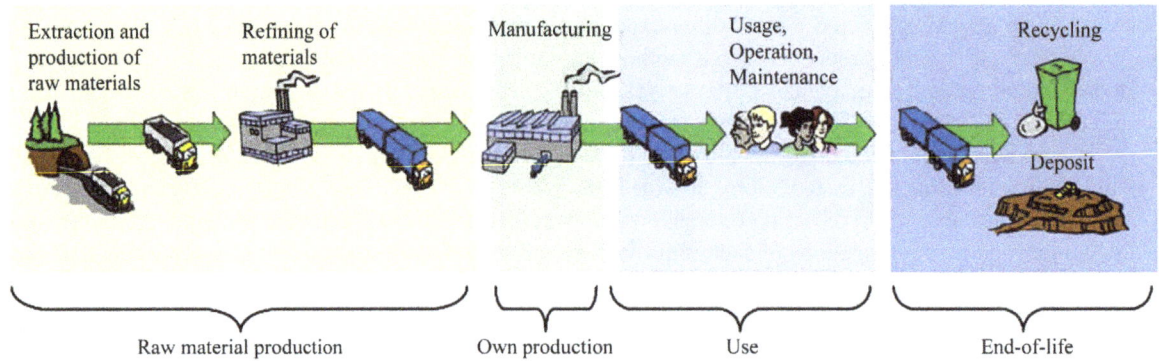

Figure 1. Life cycle assessment.

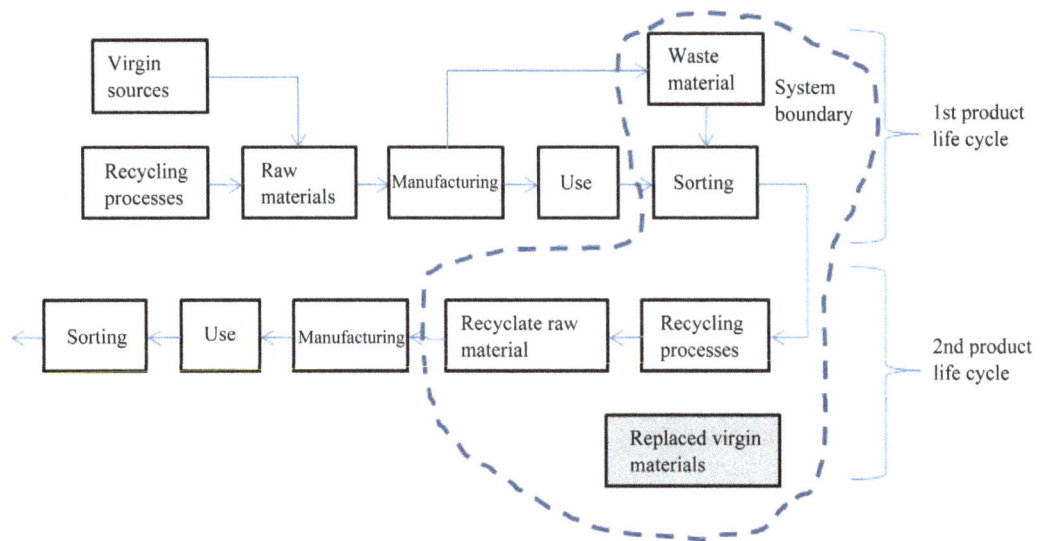

Figure 2. System boundary of LCA recycling study.

Often these product life cycles are for different products, *i.e.* cable plastic waste is rarely used to make new cables but rather to make other products. This is referred to as open-loop recycling, or, since it often entails a loss of valuable material properties, down-cycling [4].

Another way of seeing it is that the product or service under investigation is not the cable but rather the waste management of the cable, where the waste material is the input and the produced recyclate is the output, see **Figure 3**. In such a perspective the study could be compared to a cradle-to-gate LCA for a commodity from virgin origin, see **Figure 1**.

Focusing on the service required to manage the waste in the best way makes it natural to present the results per unit of cable plastic waste or per unit of cable waste, *i.e.* sometimes including the conducting metal. In LCA language these are the functional units used. A correct generic name of the *functional unit* would be *waste management per unit of waste material*. The starting point is the waste. Something has to be done about it; it cannot just be left in a pile; it has to be managed.

The waste ladder in **Figure 4** portrays the waste management options generally available. It is considered in general to be environmentally preferable to be as high on the ladder as possible. The waste ladder or waste hierarchy is encouraged by the European Union (EU) Waste Framework Directive [9], though departing from the hierarchy could be justified for, among other, reasons of technical feasibility and economic viability. In this paper, the climate impact associated with each step is calculated for the management of plastic waste from cables in order to further stimulate companies to move up the waste ladder.

The choice of system boundary and functional unit(s) means that there is no need to include the actual cable manufacturing or the use of the cables in the calculations. This is of course very advantageous since it limits

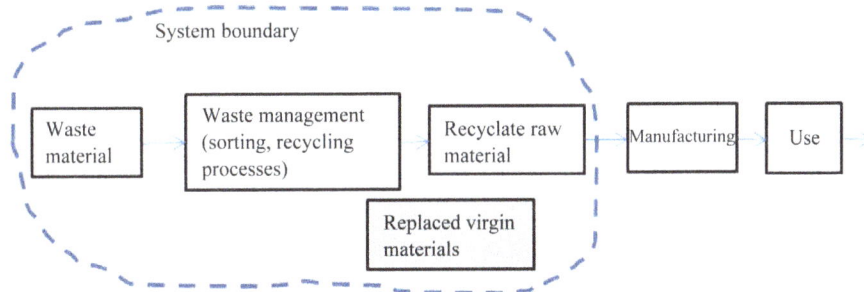

Figure 3. System boundary of cradle-to-gate waste management LCA.

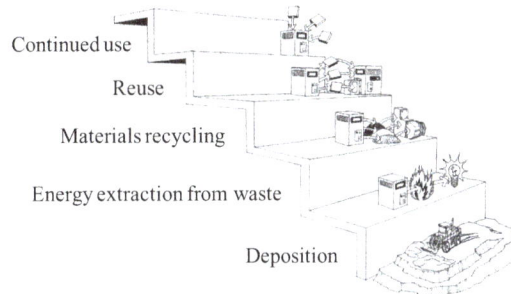

Figure 4. The waste ladder.

drastically the amount of data needed for the analysis. However, such upstream activities may only be excluded if they are not affected by any of the investigated waste management options [4]. For example, if internal recycling of production cable waste is one option, the quality must be the same (as the normally used material) so that the cable manufacturing or use is not affected.

Cable producers and recycling companies have provided site-specific data for this paper and underlying reports [3] [10]. Sometimes averages from several companies with the same process are provided. For certain processes, e.g. transports and primary metal manufacturing, generic instead of specific process data is presented. This generic data stem from public LCA databases and represent in general European or global averages. Data has mainly been drawn from the database Ecoinvent 2.0 [11].

The studied system is expanded to include *avoided processes* and the subsequent avoided environmental impacts when the recycled materials replace virgin materials. Avoided processes are shaded grey in the figures above. The choice of avoided process is very critical when system expansion is used [4] [7]. For example, since recycled copper can replace virgin copper, the environmental burdens for virgin copper manufacturing are subtracted from the studied system. Similarly, plastic recycle can replace virgin plastic in some applications. However, it is rare that plastic recyclate can actually replace virgin plastic fully, more often some form of loss of material properties occur and this makes it difficult to define the replaced material. ISO 14044 [5] contains guidance concerning deciding the avoided or replaced material. The replacement should be based on, in priority order:

- Physical properties (e.g. mass or energy content);
- Economic value (e.g. market value of the scrap material or recycled material in relation to market value of primary material).

In order to follow ISO 14044 and also to somewhat capture that recycling almost always entail down-cycling, *i.e.*, the recyclate has less good properties than the virgin resource, this paper presents two different scenarios:

- 1 to 1. In the 1 to 1 scenario 1 kg of recyclate is replacing 1 kg of the virgin material; e.g. 1 kg of recyclate polyvinyl chloride (PVC) compound replaces 1 kg of virgin PVC compound. Thus, the environmental burdens associated with manufacturing of virgin PVC compound is subtracted from the studied system (the cradle-to-gate waste management LCA) on a 1 to 1 basis.
- Market based. In the market based scenario, it is assumed that the loss of quality of the recyclate is proportional to the relation between the price paid for the recyclate and the price paid for the virgin material. Due to the loss of quality the recyclate cannot replace virgin material of the same type. What it can replace we do

not know, so we assume, that the "environmental burdens saved" are proportional to the loss of quality which we assume is proportional to the difference in price. For example, if PVC recyclate is paid at 88 euro (€) per tonne and the price of virgin PVC is 1320 € per tonne, $88/1320 = 0.07 = 7\%$ of the environmental burdens of virgin PVC manufacturing are subtracted from the studied system to account for the material that the recyclate is replacing.

Normally in LCA, at least five impact categories are used: climate change; acidification; ozone depletion; photochemical smog and eutrophication. In this paper, only climate change results are presented, in carbon dioxide equivalents (CO_2eq). Since CO_2eq is a good indicator for energy related environmental impact and most data sets related to cable waste management is dominated by energy use it has been shown [3] that CO_2eq is a good indicator for the environmental impact of cable waste recycling. For calculation of non CO_2 gases to CO_2eq the latest characterization factors from the Intergovernmental Panel on Climate Change (IPCC) [12] have been used.

2.3. Life Cycle Cost Methodology

Life cycle cost (LCC) [13] as applied in this paper refers to all costs and incomes of a particular waste incurred to the owner of the waste. For example, cost of transport (internal and external), waste processing costs, disposal costs and price of recyclates. The LCC thus shows (the waste owner) the profitability of each studied waste management option. The collection of data and identification of data required were achieved in parallel with the LCA.

3. Results

In order to calculate the benefits of different recycling routes, monetary and climate change data about the involved recycling processes, transports and avoided (replaced) products have been collected. The units are euro (€) and gram carbon dioxide equivalents (CO_2eq). In order to facilitate calculations and comparisons, data have been assembled in one sheet, see **Table 1**. The idea with the sheet is to facilitate finding and marking data that is relevant for a unique comparison, sum it up and arrive at the results.

3.1. Reduction of Climate Impacts by Moving up the Cable Plastic Waste Ladder

The potential gains of improving the cable plastic waste recycling by moving up the "waste ladder" are illustrated in **Figure 5**. All values are per kg of plastic waste. Note that the potential gains by the associated metal recycling are not shown in this figure.

The spans given reflect the different scenarios—market based or "1 to 1"—and the different polymers involved. Apart from the landfill figure, all figures are related to avoided products/processes, in the green part of **Table 1**. It is the replaced or avoided product that gives the largest climate impact contribution. Recycling processes and transports give only minor climate impacts and are therefore not included in the **Figure 5** calculations. All figures are absolute, *i.e.* relative to zero.

Avoiding plastic (and metal) cable waste completely is of course the primary target of all cable producers, but not always possible. On the other hand, moving one or two steps up the ladder is not only possible but also to a degree driven by legislation limiting landfill and energy recycling. Moving one step up the waste ladder would mean avoiding around 0.5 kg CO_2eq/kg plastic. Consumption of plastic compounds by the European cable industry in 2012 was 1.23 million tonnes [14]. Plastic waste from cable manufacturers are around 5% of their total use of plastic [3]. If the industry as a whole can move one step up the waste ladder about $1,230,000,000 \times 0.05 \times 0.5$ kg CO_2eq/kg = 30,750 tonnes of CO_2eq can be avoided annually.

3.2. Life Cycle Cost or Economic Feasibility

Recycling processes and transports may only have minor climate impacts, but they are very important from an economical point of view. Therefore indicative price information is given in **Table 1**, in order to investigate the economical feasibility of different recycling options. Results from the LCC show that moving one step up at the top of the waste ladder can increase profits by almost 2 € per kg plastic, see **Table 1**. At the bottom of the plastic waste ladder there is a landfill cost of 0.12 €/kg and additional transportation costs.

Table 1. Hot milling of PVC scrap compared to external recycling.

New recycling route	Hot milling		Compared to	Current way of recycling	Per kg cable		External recycling
Process(es)/ materials	Per kg cable		Comment/calculation	Process(es)/ materials	Per kg cable		Comment/calculation
	€	gram CO_2eq			€	gram CO_2eq	
Needed processes				Needed processes			
Production waste cable granulation	0.14	26		Production waste cable granulation	0.14	26	
Needed processes	Per kg plastic		Note that values below are given per kg polymer and may need recalculation to per kg cable by multiplication with value for kg polymer per kg cable.	Needed processes	Per kg plastic		Note that values below are given per kg polymer and may need recalculation to per kg cable by multiplication with value for kg polymer per kg cable.
	€	gram CO_2eq			€	gram CO_2eq	
Plastsep, Swedish electricity	0.055	2.2		Plastsep, Swedish electricity	0.055	2.2	
Plastsep, European electricity	0.055	8.9		Plastsep, European electricity	0.055	8.9	
Compounding PVC	0.23	6.2		Compounding PVC	0.23	6.2	
Compounding polyolefin	0.23	15		Compounding polyolefin	0.23	15	
Melt filtrating PVC	0.33	6.2		Melt filtrating PVC	0.33	6.2	
Melt filtrating polyolefin	0.33	15		Melt filtrating polyolefin	0.33	15	
Hot milling	0.004	4.3		Hot milling	0.004	4.3	
Disposal, polyvinylchloride	0.12	66		Disposal, polyvinylchloride	0.12	66	
Disposal, plastics, mixture	0.12	90		Disposal, plastics, mixture	0.12	90	
Nedeed transports	Per tonkm transport		Multiply values below for € and gram CO_2eq per tonkm transport with actual transport distance to get gram CO_2eq/tonne and €/tonne for the actual transport. Convert to gram CO_2eq /kg and €/kg by dividing with 1000.	Nedeed transports	Per tonkm transport		Multiply values below for € and gram CO_2eq per tonkm transport with actual transport distance to get gram CO_2eq/tonne and €/tonne for the actual transport. Convert to gram CO_2eq/kg and €/kg by dividing with 1000.
	€	gram CO_2eq			€	gram CO_2eq	
Lorry, Trailer 26t Euro3, NTM	0.10	51		Lorry, Trailer 26 t Euro3, NTM	0.02	10	200 km transport to granulation and compounding: $200 \times 0.1/1000 = 0.02$ euro and $200 \times 51/1000 = 10$ gram CO_2eq
Transport, lorry >16t, fleet average/ RER S	0.10	134		Transport, lorry >16 t, fleet average/RER S	0.07	94	700 km transport to user in Europe: $700 \times 0.1/1000 = 0.07$ euro and $700 \times 51/1000 = 94$ gram CO_2eq
Total needed processes	0.004	4.3		Total needed processes	0.460	137	

Continued

Avoided processes	Per kg material (€	Market gram CO$_2$eq	1 to 1)	Note that values below are given per kg material and may need recalculation to per kg cable by multiplication with value for kg material per kg cable.
Copper, primary, at refinery/GLO S	−5.5	−3160	−3160	
Copper granulate	−5.3	−3065	−3160	
Copper fluff	−4.6	−2686	−3160	
Aluminium, primary, at plant/RER S	−1.6	−12,200	−12,200	
Aluminium granulate	−1.5	−11,346	−12,200	
Aluminium fluff	−1.3	−10370	−12,200	
Heavy fuel oil, at regional storage/RER S	−1.1	−455	−455	
HFFR as oil replacement	0.058	9	−166	
Polyolefins as oil replacement	−0.022	−9	−477	
PVC compound for cable I	−1.3	−1500	−1500	
PVC recyclate I	−0.09	−100	−1500	
HFFR compound for cable	−2.0	−1170	−1170	
HFFR recyclate	−0.11	−65	−1170	
Compounding PVC	−0.23	−6.2	−6.2	
Compounding polyolefin	−0.23	−15	−15	
Avoided transports	Per ton km transport (€	gram CO$_2$eq)		Multiply values below for € and gram CO$_2$eq per tonkm transport with actual transport distance to get gram CO$_2$eq/tonne and €/tonne for the actual transport. Convert to gram CO$_2$eq /kg and €/kg by dividing with 1000.
Lorry, Trailer 26t Euro3, NTM	−0.02	−10	−10	200 km transport from PVC supplier avoided.
Transport, lorry >16t, fleet average/RER S	−0.10	−134	−134	
	Per kg material (€	Market gram CO$_2$eq	1 to 1)	
Total avoided processes	−1.550	−1516	−1516	
Total of needed and avoided processes	−1.546	−1512	−1512	

Avoided processes	Per kg material (€	Market gram CO$_2$eq	1 to 1)	Note that values below are given per kg material and may need recalculation to per kg cable by multiplication with value for kg material per kg cable.
Copper, primary, at refinery/GLO S	−5.5	−3160	−3160	
Copper granulate	−5.3	−3065	−3160	
Copper fluff	−4.6	−2686	−3160	
Aluminium, primary, at plant/RER S	−1.6	−12,200	−12,200	
Aluminium granulate	−1.5	−11,346	−12,200	
Aluminium fluff	−1.3	−10,370	−12,200	
Heavy fuel oil, at regional storage/RER S	−1.1	−455	−455	
HFFR as oil replacement	0.058	9	−166	
Polyolefins as oil replacement	−0.022	−9	−477	
PVC compound for cable I	−1.3	−1500	−1500	
PVC recyclate I	−0.09	−100	−1500	
HFFR compound for cable	−2.0	−1170	−1170	
HFFR recyclate	−0.11	−65	−1170	
Compounding PVC	−0.23	−6.2	−6.2	
Compounding polyolefin	−0.23	−15	−15	
Avoided transports	Per tonkm transport (€	gram CO$_2$eq)		Multiply values below for € and gram CO$_2$eq per tonkm transport with actual transport distance to get gram CO$_2$eq/tonne and €/tonne for the actual transport. Convert to gram CO$_2$eq /kg and €/kg by dividing with 1000.
Lorry, Trailer 26 t Euro3, NTM	−0.10	−51	−51	
Transport, lorry > 16 t, fleet average/RER S	−0.10	−134	−134	
	Per kg material (€	Market gram CO$_2$eq	1 to 1)	
Total avoided processes	−0.09	−100	−1500	
Total of needed and avoided processes	0.370	37	−1364	

Continued

Bottom line comparison of new recycling route compared to current way of recycling

Process(es)/ materials	€	Per kg		Conclusions
		Market	1 to 1	
		gram CO_2eq		
New recycling route	−1.546	−1512	−1512	For 100 tonnes the savings are 100 × 1.92 × 1000 = 192 000 euro and in between 154.8 - 14.8 tonnes CO_2eq.
Current way of recycling	0.370	37	−1364	
Difference	−1.92	−1548	−148	

Notes to the calculation of the LCA comparison between Hot milling and External recycling in **Figure 6** and **Table 1**. Boxes coloured yellow in table 1 apply! Hot milling, left columns: Hot milling costs 0.004 €/kg plastic and entails 4.3 gram CO_2eq/kg plastic emissions. Production of virgin PVC is avoided, thus 1.3 €/kg plastic and 1500 gram CO_2eq/kg plastic is avoided. Compounding of PVC is avoided, thus 0.23 €/kg plastic and 6.2 gram CO_2eq/kg plastic is avoided. Transport, 200 km, of virgin PVC is avoided, thus 0.1 × 200/1000 = 0.02 €/kg plastic and 51×200/1000=10 gram CO_2eq/kg plastic is avoided. Since there is no down–cycling of the material (no loss of quality), the market perspective and the 1 to 1 perspective yield the same results! For Hot milling, the total of needed processes minus avoided processes is a gain of 1.546 €/kg plastic and avoidance of 1512 gram CO_2eq/kg plastic. External recycling, right columns: Granulation of hardened lumps costs 0.14 €/kg plastic and entails 26 gram CO_2eq/kg plastic emissions. Compounding the granulated PVC costs 0.23 €/kg plastic and entails 6.2 gram CO_2eq/kg plastic emissions. Transport, 200 km, of PVC lumps to granulation costs 0.1 × 200/1000 = 0.02 €/kg plastic and entails 51 × 200/1000 = 10 gram CO_2eq/kg plastic emissions. Transport, 700 km, of compunded PVC recyclate to user in Europe costs 0.1 × 700/1000 = 0.07 €/kg plastic and entails 134 × 700/1000 = 94 gram CO_2eq/kg plastic emissions. Reuse of PVC recyclate bring an income of 0.09 €/kg plastic and avoids 88/1320 × 1500 = 100 gram CO_2eq/kg plastic emissions in a market perspective and 1500 CO_2eq/kg plastic in a 1 to 1 perspective. For external recycling, the total of needed processes minus avoided processes is a loss of 0.37 €/kg plastic. The climate impact range between emissions of 37 gram CO_2eq/kg plastic and avoidance of 1364 gram CO_2eq/kg plastic. Bottom line: Employing hot milling instead of external recycling saves −1.546 − 0.370 = −1.92 €/kg plastic (minus sign means savings/income/avoidance) and avoids emissions between −1511 − 37 = −1548 gram CO_2eq/kg plastic and −1512 − (−1364) = −148 gram CO_2eq/kg plastic.

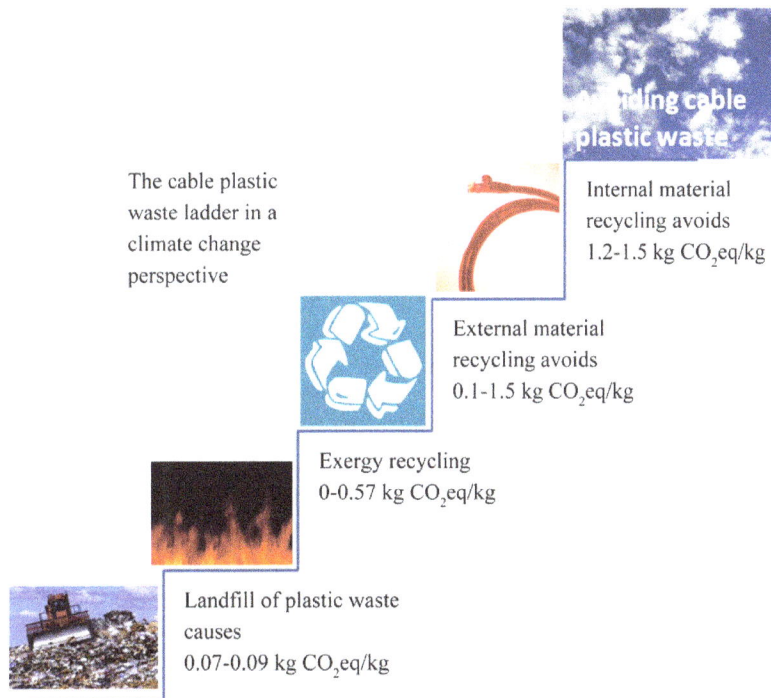

The cable plastic waste ladder in a climate change perspective

Internal material recycling avoids 1.2-1.5 kg CO_2eq/kg

External material recycling avoids 0.1-1.5 kg CO_2eq/kg

Exergy recycling 0-0.57 kg CO_2eq/kg

Landfill of plastic waste causes 0.07-0.09 kg CO_2eq/kg

Figure 5. Potential climate change gains (per kg of plastic waste) by moving up the cable plastic waste ladder.

3.3. Calculation Example

Below is given an example of how to use the data in **Table 1** to calculate the economics and the climate impact of a particular recycling case. It is always easier to understand the results if they are compared to one another. As a standard comparison, the current way of managing the cable plastic waste is used. The yellow boxes in

Table 1 are summed up in the Total boxes. A clean sheet to perform other calculations is available in the underlying report [3] that also contains other calculation examples and a step-by-step procedure on how to perform them.

Internal versus External Recycling

Start and stop PVC and HFFR (halogen free flame retardant) scrap at cable extruders can be recycled directly back into extruders via hot milling of the scrap. This might need investment in mills which is not considered in this calculation. The extra work involved could often be handled by the extruder operator, thus, normally hot milling at extruders does not entail any extra work costs. It is difficult to clean the mills. Therefore, hot milling is only relevant at extruders that run the same material all the time. For increased meaning and understanding, hot milling at extruders should be compared to an alternative. A currently used alternative is to sell the hardened scrap lumps from the extruder to an external waste handler who granulates them and pass them on for mechanical recycling in a different product. It is a good idea to make a rough drawing of the processes involved in both recycling routes, see below. Avoided processes and materials are coulored yellow in **Figure 6**. Note that upstream processes like transport and compounding of virgin PVC is avoided by hot milling the scrap at extruders. A minus sign in **Table 1** means saved or avoided € or CO_2eq. It is recommended to do the calculations both with a market perspective on the avoided burdens and with a "1 to 1" perspective.

When the table has been completed a comparison of the bottom lines for hot milling and external recycling is done. The conclusion, in the example, is that hot milling can save more than 192000 € annually and avoid between 15 - 155 tonnes of CO_2eq annually. Per kg, the figures compare well with those given for internal and external material recycling in the cable plastic waste ladder in **Figure 5**.

4. Discussion

The methodology described and applied has evolved during a four year long cooperation between industrial waste management experts, the cable industry, LCA practitioners and other stakeholders. It aims to bring to the industrial decision-maker the necessary economic and environmental facts to judge the merits of competing waste management options and thus facilitate movement up the waste ladder. Some barriers and incentives related to this aim of improving waste management of plastic cable waste in particular and other waste are discussed below.

4.1. Restricted Substances May Hinder Recycling

It should be pointed out that the use in cables of restricted phthalates as well as restricted substances such as lead, bromide and antimony may hinder the possibilities to use recycled plastic waste from used cable. The inclusion of restricted substances in cables may also hinder all forms of external recycling of plastic waste from cable production. The subject of restricted substances in cable waste will be the focus in the Wire and Cable project run by Swerea IVF, from 2014 and onwards. The quality of various waste materials and the phasing out of hazardous waste is regulated under the Waste Framework Directive [9], where End-of-Waste criteria is developed for priority waste streams, among them plastic materials. Chemical regulations like REACH (Registration, Evaluation, Authorisation and Restriction of Chemicals) [15] and product directives like RoHS (Restriction of

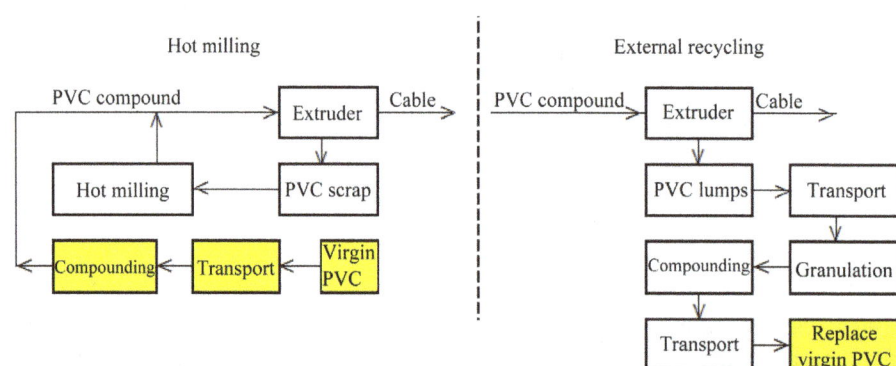

Figure 6. Hot milling of PVC scrap compared to external recycling of hardened PVC lumps.

Hazardous Substances) [16] regulates hazardous substances in products and material used in products, regardless virgin or recycled material. This potential barrier for recycling exists for all waste materials.

4.2. Identical Processes

When comparing recycling routes that contain identical processes it is not necessary to include the identical processes since they equal out each other. However, when you carry out the analysis in practice it is easier to just list all the processes involved without thinking about if they equal out or not. Once included in the analysis, the processes carry some information, so you might as well leave them in. The recommendation is therefore to include also identical processes in a comparison once you have identified them. "Identical processes" that are initially forgotten need not be taken in to the calculations.

4.3. Limitations of the System Boundary

To be able to limit the analysis to include only the waste management processes is very advantageous since it limits drastically the amount of data needed for the analysis. One reason why this is often possible for cable waste is that when producing cables, achieving the specified product quality properties is far more important than advancing on the waste ladder. Thus, in practice, it can be assumed that the manufacturing and use of the cable will in general not be affected by any of the waste management options investigated; simply because the cable properties must remain the same. So such upstream activities can most often be excluded from the LCA. Seen from another perspective, achieving the specified product properties with the use of waste materials is probably the largest barrier to (internal) material recycling.

Downstream, the recommended use of two waste perspectives (market and 1 to 1), can be seen as a LCA sensitivity analysis regarding the market for the recyclate [4]. In general, the 1 to 1 approach overestimates the CO_2eq savings, while the market approach probably in most cases underestimates the CO_2eq savings. The true climate change influence most probably is somewhere in between these two extremes. It is therefore recommended to do both calculations and present the results as a span, as is done in the example calculations.

From an economic point of view the market approach must be used in order to understand the economic implications. PVC recyclate rarely sell at the same price as virgin PVC. However, when there is no price difference between the recyclate and the virgin material, then the market approach gives the same result as the 1 to 1 approach. Put another way, the market approach is only relevant when so called down-cycling of wastes occur, *i.e.*, the waste is not reused for the same product/purpose as it was used originally or maintaining its origin quality.

4.4. Ecodesign of Cables

Within the Wire and Cable project, some attempts were made to elaborate guidelines for how to design cables in order to facilitate material recycling after their use. These attempts were however unsuccessful. One reason may be that in life cycle assessment of cables, as apart from LCA of cable recycling, the materials in the cable generally carry insignificant environmental impact compared to the environmental impact associated by the electricity losses during a cables life cycle [17]. In other words, the dominant environmental impact occurs during the use phase of a cable and therefore improvement work should focus on the use phase. Subsequently, less attention is put on the recycling stage. However, from a cable manufacturing point of view, the material costs are significant, and this should warrant some interest at least for improving the recyclability of production cable waste.

4.5. The Relevance of Climate Change

Plastsep is a technology based on sink-float separation and wet shaking table that is used to separate heavy plastics like PVC from the lighter polyolefines. Applying Plastsep to the mixed plastic waste output after granulation of used cables has proven both economically and environmentally beneficial under all conceivable circumstances [12]. It was further shown [12] that applying Plastsep not only avoided climate impact but also photochemical smog formation, eutrophication and acidification. This indicates that climate change can serve as an indicator also for these other environmental impact categories.

4.6. Prices

All prices should be seen as indicative. A variation of at least $+-$ 100% should be taken for granted. Nevertheless, having access to price information of some kind is very beneficial since it is never a potential climate impact avoidance alone that will make companies move up the waste ladder. Moving up the waste ladder will have to be economically beneficial (or enforced by law) otherwise it will not happen. With the information given companies can get a first idea which recycling routes are economically viable for them and which are not.

4.7. Applicability of Methodology for Other Waste Materials

The applicability of the applied methodology to other waste materials than plastic cable waste is discussed below for each of the five steps in the methodology.

4.7.1. Cradle-to-Gate LCA (and LCC) for Waste Material to Recyclate
Data relevant to the particular waste material is needed on:
- Recycling processes in the form of energy use, emissions, yield, price etc.
- Transports in the form of energy use, emissions and price.
- Primary material production in the form of energy use, emissions, price.
- Price of recyclates.

4.7.2. Quantifying the Climate Impact for Each Step on the Waste Ladder
To accomplish a quantified waste ladder, the particular waste material need to be well defined and the market for the waste material known, *i.e.* prices of recyclates. The quantified waste ladder is a pedagogic instrument to be used to rouse interest. The specific case should always be calculated with specific data.

4.7.3. Use of Economic and Environmental Data in Parallel
Cost data is often very sensitive. The related price information could be much easier to get access to and it is also more relevant than the cost. Fluctuation of market prices has to be considered in order to achieve long-term sustainability of waste management options.

4.7.4. Climate Impact Presented as a Span
It may be that some other environmental impact category than climate impact is more relevant to use or that several impact categories are needed. This should be checked for all waste materials.

As discussed above, to calculate and present the climate impact as a span, can be seen as an LCA sensitivity analysis regarding the market for the recyclate. The true climate change influence most probably is somewhere in between the top value and the lower value. If there is no price difference between the recyclate and the virgin material, then the top or high value of CO_2eq savings apply.

4.7.5. Do-It-Yourself Calculations
The vision of laymen doing correct life cycle calculations may be difficult to realize, but aiming there will aid in understanding and communicating the results of the life cycle calculations. This will, for example, facilitate exchange and correct understanding of LCA data between LCA practitioners. The calculations (for cable waste) can be achieved by the following steps:
1) Copy the table with all the process data and enlarge it to A3-size or larger.
2) Identify the recycling routes you want to compare. Draw simple process flow diagrams of both recycling routes, from waste to recyclate.
3) In the A3-sheet, identify and mark the (grey) processes *needed* to enable the recycling
4) In the A3-sheet, identify and mark the (green) products/materials and processes that are *avoided/replaced* by the recyclate
5) Convert data to per kg cable if necessary, see examples
6) Convert transport data according to instructions in the table
7) Examine the bottom lines of the recycling routes and calculate the difference in savings per kg
8) Calculate the difference in savings per year for your company

5. Conclusions

This paper suggests cradle-to-gate life cycle assessment as a method to analyze the pros and cons of different cable scrap recycling options at hand. Economic and environmental data about different recycling processes and other relevant processes and materials have been collected and are presented. Cable producers could use this data and the proposed method to assess the way they deal with the cable plastic waste today and compare it with available alternatives and thus illuminate the improvement potential of recycling cable plastic waste both in an environmental and in an economic sense. The methodology suggested and applied consists of:

1) Cradle-to-gate LCA for waste material to recyclate;
2) Quantifying the climate impact for each step on the waste ladder;
3) Use of economic and climate impact data in parallel;
4) Climate impact presented as a span; and
5) Do-it-yourself calculations.

In this paper, comparisons between internal and external recycling of cable plastics show that between 1.5 - 0.15 kg CO_2eq can be saved per kg of plastics when moving one step up the waste ladder from external recycling to internal recycling. In economic terms, this one step up at the top of the waste ladder can increase profits by almost 2 € per kg plastic.

The suggested method is probably applicable also for other waste materials in society in order to move towards improved use of finite resources.

References

[1]   Unger, N. and Oliver, G. (2008) Life Cycle Considerations about Optic Fibre Cable and Copper Cable Systems: A Case Study. *Journal of Cleaner Production*, **16**, 1517-1525. http://dx.doi.org/10.1016/j.jclepro.2007.08.016

[2]   JRCb (2011) Supporting Environmentally Sound Decisions for Construction and Demolition (C & D) Waste Management.

[3]   Zackrisson, M. (2013). Recycling Production Cable Waste—Environmental and Economic Aspects. Swerea IVF Report 13003, Mölndal.

[4]   JRCa (2011) Supporting Environmentally Sound Decisions for Waste Management. JRC European Commission.

[5]   ISO (2006) ISO 14044. Environmental Management—Life Cycle Assessment—Requirements and Guidelines.

[6]   Wolf, M.-A. and Rana, P. (2012) The International Reference Life Cycle Data System. http://eplca.jrc.ec.europa.eu/uploads/2014/02/JRC-Reference-Report-ILCD-Handbook-Towards-more-sustainable-production-and-consumption-for-a-resource-efficient-Europe.pdf

[7]   Björklund, A. and Finnveden, G. (2005) Recycling Revisited-Life Cycle Comparisons of Global Warming Impact and Total Energy Use of Waste Management Strategies. *Resources, Conservation and Recycling*, **44**, 309-317. http://dx.doi.org/10.1016/j.resconrec.2004.12.002

[8]   Zackrisson, M., Cristina, R., Kim, C. and Anna, J. (2008) Stepwise Environmental Product Declarations: Ten SME Case Studies. *Journal of Cleaner Production*, **16**, 1872-1886. http://dx.doi.org/10.1016/j.jclepro.2008.01.001

[9]   European Commission (2008) Directive 2008/98/EC of the European Parliament and of the Council of 19 November 2008 on Waste and Repealing Certain Directives.

[10]   Zackrisson, M. (2012) Life Cycle Assessment of Cable Recycling. Part I: Plastsep Compared to State of the Art. Swerea IVF Report, Swerea IVF AB, Mölndal.

[11]   Ecoinvent (2010) The Life Cycle Inventory Data Version 2.2. Ecoinvent Database.

[12]   IPCC (2007) Climate Change 2007 the Physical Science Basis. http://www.ipcc.ch/publications_and_data/ar4/wg1/en/ch2s2-10-2.html#table-2-14

[13]   IEC (2004) Dependability Management—Part 3-3: Application Guide—Life Cycle Costing. IEC 60300-3-3.

[14]   AMI (2012) AMI's Guide to the Cable Extrusion Industry in Europe (Edition 6). www.amiplastics.com

[15]   European Commission (2006) REGULATION (EC) No 1907/2006 of the European Parliament and of the Council of 18 December 2006 Concerning the Registration, Evaluation, Authorisation and Restriction of Chemicals (REACH).

[16]   European Commission (2011) DIRECTIVE 2011/65/EU of the European Parliament and of the Council of 8 June 2011 on the Restriction of the Use of Certain Hazardous Substances in Electrical and Electronic Equipment.

[17]   Jones, C.I. and Marcelle, C.M. (2010) Life-Cycle Assessment of 11 kV Electrical Overhead Lines and Underground Cables. *Journal of Cleaner Production*, **18**, 1464-1477. http://dx.doi.org/10.1016/j.jclepro.2010.05.008

5

Correlations of Materials Surface Properties with Biological Responses

Robert E. Baier

Industry/University Center for Biosurfaces, 110 Parker Hall, State University of New York at Buffalo, Buffalo, NY, USA
Email: baier@buffalo.edu

Abstract

More than 50 years have passed since it was first recognized that the surface properties, and predominantly the surface energies of materials controlled their interactions with all biological phases via their spontaneous acquisition of proteinaceous "conditioning films" of differing degrees of denaturation but usually of the same substances within any given system. This led to the understanding that useful engineering control of such interactions could thus be manifested through adjustments to those surface properties, giving significant control and utility to the biomaterials developer without requiring detailed discovery of the biological specifications of the components involved. Thus, effective selection of adhesive versus abhesive (non-stick, non-retention) outcomes for such useful appliances as dental implants versus substitute blood vessels, or water-resistant bonded structures versus clean, nontoxic ship bottoms is now facilitated with little biological background required. A historical overview is presented, followed by a brief survey of the forces involved and most useful analyses applied. Utility for blood-contacting materials is described in contrast to utility for bone- and tissue-contacting materials, demonstrating practical uses in controlling cell-surface interactions and preventing biofouling. New research directions being explored are noted, urging applications of this prior knowledge to replace the use of toxicants.

Keywords

Surface Property, Biological Response, Adhesive, Blood-Contacting Materials

1. Introduction

Biological responses to non-physiological surfaces are, usually, mediated by spontaneous deposits of organic films and particulate matter from all biological fluids. The earliest events follow a common pattern. Such interactions result in differing degrees of bioadhesion and can be effectively correlated usually, even controlled by the surface properties, (especially surface energies) of the substrata involved [1].

There are convincing proofs of the bioengineering utility of surface property modification to minimize biological fouling [2]. Such proofs are found across the broad range of biomaterials development, from the successful implantation of substitute human blood vessels and total artificial hearts, to the fabrication of entirely nontoxic fouling-release coatings that can replace poison paints on ship bottoms. Surfaces contacting blood have received the most careful scrutiny, and first revealed the general features of biological response later found in other circumstances [3], like milk fouling of pasteurizers [4], oceanic fouling of heat exchangers [5], cell culture propagation [6] and dental plaque formation [7]. Control of the surface properties of biomaterials is a unifying approach to control of interactions between all that which is alive and all that which is not.

2. Historical Overview

It has been a successful strategy to tailor the surface properties of biomedical organ substitutes to exhibit desired degrees of biological response. As an example, general approval of the US Food and Drug Administration was granted as long ago as January, 1979 for human peripheral vascular reconstruction using tanned umbilical cord veins meeting specific surface properties and standards [8]. The natural pavements of endothelial cells of the original living blood vessels were lost during their processing from fresh or frozen umbilical cords [9]. What was preserved was mainly the surface quality of the sub-endothelial lining, a physiologically tolerable layer, along with a basement membrane and elastic internal lamina. The primary quality control criterion utilized was the "critical surface tension" parameter [10], adjusting processing conditions to maintain this value in the mid-20 dynes/cm. The international experience included thousands of successful human implantations, saving numerous limbs from hitherto inevitable amputation [11].

In the occasional circumstances when properly surface-controlled materials do become coated with clotted blood, usually as a result of constriction of tubes at their connections or stagnation in the outflow tract, post-implantation analyses show the clots to reside innocently in the lumens of the vessels, exhibiting little to no adhesion to "biocompatible" walls [12]. Similar findings, and understanding, regarding materials surface properties in other settings will allow prediction and control of biological responses to materials in food processing units [13], in the sea [14], in tissue culture [15], in the womb [16], in the eye [17], and in the oral cavity [18].

3. Background Concepts

3.1. Control of the Forces

All inquiries about adhesive strengths and properties of biological deposits on various materials must include close control of the exposure geometries, flow fields and shear forces. The boundary hydrodynamics during both the deposition and detachment processes are crucial. It is not acceptable in biomaterials research, for example, to terminate a biological experiment (e.g. for the purpose of noting the accumulating microfouling films) by "rinsing" a sample in an unspecified or non-quantifiable manner [19]. Neither is it acceptable to equate the forces needed to detach or distract a biological deposit from a solid substratum with the thermodynamic strength of adhesion of the attached elements [20]. Cell and film separations from solids are not simply the reverse of the first attachment events. Joint failure upon separation is usually of the mixed-mode (cohesive and adhesive) type. Special devices for flow or shear-rate control should be employed during both the first exposure of test surfaces to biological media and in the final rinsing steps required to remove loosely held debris, co-adsorbed interfering salts, or entrained substances [21].

One can also incorporate transducers and electrodes into the circuits to record dynamic electrical events accompanying biological film deposition and adhesion of fouling elements. The technique of streaming potential measurement, for instance, allows direct calculation of the important boundary phase parameter of the plane of shear called zeta potential [22], cited in many theoretical models of biological responses to material samples.

3.2. Selection of Surface Analyses

Many techniques for rapid (and, in most cases, nondestructive) instrumental analysis of both synthetic material and biological surfaces are now available and in routine use. The sensitive procedure of multiple attenuated internal reflection spectroscopy [23] produces infrared and/or UV/visible absorption spectra of the exterior molecular layers of matter defining biological interfaces with connecting solids. The acquired spectra reveal compositions, rates of accumulation, and modifications of such important boundary layers.

There are several sensitive, simultaneous or sequential analytical techniques immediately applicable to probing the structure and composition of thin biological deposits without requiring their manipulation or removal from the substrata of interest [24]. These techniques are: a) as described above, acquisition of infrared absorption spectra by the internal reflection technique; b) determination of film thicknesses and refractive indexes by the method of reflected polarized light (called ellipsometry); c) measurement of surface electrical states and contact potentials; d) inference of wettabilities, strengths of adhesion, operational surface free energies and critical surface tensions by comprehensive contact angle analysis (not just measurements of water contact angles!); e) morphological inspection of such layers by direct scanning electron microscopy without requiring over-coating the samples with obscuring metallic conducting films [25]; f) simultaneous analysis by energy-dispersive X-ray techniques of the presence and relative abundances and locations of elements of atomic number higher than sodium [26], and g) identification of actual crystalline forms present, by glancing angle X-ray diffraction [27]. These methods have contributed significant information to the development of dental restorative composites, surgical adhesives, prosthetic implants, and extracorporeal circuits, as well as provided basic data on the initial events of blood clotting and marine fouling. The early addition of ESCA (Electron Spectroscopy for Chemical Analysis), also known as XPS (X-ray Photoelectron Spectroscopy) to the battery of methods was widely accepted [28].

4. Basic Precepts

4.1. Blood-Contacting Materials

The preferred materials of construction for pulsatile blood-handling devices are hardy elastomers of the polyether types of polyurethane, preferably fabricated to exhibit surface enrichment of "biocompatible" methyl groups [29]. These groups can be provided by direct alloying or admixture of polydimethylsiloxane (PDMS) in the original polymer blend [30]. All blood-contacting surfaces are preferably fabricated to be smooth and free of entrapped air prior to their first exposure to blood. Even with stringent preparation and precaution, the best synthetic materials for blood contact applications will still acquire spontaneously deposited plasma protein layers (usually dominated by fibrinogen)and support modest, though temporary, cellular adhesion [31].

Once this "conditioning" film is in place on mid-Critical Surface Tension (20 - 30 dynes/cm) materials with "passive" surface properties, arriving blood platelets that do attach typically maintain their natural discoid shapes [32]. They retain their granules and/or clotting factors. When in contact with higher or lower surface energy "active" materials, the platelets become distressed, undergoing a "viscous metamorphosis" that triggers both thrombosis and activation of the coagulation cycle. In all cases, segmented polymorphonuclear leucocytes, also called neutrophils, then stream toward the platelet- or platelet-debris-strewn foreign surfaces. On "active" incompatible surfaces, these perform as phagocytic units while white cells attach without any beneficial effect. Rather, the white cells add to the growing adherent cellular mass and accelerate the thrombotic episodes.

On "passive" compatible materials, selected by virtue of their less-retentive surface properties [33], the white cells function as a "clean-up squad", using fibrinolytic enzymes to break up the original fibrinogen films. Interfacial shear forces are then sufficient to detach the original bound mass, seldom to form again. Long-term, dynamic equilibrium with fresh flowing blood, then, is obtained through a more "native" plasma protein film acting as a "passivation" layer.

4.2. Bone- and Tissue-Contacting Materials

Basically, the operating rules for materials choices for bone- or tissue-contact devices are just the opposite of those for blood-contact devices, with a few interesting exceptions [34].

Biological mineral attachment to various materials has been demonstrated by the surface analytical techniques described earlier-including X-ray diffraction-to be dominated by calcium phosphate microcrystals of hydroxyapatite form. This basic mineral phase of bones and teeth has been grown, inadvertently, on flexible elastomeric and tissue materials used in artificial hearts and substitute heart valves [35]. Although this presents difficulties in tissue-valve and artificial heart development programs, where unwanted mineralization must be limited or overcome, one might seize upon this observation and turn adversity into virtue for use of similar materials in orthopedic or dental applications [36].

In the ongoing pursuit of improved surface conditions for dental implants of all types, including complicated structures that must interface with bone on one aspect, and overlying tissue on another, and protrude through the tissue interface into the oral cavity with an infection-free stable seal, coatings of the same elastomeric materials used in artificial hearts might be applied. Additional roughening and surface-activation would enhance cellular in-growth processes and stimulate osteogenesis [37].

The more conventional approach, using intrinsically high surface energy or bioactive materials, has meanwhile served quite well. This involved either the initial choice of self-cleansing (through surface dissolution or erosion in the host site) glass compositions [38], or calcium phosphate minerals, or the scrupulous prior cleansing (from metallic implants) of all organic debris and polishing agents. Their sterilization in the same process, and resultant activation for adhesion, is readily accomplished by exposure to radio-frequency-initiated glow discharges [39]. Apparatus to achieve such surface activation and cleansing is becoming generally available, and protocols require only a few minutes to execute [40]. Useful improvements in dental implant immobilization, encouragement of direct tissue binding, and elimination of fibrous encapsulation do result. Similar apparatus, scaled to handle larger implants such as artificial hips, can provide excellent surface cleaning and activation of these prostheses to enhance their cementation into prepared bone sites.

5. Practical Use

5.1. Controlling Cell-Surface Interactions

One of the interesting findings of microbiology has been that many hetero typically bound bacteria, isolated from oceanic films as well as from human dental plaque, often have filamentous tufts at one end specialized for preferential adhesion [41]. Internal reflection infrared spectra identifies the polar tuft material as being of mainly glycoprotein composition, at variance with the general hypothesis that lipoteichoic acid (a poly glycerol phosphate polymer) dominates the adhesive sites of those microorganisms.

Even such organisms, arguably specialized by evolution for colonization of any "foreign surface", can be denied successful adhesion (that is, there is no retention after gentle removal forces are applied). Proof of this requires only that one employ substrata exhibiting surface properties in the bioabhesive, or adhesion-resistant "biocompatible", zone identified long ago in studies dealing with mammalian cells [42]. The most utilitarian" surface energy conversion" coating of the sort required is that formed by covalent binding of methyl-silanes (often to create poly-dim ethylsiloxane layers) onto glass, silica, other mineral or metallic surfaces. Standard laboratory "siliconization" of glassware or metals can be excellent in this regard, but only when exposing new exterior chemical arrays of closely packed methyl groups to the colonizing bio-environment. Silicones not dominated by methyl group side chains do not resist bioadhesion [43]! Closely packed methyl group "lawns" exhibit a composite critical surface tension or apparent surface free energy of about 22 dynes per centimeter, in the minimally adhesive zone of the surface energy scale. The prevention of adhesion is mediated in these cases, nevertheless, by native structure persistence of deposited glycol proteinaceous films as discussed earlier [44].

Assuring specific biological responses to materials on the basis of their relative surface properties eventually will allow us to overcome still-troublesome biomaterials-centered infections. Early examples of the latter problem were those of pelvic inflammatory disease (PID) associated with Intrauterine Contraceptive Devices (IUDs) [45] having multi-filamentous nylon or frayed, micro-fibrillated polypropylene tail-strings. These structures provided sites for attachment and migration of infective microorganisms from the vaginal canal to the normally infection-free uterus, where they colonized as "biofilms" on other intrinsically bioadhesive materials used for the IUD bodies. Multimillion dollar legal judgments were levied against the major pharmaceutical firms that introduced these inappropriate biomaterials to the human reproductive system.

Unfortunately, the infection-related biomaterials problems diminished attention to beneficial features of earlier, more benign intrauterine contraceptive devices. The spontaneous coating of such devices by adsorbed glycoproteinaceous layers from the cervical mucus fluid supported speculation that such coatings may prevent the "capacitation" of sperm transiting the IUD locale. Sperm that do not experience such required changes in their initial surface (adhesive) properties do not successfully engage in another form of heterotypic adhesion, that of sperm-to-ovum, upon which fertilization of mammalian oocytes is premised. Again, the practical application of biomaterials with better-selected surface properties can make IUD regulation of fertility a useful, safe and effective option for world-wide population control.

It is abundantly clear that "strength" of biological adhesion (more accurately, "resistance to detachment" or "strength of retention") is associated with defined ranges of surface energies, or critical surface tensions, for solid substrata in extremely diverse biological circumstances [46]. Other examples abound: demonstration that the minimum binding strength (resistance to detachment) for liver cells in tissue culture is sharply in the zone between 20 and 30 dynes/cm [47]; statistically sound observations, on solid substrata ranging in surface free energy from about 10 to over 50 dynes/cm (ergs per square centimeter), that the spread areas and associated degrees of distortion of settled human cells [48] and blood platelets, freshly obtained from plasmapheres is, are minimized in the critical surface tension zone between 20 and 30dynes/cm; observation of minimal forces required to pull musselbyssus discs from solid materials having these same critical surface tensions; and more. Even the induction of hemolytic damage to circulating red blood cells is minimized when the critical surface tensions of the walls of the shearing device are adjusted to the mid-20's dynes/cm range [49]. This last observation strongly implies that biological macromolecules that arrive at, and then are displaced from, solid surfaces of varying surface properties carry away with them (back into solution or suspension) varying "messages" based on their differing degrees of surface-contact-induced "denaturation" from their original solution states [50].

5.2. Preventing Biofouling

Biofouling refers to those unwanted deposits that frequently occur on contact lenses, dentures, periscope windows, and ship bottoms, to name just a few cases. A continuing difficulty in all food processing operations is the deposition of organic matter and mineral layers on the surfaces of heat exchangers and membranes. Dairy products are near the top of the list of problem-makers in this regard [51]. After many years of study, the predominant compositions of the first most strongly bound layers are generally not known. Differences of opinion still exist, for example, on whether the earliest deposits from raw milk are mainly proteins or minerals such as the calcium-phosphate-rich "milk stone". This is a fertile territory for biomaterials specialists to enter and enrich, using the same concepts of materials-related control of bioadhesion already successful in medical-device development.

For example, special flow cells may be designed to allow control of the surface shear rates and stresses, investigating the earliest fouling events for heated surfaces in contact with homogenized whole milk [52]. Preliminary results already available endorse the concept that milk protein adsorption is the first, essentially irreversible, event in surface fouling by dairy products, with mineral deposition occurring much later.

Useful flow cell devices have already been constructed [53]. Some apply "voltage clamping" circuitry so that the primary events of biofilm deposition and secondary events of microbial attachment, polymer exudation, and mineralization can be observed on substrata surfaces controlled electronically at any desired surface potential/charge condition. Such flow cells helped in documentation of the early events of microbiological fouling of model heat exchange surfaces in the warm subtropical waters of the Gulf of Mexico. Acquisition of that fundamental knowledge on biological fouling of heat exchange devices in warm seawater was of critical importance to international efforts to extract stored solar energy from the tropical oceans, using the principles of "ocean thermal energy conversion". The Gulf of Mexico findings were compelling in their demonstration of slow deposition of humic-like conditioning films on the heat exchange surfaces. These adsorbed films were colonized in less than 3 days by pioneer bacteria of both flagellated and un-flagellated types. Within 6 days of exposure, growing deposits of polymer exudates were revealed around the pioneer microorganisms by using a special adaptation of scanning electron microscopy, eliminating the need for obscuring electro-conductive layers of sputtered metals. At the same exposure time, deposition of calcareous smatter, usually associated with diatoms and other algal forms, was noted. Patches of extreme microbiological diversity, with ecological succession already in progress, were observed on many test surfaces in the period between 3 and 6 days. It is certainly clear from these, and considerable supporting data, that biomaterials uses within that 70% of the earth's surface that is" wet" will require attention to (and control of) such biofouling phenomena [54].

Already gathered experimental data on critical inter facial layers of biological films in marine environments suggest nontoxic mechanisms by which their adhesion-or lack thereof to practical materials—can be controlled [55]. Recognizing serious drag-enhancing penalties associated with bacterial/slime fouling layers on even toxic marine paint surfaces, intriguing early results suggest that biofilms, or their synthetic analogues, maybe created to eliminate the drag effects of more-usual fouling layers and perhaps even to provide significant drag reduction

[56]. Proof of-principle testing, using natural low-drag skin of living porpoises and killer whales [57], has shown the best surfaces have low-critical-surface-tension, protein-dominated characters inconsonance with similar findings for fouling-resistant layers of human oral mucosa and blood vessel endothelium [58]. Since synthetic materials for numerous biomedical devices have been successful without the need for toxicants to prevent fouling by even concentrated biological fluids, it should not be surprising that field tests endorsed similar material surface modifications for service in seawater. Results indicated that adjustment of a material's critical surface tension to the zone between 20 and 30 dynes/cm correlated with the most facile detachment of fouling debris [59]. The coincidence of this finding with results from tests of biomedical devices is encouraging in suggesting a strong conservatism among natural mechanisms promoting or preventing biological adhesion.

It even has been shown that algae produce base attachments of significantly greater area in order to resist separation from substrata having low critical surface tensions [60]. Diatom colonization success has also been demonstrated to be dependent on the initial surface energy of test substrata [61]. Our major conclusions must be that understanding, prediction, and control of surface properties of materials in all biological settings will be crucial to achieving improved performance in practical cases. Interestingly, here is a situation where large-scale (and large volume) commercial benefits may arise from the preceding "small science" (and small volume) of biomaterials development for artificial internal organs.

6. New Research Directions

Many authors still cite the needs for further study of the events of "fouling" in medical equipment, so it is only necessary to add some new discussion of the areas of biological adhesion noted here [62]. It is especially important to note the striking similarities in the fouling processes that occur in the bloodstream and the oral cavity, as similar to events in cooling water structures and processing equipment for various dairy products.

A need certainly still exists for improved understanding, on a basic level, of the deposition of macromolecules and of living cells at solid and semisolid (hydrogel, tissue) surfaces. Lack of the required detailed knowledge of these fundamental processes limits practical control of biological adhesion on a more general basis. Specific major areas for continuing study include the effects of different material surface properties (texture, charge, chemistry, energy) on binding of deposited macromolecules, seeking evidence for selective retention of particular components or sub fractions, determining the actual degree of coverage of the "cleaned" original surfaces, assessing the orientation of the adsorbed molecular entities, and defining the longer term modifications to the original surfaces' properties and to the attached molecules. Further, it is important to learn more about the cascade of specific cells arriving at and attaching to indwelling engineering materials, particularly noting cellular exudation of bio reactants or enzyme catalysts concentrated at the interfaces [63]. There is a known tendency for attached cells to produce polymeric exudates that both permanently bind them to the "pellicles" first acquired and engulf the growing, metabolizing units in "coats of slime". These events must be better understood and controlled if biomaterials-centered infections are to be successfully combatted [64].

It is also extremely important to extend modern surface analytical methods to the problems of identifying, more completely, the nature of cellular exopolymers produced, addressing as well their mode of production and changes with time and conditions. This is a crucial topic in many emerging sub-disciplines of what is generally called "biotechnology". The reactions between "exported" cellular products and the adsorbed, usually glycolproteinaceous, films that provide early binding of the cells to the starting surfaces must be ascertained. Improved measures of the actual strengths of adhesion (retention) between the cellular polymers and the films coating solid surfaces of differing physical chemical states must be developed. A practical focus of these efforts could be toward the identification of any "weak links" in the chain of cellular colonization, growth, and binding processes, that may be subject to direct interruption or strengthening [65].

Other biological responses to materials' surface properties are of equally urgent concern, specifically as they influence the events occurring at the initial attachment interfaces. It is recognized that cellular migration as well as other transport processes in tissue or fluid phases may limit the rate of cellular attachment to different engineering surfaces. The compositions, adhesive and cohesive strengths, and densities of the attached cell layers, as well as the geometries and hydrodynamic features at specific sites, are clearly important. These factors bear heavily on issues of human health and disease through their influence on the re-suspension and removal of biological deposits from the interfacial zones. Based upon the criteria reviewed here, one important lesson for fouling limitation in natural fluids may be that complete prevention of film formation and early cellular adhesion is not necessary: appropriate adjustments of material surface properties can control-indeed enhance or limit-the rate and reliability of re-entrainment of original deposits into the adjacent aqueous phases [66]. It is not yet known whether similar adjustments in materials' surface properties can modulate the biological responses of still-attached organisms to systemic (volume) agents for therapy or antisepsis.

Another identified research priority is for improvement and introduction of uniformity in testing methods regarding biological film deposition and cellular attachment, so that more direct comparisons of different materials in different circumstances can be made [67]. What seems to be required are small standardized units that can be adjusted to experience known flow rates, shear forces, nutrient conditions and so on, the purpose being to provide bio response indices with comparable meanings under different conditions at different times and at different sites. Availability and regular use of inexpensive, standardized test units would provide a continually expanding data base concerning the effects of material surface properties and treatments on rates of deposit formation, for example, under a variety of physical, chemical and biological conditions. More effective means of cleaning and sterilizing biomaterials surfaces are also required since it is obvious that current techniques, with the possible exception of still-emerging glow-discharge-plasma processes, are not likely to completely remove cellular and adsorbed film layers closest to the material surfaces [68]. When incomplete cleanliness is accepted, despite sterility, subsequent buildup of secondary fouling deposits occurs at rates much more rapid than that observed on truly clean substrata. Obviously, immunologic, antigenic, pyrogenic responses can still be triggered by remnants of even sterile biological debris. Thus, research directed at more complete removal of adherent deposits and their anchoring organic films, by both mechanical and chemical cleaning techniques, concomitant with or as a precursor to sterilization, is essential [69].

7. Concluding Remarks

Application of ambient environment as well as high-vacuum surface analytical techniques to biomaterials reveals fundamental similarities in the primary events of bioadhesion to them that are well correlated with the initial materials' surface properties. Initial fouling film formation in milk processing equipment exhibits a similar pattern to process in the oralcavity and in subtropical ocean water heat exchangers, including subsequent microorganism attachments, polymer exudation, and mineralization in most such circumstances. Implications of the identification of preferred surface energy ranges favoring or inhibiting permanent biological adhesion are clear in results from medical device trials including the artificial heart, dental implants, artificial hips, and substitute blood vessels. Calcification of blood-contacting surfaces of flexing elastomeric heart-assist-sacs indicates directions for similar methods to promote bone formation on dental or orthopedic fixtures [70]. Research priorities in the field of biological responses to materials surfaces should include additional attention to the effects of different surfaces on macromolecular retention, to the properties of various cells colonizing immersed surfaces, to the nature of exopolymers from cells attached to surfaces, to reactions between cellular products and surface "conditioning" films, to strengths of adhesion between cellular polymers and pre-adsorbed films, to transport processes in the bulk tissue or fluid phases, to the geometries and hydrodynamics of specific systems, and to selection of surface properties that will enhance the utility of chemical and mechanical techniques for removal of fouling films, while obtaining sterilization.

References

[1]   Baier, R.E. (2006) Surface Behaviour of Biomaterials: The *Theta Surface* for Biocompatibility. *Journal of Materials Science: Materials in Medicine*, **17**, 1057-1062. http://dx.doi.org/10.1007/s10856-006-0444-8

[2]   Baier, R.E. (1982) Conditioning Surfaces to Suit the Biomedical Environment: Recent Progress. *Journal of Biomechanical Engineering*, **104**, 257-271. http://dx.doi.org/10.1115/1.3138358

[3]   Baier, R.E. and Dutton, R.C. (1969) Initial Events in Interactions of Blood with Foreign Surfaces. *Journal of Biomedical Materials Research*, **3**, 191-206. http://dx.doi.org/10.1002/jbm.820030115

[4]   Baier, R.E. (1981) Modification of Surfaces to Reduce Fouling and/or Improve Cleaning. *Proceedings of Fundmentals and Applications of Surface Phenomena Associated with Fouling and Cleaning in Food Processing*, Tylosand, 6-9 April 1981, 1-22.

[5]   Baier, R.E. (1981) Early Events of Micro-Biofouling of All Heat Transfer Equipment. In: Somerscales, E.F.C. and Knudsen, J.G., Eds., *Fouling of Heat Transfer Equipment*, Hemisphere Publishing Corp, Washington, DC, 293-304.

[6]   Baier, R.E. (1985) Cell Seeding: Biomaterial Surface Preparation. *ASAIO Journal*, **8**, 104-108.

[7] Baier, R.E. (1973) Occurrence, Nature, and Extent of Cohesive and Adhesive Forces in Dental Integuments. In: Lasslo A. and Quintana, R.P., Eds., *Surface Chemistry and Dental Integuments*, Charles C. Thomas Publisher, Springfield, 337-391.

[8] Baier, R.E., Akers, C.K., Perlmutter, S., Dardik, H., Dardik, I. and Wodka, M. (1976) Processed Human Umbilical Cord Veins for Vascular Reconstructive Surgery. *Transactions—American Society for Artificial Internal Organs*, **22**, 514-524.

[9] Baier, R.E. (1978) Physical Chemistry of the Vascular Interface: Composition, Texture, and Adhesive Quality. In: Sawyer, P.N. and Kaplitt, M.J., Eds., *Vascular Grafts*, Appleton-Century-Crofts, New York, 76-107.

[10] Baier, R.E. and Loeb, G.I. (1971) Multiple Parameters Characterizing Interfacial Films of a Protein Analogue, Poly-methylglutamate. In: Craver, C.D., Ed., *Polymer Characterization: Interdisciplinary Approaches*, Plenum Press, New York, 79-96.

[11] Dardik, H., Baier, R.E., Meenaghan, M., Natiella, J., Weinberg, S., Turner, R., Sussman, B., Kahn, M., Ibrahim, I. and Dardik, I.I. (1982) Morphologic and Biophysical Assessment of Long Term Human Umbilical Cord Vein Implants Used as Vascular Conduits. *Surgery, Gynecology & Obstetrics*, **154**, 17-26.

[12] Baier, R.E. and Abbott, W.M. (1978) Comparative Biophysical Properties of the Flow Surfaces of Contemporary Vascular Grafts. In: Dardik, H., Ed., *Grafts Materials in Vascular Surgery*, Symposia Specialists, Inc., Miami, 70-103.

[13] Baier, R.E. and Meyer, A.E. (1985) Surface Chemical Approaches to Decontamination and Disinfection. In: Lund, D., Plett, E. and Sandu, C., Eds., *Fouling & Cleaning in Food Processing*, University of Wisconsin-Madison, Madison, 336-339.

[14] Baier, R.E. (1984) Initial Events in Microbial Film Formation. In: Costlow, J.D. and Tipper, R.C., Eds., *Marine Biodeterioration: An Interdisciplinary Study*, Naval Institute Press, Annapolis, 57-62. http://dx.doi.org/10.1007/978-1-4615-9720-9_8

[15] Baier, R.E. and Weiss, L. (1975) Demonstration of the Involvement of Adsorbed Proteins in Cell Adhesion and Cell Growth on Solid Surfaces. In: *Applied Chemistry at Protein Interfaces*, Advances in Chemistry Series, Vol. 145, American Chemical Society, Washington DC, 300-307.

[16] Baier, R.E. and Lippes, J. (1975) Glycoprotein Adsorption in Intrauterine Foreign Bodies. In: *Applied Chemistry at Protein Interfaces*, Advances in Chemistry Series, Vol. 145, American Chemical Society, Washington DC, 308-318.

[17] Baier, R.E. and Thomas, E.B. (1996) The Ocean: The Eye of the Earth. Contact Lens Spectrum, 37-44.

[18] Baier, R.E., Meyer, A.E., Natiella, J.R. and Carter, J.M. (1984) Surface Properties Determine Bioadhesive Outcomes: Methods and Results. *Journal of Biomedical Materials Research*, **18**, 337-355. http://dx.doi.org/10.1002/jbm.820180404

[19] DePalma, V.A. and Baier, R.E. (1978) Microfouling of Metallic and Coated Metallic Flow Surfaces in Model Heat Exchange Cells. *Proceedings of the Ocean Thermal Energy Conversion (OTEC) Biofouling and Corrosion Symposium*, U.S. Department of Energy, PNL-SA-7115, Washington DC, 89-106.

[20] Baier, R.E. (1982) Comments on Cell Adhesion to Biomaterial Surfaces: Conflicts and Concerns. *Journal of Biomedical Materials Research*, **16**, 173-175. http://dx.doi.org/10.1002/jbm.820160210

[21] Baier, R.E. and DePalma, V.A. (1979) Flow Cell and Method for Continuously Monitoring Deposits on Flow Surfaces. 8 Claims. U.S. Patent No. 4, 175, 233.

[22] Working Group on Physicochemical Characterization of Biomaterials, National Heart, Lung, and Blood Institute, National Institutes of Health, Leading to Publication of "Guidelines for Physico-Chemical Characterization of Biomaterials", NIH Publication No. 80-2186, September 1980.

[23] Baier, R.E. and Zisman, W.A. (1970) Wettability and Multiple Attenuated Internal Reflection Infrared Spectroscopy of Solvent-Cast Thin Films of Polyamides. *Macromolecules*, **3**, 462-468. http://dx.doi.org/10.1021/ma60016a017

[24] Baier, R.E., Shafrin, E.G. and Zisman, W.A. (1968) Adhesion: Mechanisms that Assist or Impede It. *Science*, **162**, 1360-1368. http://dx.doi.org/10.1126/science.162.3860.1360

[25] Baier, R.E., Meyer, A.E., DePalma, V.A., King, R.W. and Fornalik, M.S. (1983) Surface Microfouling during the Induction Period. *Journal of Heat Transfer*, **105**, 618-624. http://dx.doi.org/10.1115/1.3245630

[26] Baier, R.E., Forsberg, R.L., Meyer, A.E. and Lundquist, D.C. (2014) Ballast Tank Biofilms Resist Water Exchange but Distribute Dominant Species. *Management of Biological Invasions*, **5**, 241-244. (Special ICAIS Issue) http://dx.doi.org/10.3391/mbi.2014.5.3.07

[27] Baier, R.E., Mack, E.J., Rogers, C.W., Pilie, R.J. and DePalma, V.A. (1981) Source Assessment of Atmospheric Aerosols: Spectroscopic Data from a Rapid Field Technique. *Optical Engineering*, **20**, 866-872. http://dx.doi.org/10.1117/12.7972828

[28] Vargo, T.G., Hook, D.J., Gardella, J.A., Eberhardt, M.A., Meyer, A.E. and Baier, R.E. (1991) A Multitechnique Sur-

face Analytical Study of a Segmented Block Copolymer Poly (Ether-Urethane) Modified through an H_2O Radio Frequency Glow Discharge. *Journal of Polymer Science Part A: Polymer Chemistry*, **29**, 535-545. http://dx.doi.org/10.1002/pola.1991.080290410

[29] Boretos, J.W., Pierce, W.S., Baier, R.E., Leroy, A.F. and Donachy, H.J. (1975) Surface and Bulk Characteristics of a Polyether Urethane for Artificial Hearts. *Journal of Biomedical Materials Research*, **9**, 237-340. http://dx.doi.org/10.1002/jbm.820090308

[30] Pierce, W.S., Donachy, J.H., Rosenberg, G. and Baier, R.E. (1980) Calcification inside Artificial Hearts: Inhibition by Warfarin-Sodium. *Science*, **208**, 601-603. http://dx.doi.org/10.1126/science.7367883

[31] Baier, R.E. and Kurusz, M. (2012) Understanding Blood/Material Interactions: Contributions from the Columbia University Biomaterials Seminar. *ASAIO Journal*, **58**, 450-454. http://dx.doi.org/10.1097/MAT.0b013e3182631e3e

[32] Baier, R.E. (1987) Selected Methods of Investigation for Blood-Contact Surfaces. In: Leonard, E.F., Turitto, V.T. and Vroman, L., Eds., *Blood in Contact with Natural and Artificial Surfaces, Annals of the New York Academy of Sciences*, **516**, 68-77. http://dx.doi.org/10.1111/j.1749-6632.1987.tb33031.x

[33] Baier, R.E., DePalma, V.A., Goupil, D.W. and Cohen, E. (1985) Human Platelet Spreading on Substrata of Known Surface Chemistry. *Journal of Biomedical Materials Research*, **19**, 1157-1167. http://dx.doi.org/10.1002/jbm.820190922

[34] Baier, R.E., Meyer, A.E. and Natiella, J.R. (1992) Implant Surface Physics and Chemistry: Improvements and Impediments to Bioadhesion. In: Laney, W.R. and Tolman, D.E., Eds., *Tissue Integration in Oral, Orthopedic, and Maxillofacial Reconstruction*, Quintessence Publishing Co., Inc., Chicago, 240-249.

[35] Banas, M.D. and Baier, R.E. (2000) Accelerated Mineralization of Prosthetic Heart Valves. *Molecular Crystals and Liquid Crystals Science and Technology*, **354**, 249-267. http://dx.doi.org/10.1080/10587250008023619

[36] Sendax, V.I. and Baier, R.E. (1992) Improved Integration Potential for Calcium-Phosphate-Coated Implants after Glow Discharge and Water Storage. *Dental Clinics of North America*, **36**, 221-224.

[37] Baier, R.E. (1981) Catheter for Long-Term Emplacement. 8 Claims. U.S. Patent No. 4, 266, 999.

[38] Baier, R.E. (2002) A Challenging Anomaly—Glass that Does Not Clot Blood! *The Glass Researcher*, **12**, 23-24.

[39] Baier, R.E. and Meyer, A.E. (1988) Implant Surface Preparation. *International Journal of Oral & Maxillofacial Implants*, **3**, 9-20.

[40] Baier, R.E., Carter, J.M., Sorenson, S.E., Meyer, A.E., McGown, B.D. and Kasprzak, S.A. (1992) Radiofrequency Gas Plasma (Glow Discharge) Disinfection of Dental Operative Instruments, Including Handpieces. *Journal of Oral Implantology*, **18**, 236-242.

[41] Glantz, P.O., Baier, R.E. and Christersson, C.E. (1996) Biochemical and Physiological Considerations for Modeling Biofilms in the Oral Cavity: A Review. *Dental Materials*, **12**, 208-214. http://dx.doi.org/10.1016/S0109-5641(96)80024-8

[42] Baier, R.E. and DePalma, V.A. (1971) The Relation of the Internal Surface of Grafts to Thrombosis. In: Dale, W.A., Ed., *Management of Arterial Occlusive Disease*, Year Book Medical Publishers, Inc., Chicago, 147-163.

[43] Gould, J.A., Liebler, B., Baier, R., Benson, J., Boretos, J., Callahan, T., Canty, E., Compton, R., Marlowe, D., O'Holla, R., Page, B., Paulson, J. and Swanson, C. (1993) Biomaterials Availability: Development of a Characterization Strategy for Interchanging Silicone Polymers in Implantable Medical Devices. *Journal of Applied Biomaterials*, **4**, 355-358. http://dx.doi.org/10.1002/jab.770040410

[44] Baier, R.E., Loeb, G.I. and Wallace, G.T. (1971) Role of an Artificial Boundary in Modifying Blood Proteins. *Federation Proceedings*, Federation of AmerSoc for Experimental Biol, Bethesda, Vol. 30, 1523-1538.

[45] Tietze, C. (1966) Contraception with Intrauterine Devices. *American Journal of Obstetrics & Gynecology*, **96**, 1043-1054.

[46] Glantz, P.O., Arnebrant, T., Nylander, T. and Baier, R.E. (1999) Bioadhesion—A Phenomenon with Multiple Dimensions. *Acta Odontologica Scandinavica*, **57**, 238-241. http://dx.doi.org/10.1080/000163599428634

[47] Baier, R.E. (1980) Substrata Influences on the Adhesion of Microorganisms and Their Resultant New Surface Properties. In: Bitton, G. and Marshall, K.S., Eds., *Adsorption of Microorganisms*, Wiley-Interscience Publishers, Hoboken, 59-104.

[48] Baier, R.E. (1970) Surface Properties Influencing Biological Adhesion. In: Manly, R.S., Ed., *Adhesion in Biological Systems*, Academic Press, New York, 15-48. http://dx.doi.org/10.1016/B978-0-12-469050-9.50007-7

[49] Baier, R.E., Dutton, R.C. and Gott, V.L. (1970) Surface Chemical Features of Blood Vessel Walls and of Synthetic Materials Exhibiting Thromboresistance. In: Blank, M., Ed., *Surface Chemistry of Biological Systems*, Plenum Press, New York, 235-260. http://dx.doi.org/10.1007/978-1-4615-9005-7_14

[50] Baier, R.E. (1975) Blood Compatibility of Synthetic Polymers: Perspective and Problems. In: Kronenthal, R.L., Oser,

Z. and Martin, E., Eds., *Polymers in Medicine and Surgery*, Plenum Press, New York, 139-159. http://dx.doi.org/10.1007/978-1-4684-7744-3_10

[51] Baier, R.E., DePalma, V.A., Meyer, A.E., King, R.W. and Fornalik, M.S. (1981) Control of Heat Exchange Surface Microfouling by Material and Process Variations. In: Chenoweth, J.M. and Impagliazzo, M., Eds., *Fouling in Heat Exchange Equipment*, HTD-Vol. 17, AmerSoc Mechanical Engineers, New York, 97-103.

[52] King, R.W., Meyer, A.E., Ziegler, R.C. and Baier, R.E. (1981) New Flow Cell Technology for Assessing Primary Biofouling in Oceanic Heat Exchangers. *Proceedings of the 8th Ocean Energy Conference*, U.S. Department of Energy, Washington DC, 431-436.

[53] Baier, R.E., Meyer, A.E. and King, R.W. (1988) Improved Flow-Cell Techniques for Assessing Marine Microfouling and Corrosion. In: Thompson, M.F., Sarojini, R. and Nagabhushanam, R., Eds., *Marine Biodeterioration*, Oxford & IBH Publishing Co., Ltd., New Delhi, 385-394.

[54] Forsberg, R.L., Baier, R.E. and Meyer, A.E. (2014) Sampling and Experiments with Biofilms in the Environment: Part 2, Sampling from Large Structures Such as Ballast Tanks. In: Dobretsov, S., Thomason, J.C. and Williams, D.N., Eds., *Biofouling Methods*, Wiley-Blackwell, Oxford.

[55] Baier, R.E., Meyer, A.E. and Forsberg, R.L. (1997) Certification of Properties of Nontoxic Fouling-Release Coatings Exposed to Abrasion and Long-Term Immersion. *Naval Research Reviews*, **49**, 60-65.

[56] Baier, R.E., Meyer, A.E., Forsberg, R.L. and Ricotta, M.S. (1997) Intrinsic Drag Reduction of Biofouling-Resistant Coatings. *Proceedings, Emerging Nonmetallic Materials for the Marine Environment*, U.S.-Pacific Rim Workshop Sponsored by the U.S. Office of Naval Research, Honolulu, 1-36 through 1-40.

[57] Baier, R.E., Gucinski, H., Meenaghan, M.A., Wirth, J. and Glantz, P.O. (1984) Biophysical Studies of Mucosal Surfaces. In: Glantz, P.O., Leach, S.A. and Ericson, T., Eds., *Oral Interfacial Reactions of Bone, Soft Tissue & Saliva*, IRL Press Ltd, Oxford, 83-95.

[58] Baier, R.E. and Meyer, A.E. (1983) Surface Energetics and Biological Adhesion. In: Mittal, K.L., Ed., *Physiochemical Aspects of Polymer Surfaces*, Vol. 2, Plenum Publishing Corporation, New York, 895-909.

[59] Baier, R.E. (1973) Influence of the Initial Surface Condition of Materials on Bioadhesion. *Proceedings, Third International Congress on Marine Corrosion and Fouling*, Northwestern University Press, Evanston, 633-639.

[60] Fletcher, R.L. and Baier, R.E. (1984) Influence of Surface Energy on the Development of the Green Alga Enteromorpha. *Marine Biology Letters*, **5**, 251-254.

[61] Meyer, A., Baier, R., Wood, C.D., Stein, J., Truby, K., Holm, E., Montemarano, J., Kavanagh, C., Nedved, B., Smith, C., Swain, G. and Wiebe, D. (2006) Contact Angle Anomalies Indicate that Surface-Active Eluates from Silicone Coatings Inhibit the Adhesive Mechanisms of Fouling Organisms. *Biofouling*, **22**, 411-423. http://dx.doi.org/10.1080/08927010601025473

[62] Glantz, P.O.J., Arnebrant, T., Nylander, T. and Baier, R.E. (1999) Bioadhesion—A Phenomenon with Multiple Dimensions. *Acta Odontologica Scandinavica*, **57**, 238-241. http://dx.doi.org/10.1080/000163599428634

[63] Baier, R.E. (1992) Influence of Surface and Fluid Conditions on Thrombus Generation. *Proceedings of the Amer Acad of Cardiovascular Perfusion*, **13**, 143-146.

[64] Dutton, R.C., Webber, A.J., Johnson, S.A. and Baier, R.E. (1969) Microstructure of Initial Thrombus Formation on Foreign Materials. *Journal of Biomedical Materials Research*, **3**, 13-23. http://dx.doi.org/10.1002/jbm.820030104

[65] Nayak, S.C., Baier, R.E., Meyer, A.E. and Abuhaimed, T. (2010) Improvement of Root Canal X-Ray Imaging by Delmopinol Pretreatment-Assisted Contrast Media Infiltration. *Northeast Bioengineering Conference Proceedings*, Columbia University, New York, 26-28 March 2010, ABS-026, 39.

[66] L'Italien, G.J., Megerman, J., Hasson, J.E., Meyer, A.E., Baier, R.E. and Abbott, W.M. (1986) Compliance Changes in Glutaraldehyde-Treated Arteries. *Journal of Surgical Research*, **41**, 182-188. http://dx.doi.org/10.1016/0022-4804(86)90023-5

[67] Meyer, A.E., King, R.W., Baier, R.E. and Fornalik, M.S. (1985) A Field Study of Fouling of Test Surfaces Exposed to Flowing Brackish River Water. *Proceedings, Condenser Biofouling Control Symposium*, Electric Power Research Institute.

[68] Baier, R.E., Meyer, A.E., Akers, C.K., Natiella, J.R., Meenaghan, M.A. and Carter, J.M. (1982) Degradation Effects of Conventional Steam Sterilization on Biomaterial Surfaces. *Biomaterials*, **3**, 241-245. http://dx.doi.org/10.1016/0142-9612(82)90027-8

[69] Park, J.H., Olivares-Navarrete, R., Baier, R.E., Meyer, A.E., Tannenbaum, R., Boyan, B.D. and Schwartz, Z. (2012) Effect of Cleaning and Sterilization on Titanium Implant Surface Properties and Cellular Response. *Acta Biomaterialia*, **8**, 1966-1975. http://dx.doi.org/10.1016/j.actbio.2011.11.026

[70] White, J.A., Baier, R.E., Meyer, A.E., Burke, R.P. and Hausmann, E.M. (2000) Biomechanical and Biochemical Paths to Dystrophic Mineralization of Stented Cardiovascular Tissues. In: Vossoughi, J., Kipshidze, N. and Karanian, J.W., Eds., *Stent Graft Update*, Chapter 5, Medical and Engineering Publishers, Inc., Washington DC, 37-65.

6

A Robust Indicator for Promoting Circular Economy through Recycling

Francesco Di Maio, Peter Carlo Rem

Delft University of Technology, Delft, The Netherlands
Email: f.dimaio@tudelft.nl

Abstract

In order to move towards a more sustainable development, it is necessary not only to minimize the use of materials in the design stage and to find new materials as alternatives to nonrenewable ones (e.g. optical fiber instead of copper, biopolymers instead of polymers from oil) but also to reclaim as much as possible material value through effective recycling. To this extent, recycling can play a key role in multiple dimensions, while providing new business opportunities for innovative companies, having positive impacts on the society and the environment and fostering an effective circular economy as well. Because of the advanced waste management infrastructures available in developed countries, it is possible to achieve an almost complete collection of solid wastes into a variety of controlled bulk material flows. However, the picture for the follow-up step, the recycling of raw materials such as steel, non-ferrous metals, polymers and glass from these flows, is less positive. Materials value recovered from waste represents a very small fraction of European GDP. The fundamental issue is that policymakers still lack an effective key performance indicator for stimulating the recycling industry. Therefore although recycling plays an important role in the circular economy perspective, it is necessary to radically change the metric used so far to compute the recycling rate. Nowadays, the recycling rate is computed measuring the amount of material entering the recycling facilities. This approach has brought about an inaccurate and somehow misleading indicator (the recycling rate) which contributed to wrong decision making and to poor innovation in the industry. The new approach proposed in this paper considers the use of a Circular Economy Index (CEI) as the ratio of the material value produced by the recycler (market value) by the intrinsic material value[1] entering the recycling facility. It is argued that this index is related to strategic, economic and environmental aspects of recycling and it has very important implications as decision making tool. To compute the CEI it is necessary to know detailed information of the components and materials contained in each end of life (EOL) product entering the recycling facilities and how they end up in the recycled raw materials. Therefore an accurate accounting of materials (with standards if available), mass, chemical composition and smallest dimension (e.g. a screw, a plastic foil) is proposed.

---

[1]The present market value of all materials that would be needed to re-produce the EoL products that make up the waste.

Keywords

Recycling, Recycling Rate, Innovation, Policy, Resource Efficiency, Indicators

1. Introduction

The linear take-make-dispose economic model relies on large quantities of easily accessible resources, and as such is increasingly unfit for the reality in which it operates [1]. A reduction of resources consumed per unit of manufacturing output can only slowdown the depletion of those resources as it cannot modify the finite nature of their stocks. Demand and competition for limited resources increase price volatility, cause environmental degradation and threaten the competitiveness of countries. That is why the European Union as well as other countries worldwide are striving to move toward a circular economy model [2].

In order to move towards a more sustainable development and at the same time create opportunities for economic growth, a fundamental transformation in producer and consumer behavior is needed. It is crucial to increase the resource efficiency of production optimizing the use of materials in the design stage and to find new materials as alternatives to nonrenewable ones (e.g. optical fiber instead of copper, biopolymers instead of polymers from oil). In many countries, it is also necessary to improve the resource efficiency of collection. However, what is most urgent is to minimize the amount of materials which are currently disposed of, through effective recycling (cf. **Figure 1**).

The waste management infrastructure of Europe is already well developed so that it achieves an almost complete collection of solid wastes into a variety of controlled bulk material flows [3]. The picture for the follow-up step, the recycling of raw materials such as steel, non-ferrous metals, polymers and glass from these flows, is less positive. Materials value recovered from waste represents less than 0.5% of European GDP, even at the most favorable of economic conditions (EEA report). In most EU countries, recycling provides only between 5% and 15% in value of the raw materials used in manufacturing and construction [4]. The fundamental issue is that policymakers still lack an effective key performance indicator for stimulating the recycling industry.

Therefore although recycling is currently playing an important role in the circular economy perspective, it is necessary to radically change the metric used so far to compute the recycling rate. Nowadays the recycling rate is computed measuring the amount of material entering the recycling facilities. This approach has brought about an inaccurate and somehow misleading indicator (the recycling rate) which contributed to wrong decision making and to poor innovation in the industry [5] [6].

It is well known that material recycling is beneficial not only for the environment but also for the economy and the society at large. Every kilogram of recycled material can replace primary material and therefore displaces the activities that are needed to locate, mine and process it. All such activities use energy, release pollution and alter the landscape in ways that are perceived as a danger to the environment. Some raw materials (e.g. cobalt, copper, platinum, neodymium, etc.) involve also relevant social issues such as conflicts to access minerals, human rights violations, black market, etc. [7].

Because of the advanced waste management infrastructures available in developed countries, large amount of end-of-life (EoL) industrial and consumer products are available. Despite their potential value, these EoL products are still called waste residuals instead of "surface mines" waiting to be exploited [8]. However it is well es-

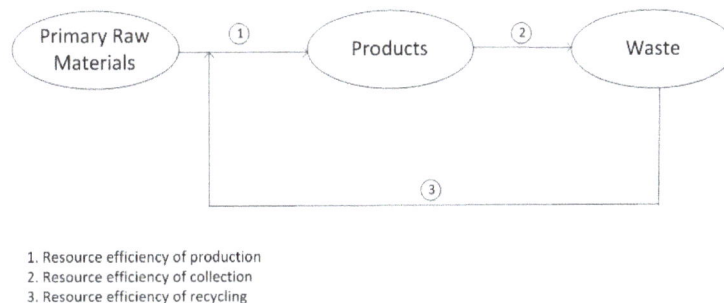

1. Resource efficiency of production
2. Resource efficiency of collection
3. Resource efficiency of recycling

Figure 1. Material flow and resource efficiencies.

tablished that increasing the efficient use of resources creates economic value (at firm, national and European level) and that the production of secondary materials is inherently more labor-intensive and less energy intensive than the production of primary materials [4]-[9]. Thus it creates significantly more jobs and requires higher levels of skills, facilitating the entry of women into the labor [10].

2. From a Linear to a Circular Economy

Europe has the world's highest net imports of resources per person, and its open economy relies heavily on imported raw materials and energy. Secure access to resources has become an increasingly strategic economic issue, while possible negative social and environmental impacts on third countries constitute an additional concern [11]. In 2013, a total amount of 5.7 billion tonnes [12] of materials has been used by the EU economy to provide its citizens with the goods and services they needed. In terms of value, the above total amount accounts to about 400 bn euro[2].

Considering that by 2050 the world population will hit 10 billion people, the increase in material use that would occur even at the current development levels and resource consumption patterns will reach about 180% of the 1990's level [13]. However economic development (in the sense of GDP increase) will take place as it is the main objective of governments in the developing as well as developed part of the world. Combining the effect of population growth and GDP increase in the developing countries results in a hefty increase of the consumption of natural resources (up to 800% of the 1990's value) [13].

Besides the implications of the fact that materials extracted from the earth and utilized for economic purposes are not literally "consumed" but become waste residuals that do not disappear and may cause environmental damage and result in unpaid social costs [14], experts have calculated that without a rethink of how materials are used in the current linear "take-make-dispose" economy, the virgin stocks of several key materials appear inadequate to sustain the modern "developed world" quality of life for all earth's peoples under contemporary technology [15]. Therefore it is necessary to move towards an industrial model that decouples economic growth from material input: The Circular Economy (CE). CE models maintain the added value in products for as long as possible and minimize waste. They keep resources within the economy when products no longer serve their functions so that materials can be used again and therefore generate more value. Thus circular business models create more value from each unit of resource than traditional linear models.

Although the CE approach contrasts with the mind-set embedded in most current industrial operations where even the terminology (value chain, supply chain, end user) expresses a linear view, several benefits may rise from the shift to the Circular Economy model and to a more resource-efficient path.

Since the early days of industrialization, companies mine and extract materials, use them to manufacture goods and sell the goods to customers (or end users) who dispose of them when they become obsolete or no longer useful. Some 65 billion tons of raw materials entered the economic system in 2010 and this figure is expected to grow to around 82 billion tons in 2020 [16].

The material saving potential arising from the transition to a CE model and to a more resource efficient path is estimated to 500 billion € per year for the European industry [17]. The job creation potential of remanufacturing and recycling in Europe is estimated at one million [1]. From the strategic point of view the benefits of the CE approach arise from the reduced risk of supply disruption and price volatility as well as from the huge potential for innovation related to new technologies (needed to increase resource productivity, material substitution, waste management and recycling), improvements of the forward and reverse cycles (optimization of the supply chain and logistics) and business models.

3. The Need of New Indexes

Taking into account the facts outlined above, we do believe that it is necessary to stimulate recycling through proper legislation and effective financial incentives. To this extent we assume that the robust and intuitive CE index proposed in this paper can help society to achieve the social, environmental, economic and strategic goalsit pursues.

To assess the environmental impact of any product throughout its life cycle, Life Cycle Assessment (LCA) is currently used [18]. Although the LCA method provides good insights about the environmental burden of each

---

[2]The value of materials at the point where they are in their final chemical composition, but not yet manufactured as a part or component.

product/industry, it is not always cost effective because a detailed LCA requires large amount of data and therefore it is time consuming [18]-[20].

Moreover the LCA studies provide information only on the environmental domain of sustainability, neglecting the economic and social ones which should be addressed simultaneously [19] [21] [22].

Also material flow accounting and analysis (MFA) is at present used to assess environmental as well as economic and other policies [23].To address in particular the economic policies, a methodological guide for economy-wide MFA (EW-MFA) has been developed by the Statistical Office of the European Commission (EUROSTAT) in 2001 [12]. However expert experiences in reviewing progress of these indicators toward numerical targets, have revealed several practical problems with the calculated indicators. For instance the time-lag in the availability of data and the fact that EW-MFA indicators are inherently macroscopic so that it is difficult to observe the direct effects of individual efforts to achieve CE [23].

4. The Driving Forces for the New Approach

The Circular Economy Index (CEI) proposed in this article uses a different approach. It aims at introducing the economic value of the materials embedded in consumers products as the property to be measured and accounted. It is argued that this index is related to a wide range of strategic, economic, social and environmental aspects of recycling and is therefore a proper instrument for decision making.

An important aspect of the CEI is that it does need data that are available in the companies' financial reports and in the bureaus of statistics so that the analysis of the performance is possible at firm as well as sector level (local, national or European).

Because the CEI intuitively represents the effectiveness of recycling firms at extracting value from the processed materials, it represents a decision making tool which will help management and policy makers to steer decision towards value creation and technological innovation.

A successful indicator for policymaking is always a compromise between the need for conceptual simplicity, the cost of evaluation and the degree to which the indicator is in parallel with current policy targets. Mass recycling rates, for example, are vastly popular for their conceptual simplicity and the relative ease of computing it for specific (categories of) EoL products (cf. **Figure 2**). Yet, the relation between the indicator and primary economic and environmental policy targets such as job creation and reduction of greenhouse gases is not very clear. Also, the indicator cannot be tuned to changes in the focus of policies. In contrast, (Social) Life Cycle Assessment can provide indicators relating to many aspects of policy, and therefore can also be tuned to changes in policy, but it is a very expensive tool and it is so complex that different LCA studies may deliver vastly different results for the same subject (cf. **Figure 2**).

Conceptual simplicity provides an indicator with robustness and reliability: different studies will reach similar conclusions, as there is little room for alternative interpretations. For the cost of evaluation, a critical point apart from conceptual simplicity is whether the computation of the indicator requires anything beyond readily available data. In regard to these two requirements, the best indicators are those that can be computed automatically

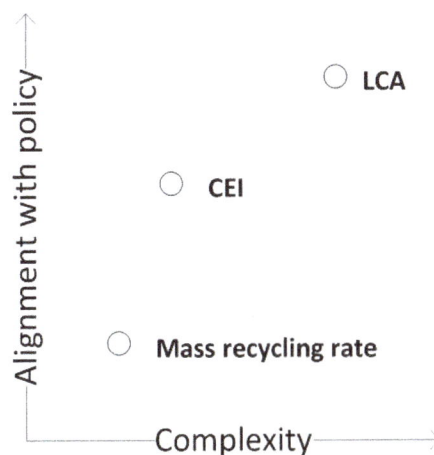

Figure 2. Alignment with policy vs complexity.

(*i.e.*, without issues requiring human interpretation) on the basis of data from standard, e.g. financial, reports.

Simple and cheap indicators typically point at a specific policy aspect, while they correlate, indirectly, with a number of other aspects. Mass recycling rates, for example, are directly related with the amount of waste that is deflected from landfills or incineration. Indirectly, mass recycling rates also correlate with the creation of jobs and the reduction of greenhouse gases. However, the relation between deflecting waste from landfills and creating jobs is so indirect that if creation of jobs becomes more important, at one point in time, than deflection from landfills, it makes sense to move to a different indicator.

The CEI proposed in this paper aims to be as simple as the mass recycling rate and better aligned with social, environmental and economic policies than the mass recycling rate while at the same moment, being simpler than LCA to be computed (cf. **Figure 2**).

5. The Circular Economy Index

Along with reuse and refurbish/remanufacturing, recycling plays an important role within the circular economy model and is often considered a cornerstone of a broader vision for the sustainability of a closed-loop society [6].

As material resources such as metal ores are becoming scarcer, there is an increasing incentive for upstream industries, such as smelters, to look for a secondary supply of resources from recycling activities. By strengthening the links between the primary resource supply sector and the recycling and waste management sector, both resource supply issues and waste management issues can be better addressed. Therefore indicators representing the recycling rate of materials and products have been used in the past. However, the definition of recycling rate varies considerably, mainly because both the numerator and denominator of the fraction that represent a recycling rate have been inconsistently chosen. The numerator often represents the amount of waste separated from waste streams for subsequent recycling, but usually the amount actually recycled is less than that separated because of the generation of residues in downstream recycling processes. This approach has brought about an inaccurate and somehow misleading indicator (the recycling rate) which might have contributed to wrong decision making and to poor innovation in the recycling industry.

The Circular Economy Index (CEI) proposed in this paper is the ratio of the material value produced by the recycler (market value) by the material value entering the recycling facility. In other words:

$$\text{CEI} = \frac{\text{Material value recycled from EOL product(s)}}{\text{Material value needed for (re-) producing EOL product(s)}}$$

Although several units to measure the resource efficiency are available (e.g. mass, volume, embedded energy, carbon footprint), the economic value (e.g. €, $) has been selected because is aligns best with the present EU policies which aim at fostering social and environmental benefits for citizens. To avoid any inconsistency, the values should be measured as soon as EoL products are collected (at the beginning of Arrow 3 in **Figure 1**) and just before the materials enter the production process (at the end of Arrow 1in **Figure 1**).

The CEI solves some problems of LCA and mass recycling rates in a very elegant way. Unlike mass recycling rates, the CEI adjusts itself automatically if some specific material becomes more expensive because it is less available, e.g. as a consequence of strategic issues, or if a material becomes cheaper because of a very efficient recycling technology.

Unlike LCA's, the computation of the CEI does not become more complex because a material is produced in alternative ways (e.g. by the primary industry or by recycling) or at different qualities. Producers will always use the cheapest material that fit the required quality, and so the CEI will automatically adjust.

As recycling rate does, CEI is easy to be computed and it uses data which are easily made available. From management point of view it is easy to understand and compute it without doing complex calculations or calculations which use estimates to be performed. However from the scientific point of view, the recycling rate is not a suitable indicator because it is not directly related to what we want to achieve (it is not clear how it correlates to the societal impacts (economics, job creation, environment, strategic, ethical issues)). On the other hand, the LCA based thinking is rather good from the scientific point of view the, but from the management point of view is not feasible because it requires complex calculations which sometimes need estimate numbers to be performed.

While it won't be possible to enforce a law requiring that all companies should perform an LCA, it will be possible to do it with the CEI because it is easy to be interpreted and computed and does not require extra human resources to be done. These two points are extremely important for successful policies. We can only have successful policies if we have a driver (the index), if we can measure it, if it is clear to everybody what it means, and if it is related to what we want to achieve. Furthermore it should be noticed the need of policies which speed up circularity. Although current policies stimulate waste collection, most materials are not properly re-used or recycled (e.g. plastics).

To compute the CEI for a specific EoL product, it is necessary to know detailed information of the components and materials contained in each end of life (EOL) product entering the recycling facilities. Therefore an accurate accounting process taking into consideration the materials (with standards if available), mass, chemical composition and smallest dimension (e.g. a screw) is proposed.

An important aspect of the CEI is the possibility to compute it also at sector or company levels by making use of the financial data contained in the financial reports at company level or in the database of the institutions which collect data at national level (e.g. bureau of statistics, chamber of commerce, etc.).

6. Experimental (KRI) and Financial (GVA) Calculation Methods

To experimentally compute the CEI it is necessary to know detailed information of the components and materials contained in each end of life (EoL) product entering the recycling facilities. Therefore the development and demonstration of Key Recycling Info (KRI) system which takes into consideration the materials (with standards if available), mass, chemical composition and smallest dimension (e.g. a screw) is necessary. EOL products should be taken apart in all components and dismantled (*i.e.* a model of flat panel display, a model of washing machine, five regular plastic packaging). Each EOL's component should be disassembled and documented so that the precise number of indivisible components as well as amount of each material contained in the selected product goods is available and collected in the Key Recycling Info sheet. From this sheet the precise market value embedded in the EOL products can be calculated at any time.

Industry plays an important role for innovation and efficiency in a market economy and is particularly prone to generate value from scarce resources. It is generally accepted that without corporate support, society will not achieve environmental sustainability, as firms represent the productive resources of the economy [24]. Therefore an index that is easy to interpret and cheap to be computed using existing information is very important to make the green business case more appealing. This way industry will likely increase their contribution to minimize environmental challenges and pressures.

The Gross Value Added method takes into consideration the value of the materials contained in each functional component a product is made of. Although this information may be considered as confidential when the product is launched on the market, it is no longer the case after a predictable period of time (which depends from the specific consumer product characteristics). As soon as the product reaches its end of life and enters the recycling facilities, it will be possible to calculate the GVA generated by processing that product.

The GVA can be computed also using financial data. The general formula of GVA is the following:

GVA = Recycling firm revenues—non factor costs.

Recycling firm revenues include mainly revenues from selling secondary materials.

Non-factor costs for recycling companies include costs for energy and input materials.

Therefore the GVA of recycling companies consists mainly of the resource added value of the recycling activities.

The CEI will be the ratio of the GVA over the material input value (cf. **Figure 3**).

7. How CEI Relates to Economic Social and Environmental Issues

The CEI is linked to several targets we want to achieve in society (e.g. economic growth) and from management point of view requires data that can actually be easily generated (GVA). From the science point of view it relates to the societal and environmental impacts of economic growth.

Figure 4 shows how CEI correlates with the carbon footprint avoided by recycling.

From **Figure 4** it is clear that the use of current recycling rate (based on mass) as an index to stimulate recycling is not as effective as the CEI. The graph shows that reclaiming material value from waste strongly correlates with reducing carbon footprint. For example if we suppose for convenience that a car is made of just four materials (steel, aluminum, copper and plastics), the current target recycling rate of 85% set by the European

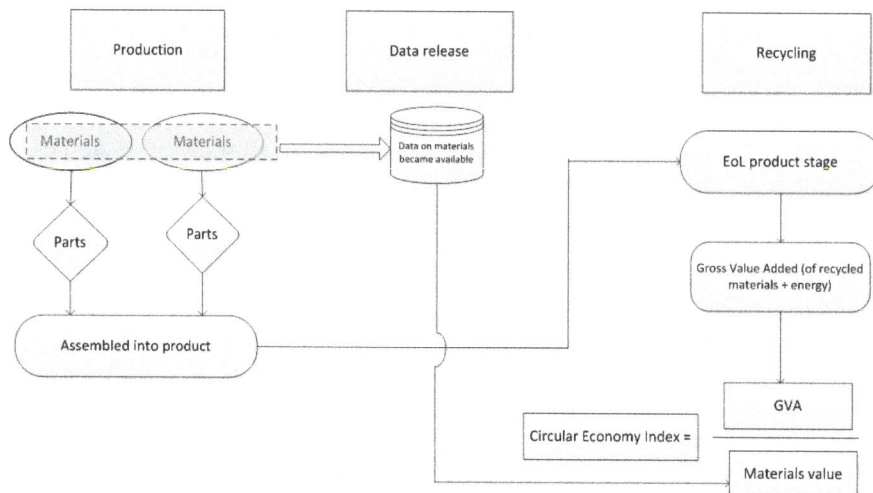

Figure 3. The circular economy index at EoL product level and at sector level.

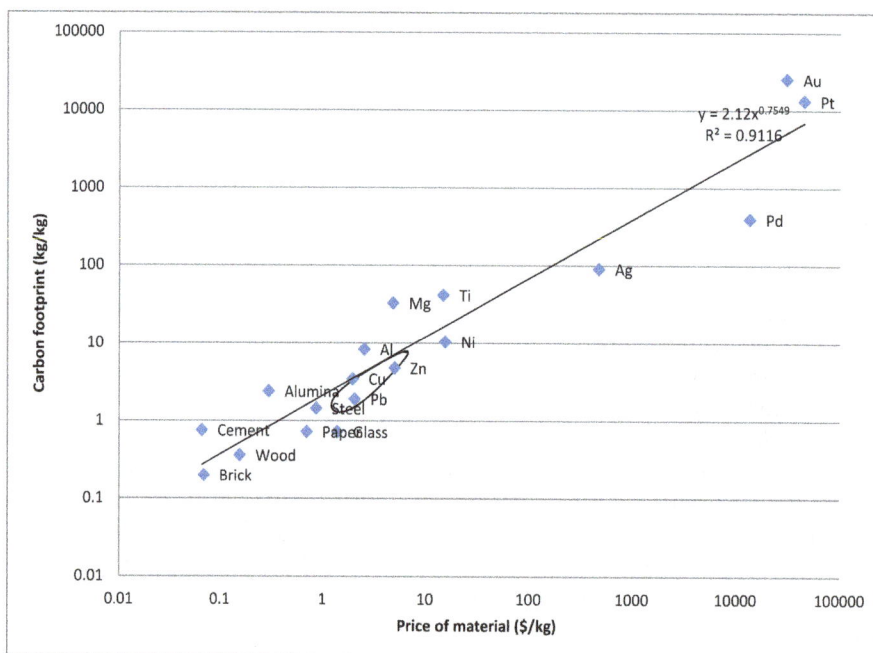

Figure 4. The carbon emission in kilograms of CO_2 per kilogram of material produced versus the price of materials (data from [25]).

Commission directive [26], is driving recyclers to maximize the separation of steel rather than aluminum, copper or plastics. This is due to the available state of art technologies that can effectively separate steel scraps from the automotive shredder residues (ASR). However from **Figure 4** appears that separating the more valuable Al and plastics materials provides considerable environmental benefits along with an economic one. Considering that recycling plastics from ASR require advanced technologies, if policy makers set the recycling target based on mass rather than on value (so that there will be no difference between a kg of steel and one of Al or plastics), the recycling industry will maximize steel recycling and will not develop new technologies. Therefore these policies do not foster the recycling of materials that can generate more revenues, create more jobs and have higher footprint than steel.

The following example will make the above point even more outstanding. If the composition of the car is the one presented in **Table 1**.

Three scenarios can be envisaged:

Table 1. Car composition in kg.

Materials	Steel	Plastics	Al	Cu
Mass (kg)	600	130	50	20

1) Use of standard technology to extract 620 kg of steel contaminated by Cu from the ASR.
2) Use of advanced technology to extract 130 kg plastics plus 50 kg Al from the ASR.
3) Use of advanced technology to separate steel from Cu.

From the mass recycling rate point of view, the largest contribution is in Scenario 1 because the larger amount of material (low grade steel) is recycled from the EoL car (620 kg). However if we use a CEI approach, the separation of the 120 kg of plastics and the 50 kg of Al, produces a higher value and at the same time, has a higher positive impact on the environment (cf. **Figure 4**). It is clear that the mass recycling rate approach provides a smaller incentive to valorize the plastics and Al contained in the ASR. This is even more evident looking at Scenario 3. In this case, the mass recycling rate approach produces no outcome at all. Therefore there is no driving force for separating steel from Cu. On the other hand, the CEI approach encourages the separation of these metals because the value of 600 kg high grade steel and 20 kg Cu is much higher than the 620 kg of low grade steel coming from Scenario 1. Therefore the use of the CEI approach is a driver for the use of innovative technologies, creates economic value and jobs and it has positive effects on the environment.

8. Science and Policy Requirements

Concerns about sustainable management of natural resources and efficient use of raw materials are growing worldwide. In particular materials and energy security. At the same time, the need to minimize environmental impact is more critical than ever. Against this background, many governments have increased efforts to promote deployment of resources efficient technologies that can strengthen resources security. This may stimulate an interest in innovative green technologies so that they are now a growing sector of the sustainable technology field.

Transforming the economy onto a resource-efficient path will bring increased competitiveness and new sources of growth and jobs through cost savings from improved efficiency, commercialisation of innovations and better management of resources over their whole life cycle. This requires policies that recognize the interdependencies between the economy, wellbeing and natural capital and seeks to remove barriers to improved resource efficiency, whilst providing a fair, flexible, predictable and coherent basis for business to operate [11].

In order for decision makers to be effective, they need robust and clear indicators that show the links between social, environmental and economic goals so that they can better understand how to achieve economic growth that is in harmony with the natural systems within which we live and work [27] [28].

Indicators can be used i) individually, ii) as part of a set, or iii) in the form of a composite index that combines individual indicator scores into a single number. Such a single aggregated number can be very useful in communicating information on general sustainability to the public and to decision makers [27]. Possible disadvantages are that the methods to achieve an aggregation are often subjective [29] [30] and that every index contains hidden assumptions and simplifications [31]. Therefore, such combined indicators need to be used judiciously. Farrell and Hart (1998) state that in many cases, indicators to measure sustainability are no more than combined lists of traditional economic, environmental and social indicators with the word 'sustainable' added to the title. Nevertheless, such combination is a first significant step because it recognizes that all three areas (economic, ecological and social) matter: sustainable development is a holistic concept and ideally one should strive to consider all three pillars of sustainability simultaneously [28]. Therefore, it is important that the development of indicators does not stop at this stage [27]. Economic and ecological analysis need to be combined [32] and one should concentrate on the interaction rather than on just the environment itself [33]. The advantage of aggregate indicators is that the information is presented in a format tailored to decision makers [33]-[35]. However, we need to be careful and informed about the way of aggregation, the uncertainties, the weights and the data source. Decision makers are too busy to deal with these details and the beauty of the aggregate indicator is the fact that it does the job for them [34]. But, no single indicator can possibly answer all questions and therefore multi-dimensional indicators can be needed [36] [37].

The advantage of the CEI proposed here is that it provides a clear indication of how good a recycling company is in valorizing the materials it processes and in combination with the KRI give environmental, technical and economic information which are useful to better perform LCA, MFA, I/O analysis, to design recycling processes and to predict the amount of stock-in-use materials.

Moreover the CEI and KRI may be used by policy makers to foster innovation by properly designing environmental standards. Such standards can trigger innovation that may partially or more than fully offset the costs of complying with them. Such "innovation offsets," [38], can not only lower the net cost of meeting environmental regulations, but can even lead to absolute advantages over firms in foreign countries not subject to similar regulations. Innovation offsets will be common because reducing pollution and increase resources efficiency is often coincident with improving the productivity with which resources are used. In short, firms can actually benefit from properly crafted environmental regulations that are more stringent (or are imposed earlier) than those faced by their competitors in other countries. By stimulating innovation, strict environmental regulations can actually enhance competitiveness.

9. Conclusions

In this paper, we argue that in order to move towards a more sustainable development, it is necessary to reclaim as much as possible material value from waste through effective recycling. To date this is not happening because the current indicators are not properly related to socio-economic goals. Therefore we propose a simple and robust indicator (the CEI) that is easy to be calculated and interpreted.

We show that the CEI is related to societal targets as well as the environmental and economic ones. Extracting value from EOL products brings about increased revenues (more jobs) and at the same time, decreases the impact on the environment.

The indicator we proposed fosters the use of innovative technologies to valorize the secondary resources contained in waste streams.

References

[1] Ellen Mac Arthur Foundation (2014) Toward the Circular Economy—Accelerating the Scale-Up across Global Supply Chains.

[2] European Commission (2014) Towards a Circular Economy: A Zero Waste Programme for Europe. Communication from the Commission to the European Parliament, the Council, the European Economic and Social Committee and the Committee of the Regions, COM(2014) 398 Final/2.

[3] Golder Europe EEIG (2004) Implementation of the Landfill Directive in the 15 Member States of the European Union, ENV.A.2/ETU/2004/0016.

[4] Hüther, M. (2010) Volkswirtschaftliche Bedeutung der Entsorgungs—und Rohstoffwirtschaft, Institut der Deutschen Wirtschaft Köln.

[5] European Parliament and Council Directive 94/62/EC, 1994, on Packaging and Packaging Waste.

[6] Gutowski, T.G., Allwood, J.M., Herrmann, C. and Sahni, S. (2013) A Global Assessment of Manufacturing: Economic Development, Energy Use, Carbon Emissions, and the Potential for Energy Efficiency and Materials Recycling. *Annual Review of Environmental Resources*, **38**, 81-106. http://dx.doi.org/10.1146/annurev-environ-041112-110510

[7] Whitney, T. (2015) Conflict Minerals, Black Markets, and Transparency: The Legislative Background of Dodd-Frank Section 1502 and Its Historical Lessons. *Journal of Human Rights*, **14**, 183-200. http://dx.doi.org/10.1080/14754835.2015.1021036

[8] UNEP (2011) Recycling Rates of Metals, A Status Report. United Nations Environmental Programme.

[9] EEA Report| No 8/2011, Earnings, Jobs and Innovation, the Role of Recycling in a Green Economy.

[10] Blake, M.K. and Hanson, S. (2005) Rethinking Innovation: Context and Gender. *Environment and Planning A*, **37**, 681-701. http://dx.doi.org/10.1068/a3710

[11] European Commission (2011) Roadmap to a Resource Efficient Europe. Communication from the Commission to the European Parliament, the Council, the European Economic and Social Committee and the Committee of the Regions, COM(2011) 571 Final.

[12] EUROSTAT (2001) Economy-Wide Material Flow Accounts and Derived Indicators—A Methodological Guide. Office for Official Publications of the European Communities, Luxemburg.

[13] Ayres, R.U. (2008) Sustainability Economics: Where Do We Stand? *Ecological Economics*, **67**, 281-310. http://dx.doi.org/10.1016/j.ecolecon.2007.12.009

[14] Ayres, R.U. and Kneese, A.V. (1969) Production, Consumption & Externalities. *American Economic Review*, **59**, 282-

296.

[15] Gordon, R.B., Bertram, M. and Graedel, T.E. (2005) Metal Stock and Sustainability. *Proceedings of the National Academy of Sciences of the United States of America*, **103**, 1209-1214.

[16] Ellen Mac Arthur Foundation (2014) Toward the Circular Economy—Economic and Business Rationale for an Accelerated Transition.

[17] Europe INNOVA (2012) Guide to Resource Efficiency in Manufacturing.

[18] Guinée, J.B., Gorrée, M., Heijungs, R., Huppes, G., Kleijn, R., de Koning, A., van Oers, L., Wegener Sleeswijk, A., Suh, S., Udo de Haes, H.A., de Bruijn, H., van Duin, R., Huijbregts, M.A.J., Lindeijer, E., Roorda, A.A.H., van der Ven, B.L. and Weidema, B.P., Eds. (2002) Handbook on Life Cycle Assessment. Operational Guide to the ISO Standards. Institute for Environmental Sciences. Leiden University, Leiden.

[19] Thomassen, M.A., Dolman, M.A., van Calker, K.J. and de Boer, I.J.M. (2009) Relating Life Cycle Assessment Indicators to Gross Value Added for Dutch Dairy Farms. *Ecological Economics*, **68**, 2278-2284. http://dx.doi.org/10.1016/j.ecolecon.2009.02.011

[20] Wäger, P.A. and Hischier, R. (2015) Life Cycle Assessment of Post-Consumer Plastics Production from Waste Electrical and Electronic Equipment (WEEE) Treatment Residues in a Central European Plastics Recycling Plant. *Science of the Total Environment*, **529**, 158-167. http://dx.doi.org/10.1016/j.scitotenv.2015.05.043

[21] Glavič, P. and Lukman, R. (2007) Review of Sustainability Terms and Their Definitions. *Journal of Cleaner Production*, **15**, 1875-1885. http://dx.doi.org/10.1016/j.jclepro.2006.12.006

[22] Ness, B., Urbel-Piirsalu, E., Anderberg, S. and Olsson, L. (2007) Categorising Tools for Sustainability Assessment. *Ecological Economics*, **60**, 498-508. http://dx.doi.org/10.1016/j.ecolecon.2006.07.023

[23] Moriguchi, Y. (2007) Material Flow Indicators to Measure Progress toward a Sound Material-Cycle Society. *Waste Management*, **9**, 112-120. http://dx.doi.org/10.1007/s10163-007-0182-0

[24] Bansal, P. (2002) The Corporate Challenges of Sustainable Development. *Academy of Management Executive*, **16**, 122-131. http://dx.doi.org/10.5465/AME.2002.7173572

[25] Gutowski, T.G., Sahni, S., Allwood, J.M., Ashby, M.F. and Worrell, E. (2013) The Energy Required to Produce Materials: Constraints on Energy-Intensity Improvements, Parameters of Demand. *Philosophical Transactions of the Royal Society A*, **371**, Article ID: 20120003. http://dx.doi.org/10.1098/rsta.2012.0003

[26] European Commission (2000) Directive 2000/53/EC of the European Parliament and of the Council on End-of Life Vehicles.

[27] Farrell, A. and Hart, M. (1998) What Does Sustainability Really Mean? The Search for Useful Indicators. *Environment*, **40**, 4-9. http://dx.doi.org/10.1080/00139159809605096

[28] Van Passel, S., Nevensa, F., Mathijsb, E. and Van Huylenbroeck, G. (2007) Measuring Farm Sustainability and Explaining Differences in Sustainable Efficiency. *Ecological Economics*, **62**, 149-161. http://dx.doi.org/10.1016/j.ecolecon.2006.06.008

[29] Becker, B. (1997) Sustainability Assessment: A Review of Values, Concepts and Methodological Approaches, Issues in Agriculture. *CGIAR World Bank*, **10**, 1-63.

[30] Hueting, R. and Reijnders, L. (2004) Broad Sustainability Contra Sustainability: The Proper Construction of Sustainability Indicators. *Ecological Economics*, **50**, 249-260. http://dx.doi.org/10.1016/j.ecolecon.2004.03.031

[31] Hanley, N., Moffatt, I., Faichney, R. and Wilson, M. (1999) Measuring Sustainability: A Time Series of Alternative Indicators for Scotland. *Ecological Economics*, **28**, 55-73. http://dx.doi.org/10.1016/S0921-8009(98)00027-5

[32] Kaufmann, R.K. and Cleveland, C.J. (1995) Measuring Sustainability: Needed—An Interdisciplinary Approach to an Interdisciplinary Concept. *Ecological Economics*, **15**, 109-112. http://dx.doi.org/10.1016/0921-8009(95)00062-3

[33] Jollands, N., Lermit, J. and Patterson, M. (2003) The Usefulness of Aggregate Indicators in Policy Making and Evaluation: A Discussion with Application to Eco-Efficiency Indicators in New Zealand. ISEE Working Paper.

[34] Constanza, R. (2000) The Dynamics of the Ecological Footprint Concept. *Ecological Economics*, **32**, 341-345.

[35] Azapagic, A. (2004) Developing a Framework for Sustainable Development Indicators for the Mining and Minerals Industry. *Journal of Cleaner Production*, **12**, 639-662. http://dx.doi.org/10.1016/S0959-6526(03)00075-1

[36] Opschoor, H. (2000) The Ecological Footprint: Measuring Rod or Metaphor. *Ecological Economics*, **32**, 363-365.

[37] Veleva, V. and Ellenbecker, M. (2000) A Proposal for Measuring Business Sustainability: Addressing Shortcomings in Existing Frameworks. *Greener Management International*, **31**, 101-120.

[38] Porter, M.E. and van der Linde, C. (1995) Toward a New Conception of the Environment-Competitiveness Relationship. *The Journal of Economic Perspectives*, **9**, 97-118. http://dx.doi.org/10.1257/jep.9.4.97

7

Development of Eco-Friendly Deconstruction Technologies for Recycling Construction Waste

Myungkwan Lim[1,2], Heesup Choi[1,2]*, Hyeonggil Choi[3], Ryoma Kitagaki[3], Takahumi Noguchi[3]

[1]Department of Archtecture Engineering, Hankyong National University, Gyeonggi-do, Korea
[2]Department of Civil Engineering, Kitami Institute of Technology, Hokkaido, Japan
[3]Department of Architecture, The University of Tokyo, Tokyo, Japan
Email: *choiheesup@gmail.com

Abstract

Recently, the intended use and required performance of buildings are rapidly changing with advances in scientific technology and globalization. Furthermore, given the population growth in semi-developed and developing countries, economic growth, increasing waste, and increasing amounts of energy consumption, the industry requires the development of efficient methods to deconstruct old buildings by reducing waste and saving as much energy as possible during periods of urban redevelopment and maintenance. In general, either an ordinary breaker or a hydraulic breaker is used to deconstruct reinforced concrete buildings. This deconstruction method has the ability to shorten the construction period compared to the other methods, but it is difficult to separate the debris that gets mixed when the deconstruction is completed, as it is a rough construction method that uses large equipment. This study develops a technology that can be used to selectively heat, separate, and deconstruct the steel reinforcement inside reinforced concrete, treating the reinforcement as a conductive resistor and applying high-frequency induction heating to the reinforced concrete structure. Specifically, this study verifies the temperature characteristics of deformed bars inside reinforced concrete, the occurrence of cracks due to thermal fracture of the deformed bars, and chemical and physical weakening of concrete by thermal conduction on the surface of steel reinforcement using the high-frequency induction heating technology. Furthermore, this study considers the extent of concrete weakening in the heating range of appropriate energy and carries out a technical review of the stages that would be actually applied. This technology involves low noise and low pollution levels, and it increases the collection rate of steel reinforcement inside separated reinforced concrete members and the recycling rate of construction wastes; thus,

---

*Corresponding author.

its use is expected to reduce the energy consumption by minimizing secondary processing.

Keywords

Recycling, Deconstruction, Heat Induction, Ease to Scrap, Weakening

---

1. Introduction

Despite the annually increasing number of buildings, there is little literature on deconstruction methods. Deconstruction is strictly constrained by the surrounding environments, especially in downtown areas, and environmental pollution and safety issues such as the noise, vibration, and dust that occur during the deconstruction process also cause problems [1] [2]. The existing deconstruction methods rely largely on labor owing to the lack of deconstruction technology and experience, which causes waste of labor and loss of available materials as well as extended construction period [3]. The existing deconstruction methods do not consider the recycling of debris that occurs during the deconstruction process [4]. The reason why the actual recycling rate is problematic despite continuous recycling of wasted concrete lies in the existing deconstruction methods [5], in which various problems related to the reuse and recycling of wastes that occur owing to indiscriminate deconstruction have been pointed out [6] [7]. In the case of a hydraulic breaker or large breaker, the representative method for the current separating dismantlement, it is necessary to separate concrete members and waste. In particular, in the case of reinforced concrete members dismantled using heavy equipment, various problems can occur with regard to the reuse of steel reinforcement and recycling of waste concrete [8].

- It is necessary to convert the current "first dismantle and then separate" method to a "separate first and then dismantle" method and develop technology to improve the recycling rate that considers the efficient collect of demolished debris in the dismantlement stage.
- It is necessary to develop a partial-dismantlement technology that is effective for remodeling and repair reinforcement. Partial dismantlement, which is on the rise with the increase in remodeling and repair reinforcement, requires various technical applications and developments that are different from the existing dismantlement in terms of prevention of damage to the remaining structure or improvement in safety.
- It is necessary to propose a comprehensive management system that systemizes separating dismantlement and partial dismantlement to promote the development of the current dismantlement industry and of the entire architectural industry. This secures safety in dismantling construction, reduction in environment-polluting factors, proper handling of generated waste, and improvement in the recycling rate.

2. Complete Deconstruction Technology for Reinforced Concrete Members Using High-Frequency Induction Heating Method

2.1. Steel Reinforcement Heating Mechanism Depending on High-Frequency Induction Heating (Joule Heat)

Induction heating is a method that is used to heat conductive resistors such as metals by electrical energy that is converted from a high-frequency current-carrying conductor known as an induction coil. If alternating current flows in the lead wire, a line of magnetic force with changing direction and intensity in the surroundings occurs. If conductive materials (usually metal) are placed in the surroundings, as shown in **Figure 1**, an eddy current flows inside the metal, influenced by this changing line of magnetic force. As metal usually has an electrical resistance, Joule heating [electricity = current^2] occurs, and the metal is heated if the current flows in the metal. This phenomenon is called *induction heating*. Induction heating is less likely to be a risk, and the heat loss is also less, although the temperature rises in the non-heated areas, because only the metal is heated [9] [10].

2.2. Skin Effect and Depth of Penetration δ

In general, most currents are concentrated on the surface of a conductive resistor because the current density drops at the center of the conductive resistor if alternating current flows through it. This phenomenon is called the *skin effect*. When it gets closer to the center of the conductive resistor, the cross velocity of the magnetic

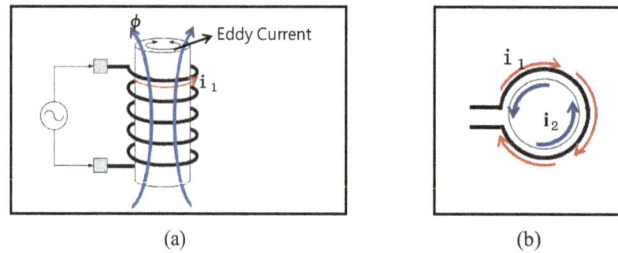

(a)   (b)

Figure 1. High-frequency induction heating mechanism. (a) Magnetic flux φ depending on alternating current i_1; (b) Induced current (Eddy Current i_2).

field speeds up. It can also be caused by the difficult flow of alternating current when the inductance increases. In Equation (1), δ [mm], the penetration depth of a high-frequency current is determined by the material permeability μ [H/m], the conductivity of the heated object σ [℧/m], and the frequency f [Hz].

$$\delta = \frac{1}{\sqrt{\pi f \mu \sigma}} \tag{1}$$

Equation (1) shows the penetration depth increases when the frequency decreases or when the material permeability and conductivity of the heated object decrease. As 90% of heating occurs from the surface to δ in terms of current and power distribution, it can be said that all eddy currents are concentrated to the depth of penetration δ from the surface of a conductive resistor [11] [12].

2.3. Thermal Model of Steel Reinforcement Using High-Frequency Induction Heating

If high-frequency alternating-current flows into the coil, an eddy current is induced and acts as source of heat that depends on the resistance of metal; here, resistance causes heat in the metal. In the case of high-frequency induction heating, there is a large difference in heating efficiency depending on the magnetic characteristics of the heated object, but a larger eddy current is induced with a magnetic substance, which is significantly affected by the magnetic field, depending on the changes in time [13].

As shown in Equation (2), the hysteresis constant η varies depending on the magnetic quality of a material; unlike the general transformer, it is easy to heat material with larger η in induction heating. However, as the heated objects are not usually made in a closed circuit like a transformer, the density of the magnetic field velocity and the effective permeability are small and η is also small, even if the heated object is a magnetic substance like cast iron. In addition, if the frequency is over about 10 [kHz], the hysteresis loss depending on the induction heating can be neglected because the eddy current loss is overwhelmingly greater than the square of frequency (as shown in **Figure 2**).

In the case of a magnetic substance, the efficiency of the heating surface increases because the depth of penetration decreases with increasing relative permeability. For steel reinforcement, with a relatively high permeability, localized heating on the surface is possible because the induced current is concentrated on the areas facing the heating coil when the magnetic field occurred from the coil is absorbed into the surface of the metal. In addition, selectively localized heating is possible because the scope of the magnetic field can be adjusted according to the diameter of the heating coil. **Figure 3** represents the heating model of steel reinforcement, which depends on the high-frequency induction heating [13] [14].

$$P_k = \eta f (B_m)^2 V \tag{2}$$

η: hysteresis constant.
V: central volume of steel reinforcement (m^3).

The principle of the magnetic field generated by the coil is summarized in accordance with Maxwell's equations. As shown in **Figure 3**, coils are distributed within a particular space; if current $i(t)$ is flowing, the magnetic vector potential A occurs at any point within the space $P(x, y, z)$. Equation (3) is the interaction formula for both the current density J and the magnetic vector potential A.

$$A = \frac{\mu}{4\pi} \int_V \frac{J}{\gamma} dV \tag{3}$$

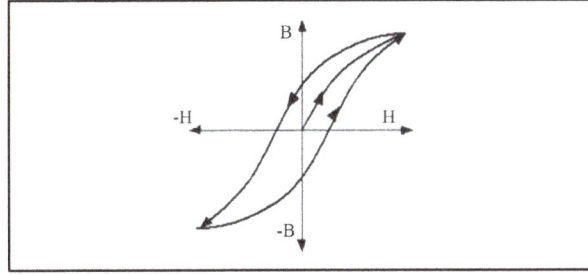

Figure 2. Hysteresis loop [15].

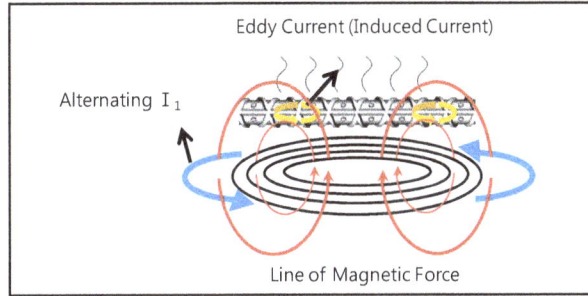

Figure 3. Thermal Model of Steel Reinforcement Depending on High-Frequency Induction Heating.

If current is alternating, the magnetic vector potential A is also alternating. Equation (4) is the relationship between the magnetic vector potential and the magnetic flux density (magnetic field velocity) B.

$$B = \nabla \times A \tag{4}$$

If the magnetic vector potential A is alternating, the magnetic flux density B is also alternating. Therefore, if a conductive resistor is used, power failure E occurs depending on Equation (5). From this current, an eddy current J occurs.

$$\nabla \times E = -\frac{\partial B}{\partial t} \tag{5}$$

$$J = \sigma E \tag{6}$$

σ: conductivity of a conductive resistor.

The entire magnetic field is determined mutually by the magnetic field due to the current that flows in the coil and the magnetic field due to the Foucault current. This causes an electromotive force in the coil. As shown in Equation (7), the electromotive force in the coil is the electric field that occurs when heating the coil.

$$V = \oint_c E \cdot d\ell = \oint_c \left(-\frac{\partial A}{\partial t} \right) d\ell \tag{7}$$

Therefore, a magnetic field in the center of the steel reinforcement, a conductive resistor, occurs that depends on the distribution of the eddy current, if coils are distributed closely to the steel reinforcement [15].

3. Temperature Characteristics of Single Steel Reinforcement Depending on High-Frequency Induction Heating

3.1. Experimental Overview

This study carries out an experiment to analyze the temperature characteristics of a deformed bar, commonly used as a conductive resistor in high-frequency induction heating, and the temperature distribution of steel reinforcement accordingly. This study sets the conditions for the diameter and length of steel reinforcement, especially of the deformed bar widely used in the field, the temperature rise characteristics and temperature distribution depending on the distance from the heating coil to the surface of steel reinforcement, and the quantity of output as experimental factors. This study identifies the temperature rise characteristics and temperature distribution at various conditions because there are large differences in heating efficiency depending on the differences in heat transfer coefficients by cross-sectional area and distance of steel reinforcement within the range of the magnetic field in high-frequency induction heating. In addition, the main purpose is to predict the temperature characteristics as basic information for various arrangements of steel reinforcement. Furthermore, this study determines the optimal conditions for output and heating positions because the depth of penetration varies with the frequency.

3.2. Experimental Method

3.2.1. Induction Heating Device

For induction heating, this study uses an experimental device with a basic frequency of 120 kHz (operating frequency of 60 - 120 kHz) and a maximum high-frequency output of 6 kW. The output stability is ±2%, and the output can be adjusted depending on the DC voltage [16]. High-frequency induction heating is a characteristic of the changed operating frequency when changing the coils, because the heating coil and heated object become one component of the circuit, and the high-frequency power is set by a resonant condenser inside the matching circuit, output lead, inductance, and inductance of heating coil. Resonant frequency employs an automatic tracking method. The heating coil is a fan-cake type and uses a coil with dimension of φ 120, rev count of 3, and coil thickness of φ 10.

3.2.2. Steel Reinforcement Heating Experiment

For the single steel reinforcement heating experiment, this study uses 150-mm steel reinforcement samples of D6, D10, D19, D25, and D32 in SD345 to identify the temperature distribution of steel reinforcement depending on the optimum output and heating. The maximum output is either 5 kW or 6 kW, and the experiment is carried out by changing the distance from the surface of the steel reinforcement to the lower bottom of the induction heating coil and measuring the temperature at the central point on the surface of steel reinforcement.

To form an arrangement similar to that of reinforced concrete specimens, this study carries out an experiment by cutting every steel reinforcement sample to 430 mm. This study measures the temperature at three positions using a heat-resistant camera: at the center of the surface of steel reinforcement, at the part facing the coil diameter, and at the part 30 cm away from the coil diameter. The heating experiment method for single steel reinforcement is show in **Figure 4**.

3.3. Temperature Rise Characteristics of Single Steel Reinforcement Depending on High-Frequency Induction Heating

The temperature rise characteristics of steel reinforcement with induction heating (length: 150 mm) are presented in **Figure 5** for output powers of 5 kW and 6 kW. As temperatures of 800°C or above are difficult to measure, any temperature above this mark is regarded as "being at 800°C". There are no large differences between 5 kW and 6 kW. Over a certain output, the depth of penetration with induction heating increases depending on the quantity of output, but if the temperature is over the Curie temperature at the surface of the steel reinforcement, the relative permeability falls to 1 [17]. The output of the electric field drops and loss occurs because the electric field reaches the critical temperature, even in the inside, without being further increased.

In the relationship between time and temperature, the targeted temperature (concrete-weakening temperature) reached up to 300°C within 60 seconds when the distance from the heating coil to the surface of steel reinforcement was 10 mm, 20 mm, or 30 mm. As the distance decreased, the temperature rose rapidly, to 600°C or above. In addition, at distances of 10 mm or 20 mm, there is a thermal equilibrium between 600°C - 800°C, and at a distance of 30 mm, there is a thermal equilibrium between 500°C - 700°C. At 50 mm, the temperature reduces to 300°C in about 300 s.

D6, D10, D19, and D25 showed a rapid temperature rise characteristic, but D32 showed reduced temperature rise trend compared to the other types of steel reinforcement. This may occur because as the induction heating heats the surface of steel reinforcement rapidly, the steel reinforcement with a thick diameter requires time to obtain a temperature difference by heating until the molecular motion can become constant depending on the heat within the steel reinforcement.

The results of heating the 430-mm-long steel reinforcement using the high-frequency induction heating

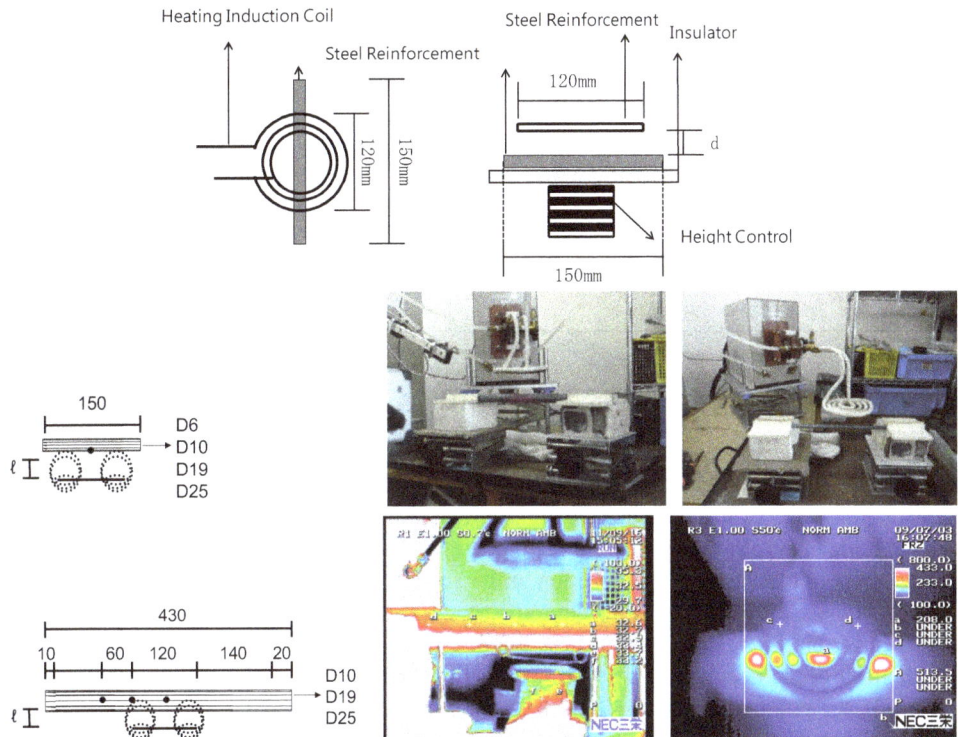

Figure 4. Experimental method of single steel reinforcement heating using high-frequency induction heating method.

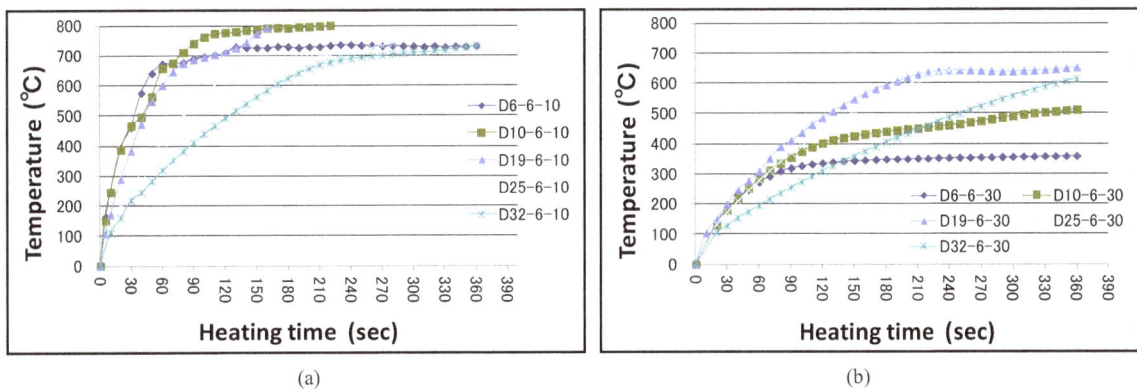

(a)

(b)

Figure 5. Temperature rise characteristics of steel reinforcement depending on high-frequency induction heating (measured at 150 mm-center). (a) 5 kW-10 mm; (b) 6 kW-30 mm. Note: D00-0-0: D (reinforced type)-output (kW)-heating distance (mm).

method are shown in **Figure 6**. Heating with an output of 5 kW showed temperature rise characteristics similar to that of the 150-mm-long steel reinforcement. This may be because little heat conduction loss occurs depending on the length of steel reinforcement and because selective heating is possible depending on the heating location.

3.4. Temperature Distribution Characteristics of Single Steel Reinforcement Depending on High-Frequency Induction Heating

The steel reinforcement heating process using the induction heater can be problematic owing to unnecessary heating, lack of heating, and concrete weakening depending on the temperature rise. This study analyzes the temperature distribution of steel reinforcement at the induction heating conditions of 5 kW and 6 kW using infrared radiation temperature measurement at 10 mm, where the highest temperature rise occurred; the results

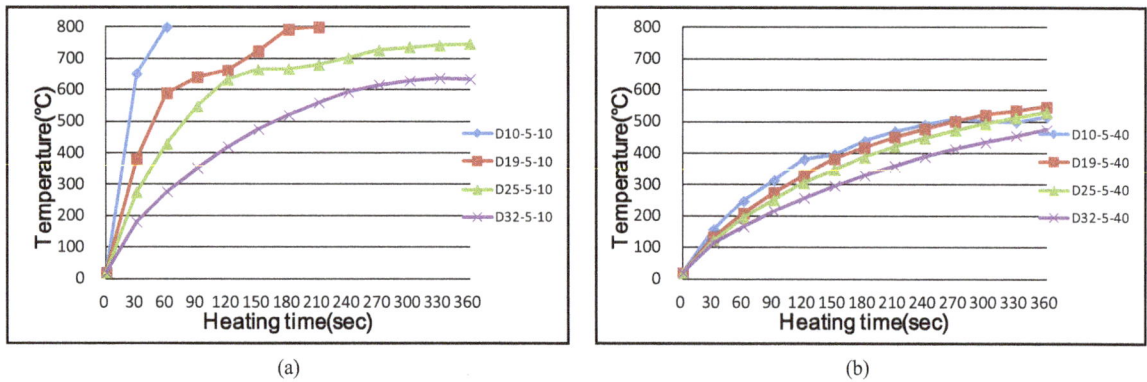

Figure 6. Temperature rise characteristics of steel reinforcement depending on high frequency induction heating (measured at 450 mm-center). (a) 5 kW-10 mm; (b) 5 kW-30 mm.

are shown in **Figure 7**. The horizontal axis of the figure is 150 mm, *i.e.*, the length of the specimen, but there are some errors in the range of temperatures owing to the radiant heat when measuring using the infrared ray temperature. The horizontal axis is 160 ± 5 mm on average, and the error of 10 ± 5 mm due to radiant heat does not have a large effect on the heating phenomenon at the center or at the end. Most of the specimens demonstrated that heating was more concentrated at the central part compared to at the end. The 120-mm-diameter steel reinforcement that was out of the heating coil showed a difference of 150°C - 450°C compared to the central part.

As the diameter of the heating coil was 120 mm in the experiment wherein the 430-mm steel reinforcement is heated at 5 kW, the point was measured with a distance of 30 mm from the center of the coil diameter toward the longitudinal direction of the steel reinforcement; the result is shown in **Figure 8**. There was not a large difference in temperature within the coil diameter, but the inside of the coil showed a temperature difference of 100°C or more when the external coil diameter was 30 mm or above. Most of the steel reinforcement showed a temperature rise by thermal conductivity of steel reinforcement outside of the magnetic field with an external coil diameter of 60 mm or above, which suggested that this was a localized heating phenomenon.

A high-temperature phenomenon on the surface of the steel reinforcement was identified at 400°C - 500°C in many cases. This high-temperature phenomenon was found only inside the heating coil, and the high-temperature phenomenon due to thermal conduction did not appear, which suggested that there was a localized heating phenomenon due to selective heating.

4. Temperature Characteristics of Cross-Steel Reinforcement Depending on High-Frequency Induction Heating

4.1. Experimental Overview

This study carries out an experiment to analyze the temperature rise characteristics and temperature distribution depending on high-frequency induction heating using SD345 steel reinforcement. This study identifies the temperature rise characteristics that can be found when using the single steel reinforcement and the changes in temperature distribution with cross-arrangement similar to that in an actual structure. If the steel reinforcement crosses, there may be a difference in heat characteristics caused by heating and conduction caused by resistance depending on the characteristics of the magnetic field. In addition, as the surface facing the coil increases and there is a large difference in heat transfer rate depending on the distance from the major steel reinforcement arranged at the bottom and the resistance area, it is expected that there will be a difference in temperature rising slope and the positioning of the heating coil.

This study identifies temperature rise characteristics, temperature gradient, and temperature distribution depending on the arrangement of steel reinforcement and distance heating calculated by thickness of covering.

4.2. Experimental Method

To identify the appropriate quantity of output or temperature distribution of steel reinforcement in the cross-steel reinforcement heating experiment, this study uses steel reinforcement of D10, D19, D25, and D32 in the case of

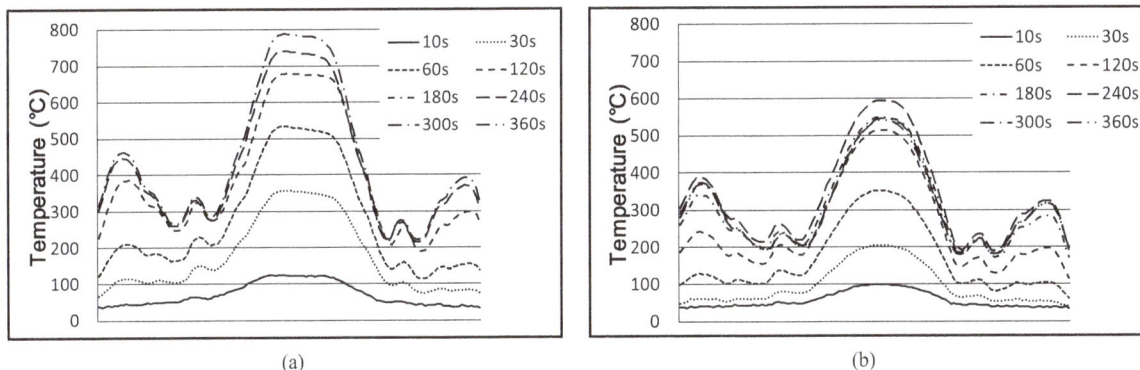

Figure 7. Temperature distribution of steel reinforcement depending on high frequency induction heating (5 kW-150 mm). (a) 5 kW-D19-10 mm; (b) 5 kW-D25-10 mm.

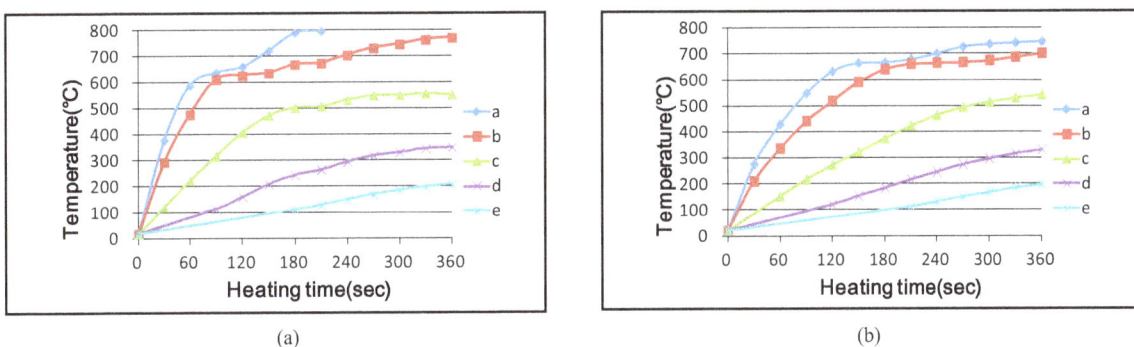

Figure 8. Temperature distribution of steel reinforcement depending on high frequency induction heating (5 kW-450 mm). (a) 5 kW-D19-10 mm; (b) 5 kW-D25-10 mm. Note: a: center of steel reinforcement, b, c, d, e: toward the end from the center of steel reinforcement 30, 60, 90, 120 mm.

SD345. To perform the experiment in the same conditions as those of the reinforced concrete members, this study cuts the length to 355 mm before use. This study assumes the steel reinforcement of D19, D25, and D32 as major steel reinforcements and carries out an experiment by cross-arranging the D10 steel reinforcement at the bottom.

As shown in **Figure 9**, this study places three steel reinforcements at intervals of 60 mm from surface to surface before heating. To select the optimal heating coil position, this study attempts a method to distribute the heating coil at the cross point of steel reinforcement (center heating) and also a method to distribute the heating coil between major steel reinforcements (side heating); the temperature characteristics depending on type and distance of steel reinforcement are then evaluated. The temperature is measured at five positions at a distance of 30 mm from the center of the upper and lower steel reinforcement using a heat-resistant camera, similar to the single steel reinforcement heating experiment.

4.3. Temperature Rise Characteristics of Cross-Steel Reinforcement Depending on High-Frequency Induction Heating

With a high-frequency output of 5 kW, the initial temperature rise curve showed the same tendency as that of the single steel reinforcement, with a sharp slope. However, it reached thermal equilibrium rapidly, showing the temperature reduction of up to 100°C compared to the single steel reinforcement. If the distance to the heating source was closer, there was no large temperature difference in the upper steel reinforcement of D10, but the upper and lower steel reinforcement showed a temperature difference between 80°C and 180°C when the heating distance was 40 mm or above. At the shortest distance, there occurs a heating phenomenon wherein a magnetic field is formed that reaches the lower steel reinforcement; at larger distances, a magnetic field is formed around the upper steel reinforcement before it reaches the lower steel reinforcement and the lower steel reinforcement is heated by conduction (as shown in **Figure 10**).

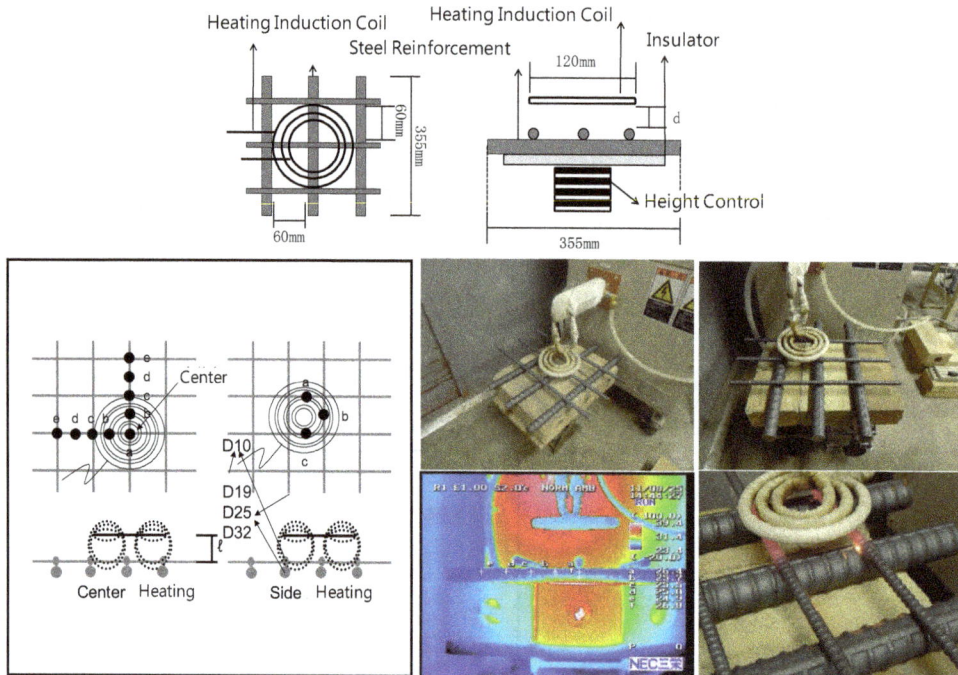

Figure 9. Experimental method of heating the cross-steel reinforcement using high-frequency induction heating.

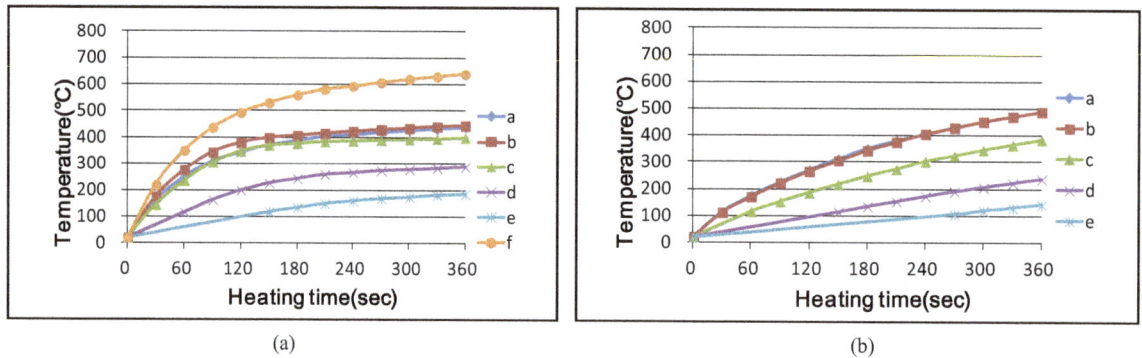

(a)   (b)

*Note: a: D00-0-0: D reinforced type—type of lower steel reinforcement—heating distance (mm).

Figure 10. Temperature rise characteristics of steel reinforcement depending on high-frequency induction heating (Center). (a) Measurement of D25-10-30 mm from the upper side; (b) Measurement of D25-10-30 mm from the lower side.

The temperature distribution shows almost the same tendency as the single steel reinforcement, but if the diameter of the heating coil is 120 mm, the temperature differential between steel reinforcement inside the diameter and steel reinforcement outside the diameter shows a large temperature differential in the upper steel reinforcement when the heating range is between 50°C and 200°C. This result can help identify the possibility of selective heating depending on the area of the magnetic field (as shown in **Figure 11**).

When the heating coil was placed between major steel reinforcements, the temperature rise characteristics showed a maximum difference of 100°C. When the heating coil was placed at the point of intersection and heated, each one of the upper and lower steel reinforcements located at the center of the coil diameter showed the same heating tendency. When the heating coil was placed between the major steel reinforcements, the heating efficiency decreased slightly, but the area of heating expanded when two upper steel reinforcements and two lower steel reinforcements in the range of the magnetic field were heated at the same time, which led to an increase in the width of the heating range. This seems to be attributable to the impact of coupling efficiency on the steel reinforcement and heating coil. As shown in Equation (8) and Equation (9), the heating coil is basically a resistor (inductance) and thus produces heat depending on [*material·length·diameter (area of surface) with the current that flows in the coil * coil resistance*]. In addition, the intensity of the magnetic force is inversely proportional to 2 - 3 square of distance and proportional to the current that occurs in the coil and coil differential (relative changes in two coils). Therefore, if the coupling efficiency of steel reinforcement and coil is bad, a high current is required. There are condensers and transformers within the conditioning unit (circuit), and they become a resistance [18] [19].

$$F = G \frac{m_1 m_2}{r^2} \qquad (8)$$

$G = 6.672 \times 10^{-11} \ \mathrm{m^3/kg \cdot s^2}$;
m = mass;
r = distance;
F = output.

$$output \ (intensity \ of \ magnetic \ field) = A \ (current) \times T \ (coil \ differential) \qquad (9)$$

In other words, as in **Figure 12**, which shows the coupling of steel reinforcement and coil, in (b), where the distance from the heating coil is the same as the distance between the coils to the surface of the steel reinforcement, two steel reinforcements are included within similar magnetic fields and thus resistance heat occurs up to a temperature like (c). Therefore, if the voltage is the same, loss within the circuit increases and the temperature of the heated object decreases with increasing current (or raised voltage).

5. Deconstruction Assessment of Reinforced Concrete Using Hitting Method

5.1. Experimental Overview

The purpose of this experiment is to evaluate the dynamic dismantlement of reinforced concrete members weakened by high-frequency induction heating. As there are no previous studies on the methods for evaluating the dismantlement of concrete members quantitatively, in most cases, qualitative evaluations such as "difficult to dismantle" are defined without evaluating the time or energy required for dismantling. Therefore, the data is not reliable as reproductive data in reviewing the expenses for dismantling or environmental load [20].

This study measures the time and energy required for unit dismantlement and makes a quantitative evaluation of dynamic dismantlement of reinforced concrete members weakened by high-frequency induction heating.

5.2. Experimental Method

The reinforced concrete members weakened by high-frequency induction heating were separated to the extent that steel reinforcement cutting can be done in the field. By installing 3-axis acceleration sensors in the hammer, the dismantling tool, this study measures the dismantling location and the time consumed by each experimental group. To calculate the time and energy taken for dismantling for single reinforced concrete specimens, the acceleration and direction of dismantling were measured until the steel reinforcement could be exposed by hitting a heated specimen that was heated in the longitudinal direction continuously. To calculate the time and energy taken for dismantling in the case of cross-steel reinforcement, the acceleration and direction of dismantling were measured until the steel reinforcement could be exposed by hitting a heated specimen that was heated by induction heating of both the center and side.

5.3. Deconstruction Assessment of Single Reinforced Concrete

The separating experiment results that compare the specimen with 30 mm steel reinforcement covering depth (D19), which requires the least electricity for the separation, before and after heating, are represented in **Figure 13** and **Figure 14**.

In most of the specimens, the destruction power showed a reducing tendency with increasing diameter of steel reinforcement, and the power required for destruction decreased with decreasing distance from the induction heating coil. In case of the specimen using D10, with the least heating efficiency, the power depending on the induction heating was the smallest. However, it showed a tendency that it reduced more greatly at 30 mm—a smaller heating distance—than at 40 mm or 50 mm. When the heating distance was smaller, the temperature of

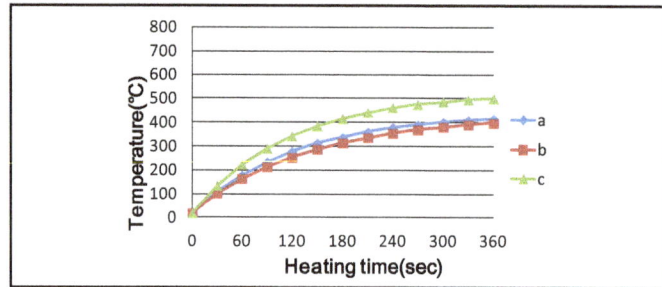

Figure 11. Temperature rise characteristics of steel reinforcement depending on high-frequency induction heating (Side).

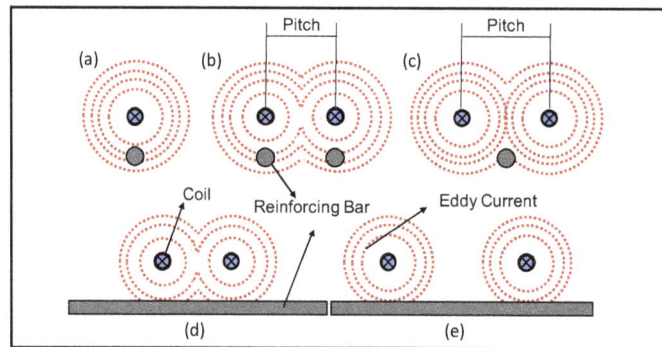

Figure 12. Improved coupling efficiency in steel reinforcement and heating coil.

steel reinforcement rose rapidly; when the diameter of steel reinforcement was small, thermal equilibrium was reached immediately after heating and the heat conduction velocity for concrete increased and then weakened immediately. This is why the breaking stress was smaller than it was for other specimens with relatively thicker covering depths.

In case of D19 and D25, with higher heating efficiency, a reduction in destruction stress with heating distance was evident. In the case of D19, the reducing rate at covering depths of 50 mm and 30 mm had about a 69% difference; for D25, there was about a 71% difference. For D32, with the smallest destruction stress, there was about an 83% reducing rate compared to the standard specimen. There may be some errors in the destruction stress caused by the increased covering thickness, but D19 showed about a 77% reducing rate, compared to the standard specimen.

5.4. Deconstruction Assessment of Cross-Reinforced Concrete

The specimen that was heated at the center showed a decreased tendency for destruction stress compared to the specimen that was heated at the side. With a covering thickness of 30 mm, the center-heated sample showed about a 39% reducing rate and the side-heated sample showed about a 31% reducing rate, which suggested that there was an 8% difference. Compared to the center heating, crack propagation showed a high efficiency when heating the side, but the center heating showed a high heating efficiency in the crack width and internal temperature rise ratio. Therefore, center heating may be influential on the stress required to separate steel reinforcement from concrete (as shown in **Figure 15** and **Figure 16**).

The overall tendency before and after high-frequency induction heating increased according to the covering thickness. The crack width caused by the expansion pressure depending on heating is more influential to the steel reinforcement than is chemical weakening, which depends on thermal conduction.

6. Conclusions

This study carries out an experiment to separate steel reinforcement from concrete as completely as possible through indirect heating using high-frequency induction heating, especially of the steel reinforcement inside the

(a)

(b)

(c)

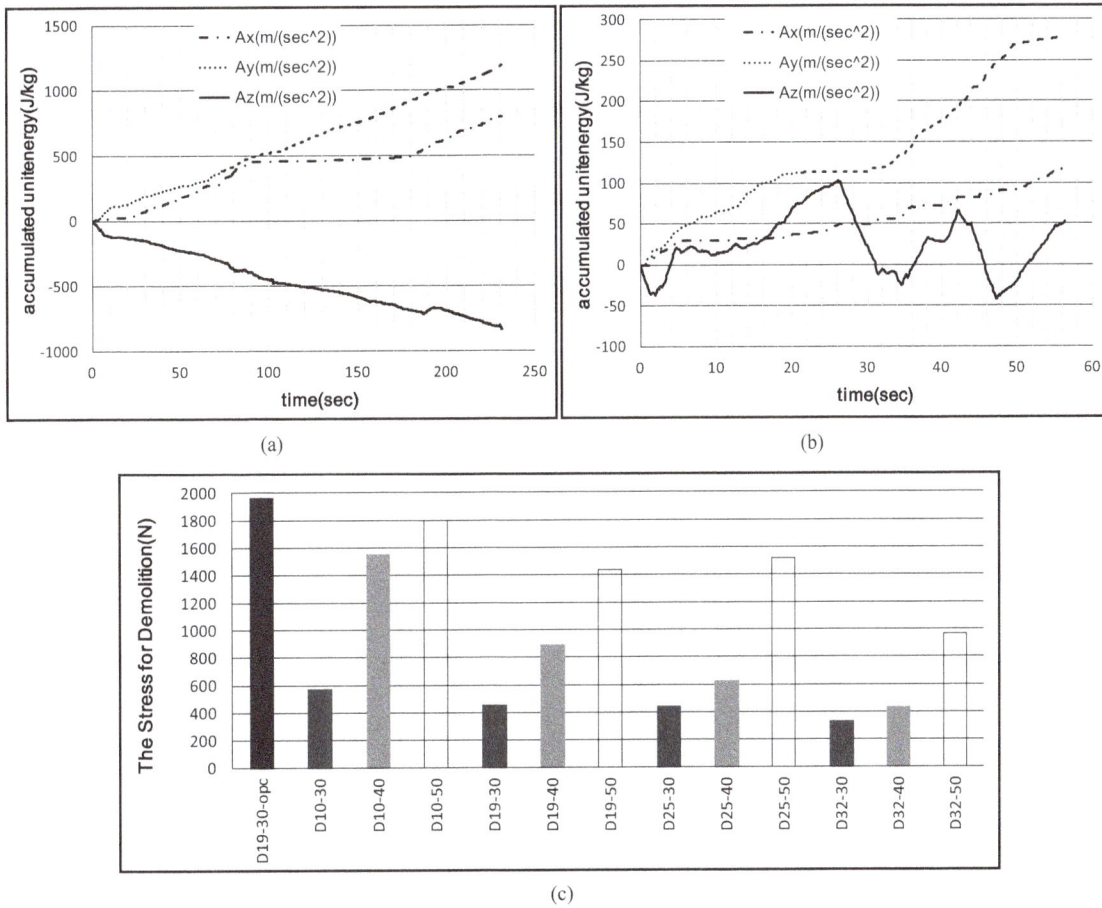

Figure 13. Experiment result of deconstruction of single reinforced concrete member after high frequency induction heating. (a) Specimen before heating (D19 - 30 mm); (b) Specimen after heating (D19 - 30 mm); (c) Resolved stress of reinforced concrete member.

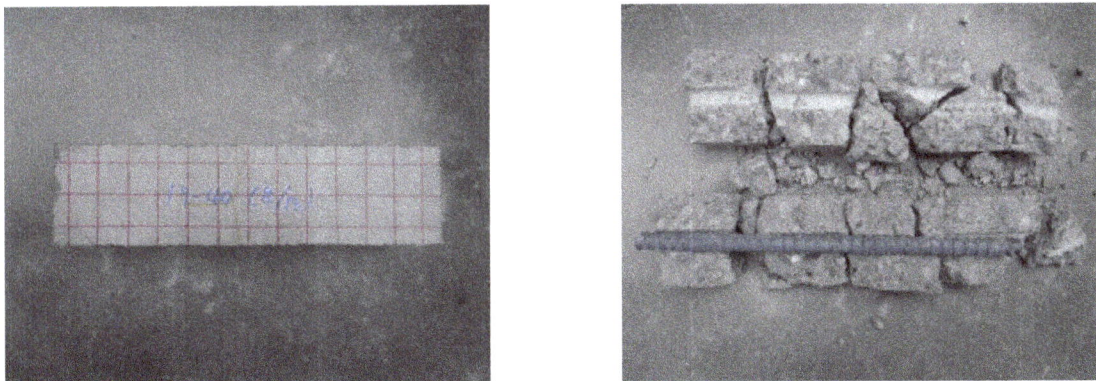

Figure 14. Experiment result of complete deconstruction of single reinforced concrete member after heating.

reinforced concrete, and the findings are as follows:

1) The frequency with the maximum output determines the depth of penetration in the induction heating. If the output increases enough to heat the steel reinforcement that is far from the heating coil, current density and intensity of magnetic field increase in general, but the depth of penetration decreases. Thus, it is necessary to take into account the appropriate output at a certain distance. As the appropriate output does not show any large difference in temperature rise with time at a distance of less than 40 mm, the output of 5 kW or above leads to heat loss at temperatures higher than the Curie point and there is a problem in power dissipation.

(a)

(b)

(c)

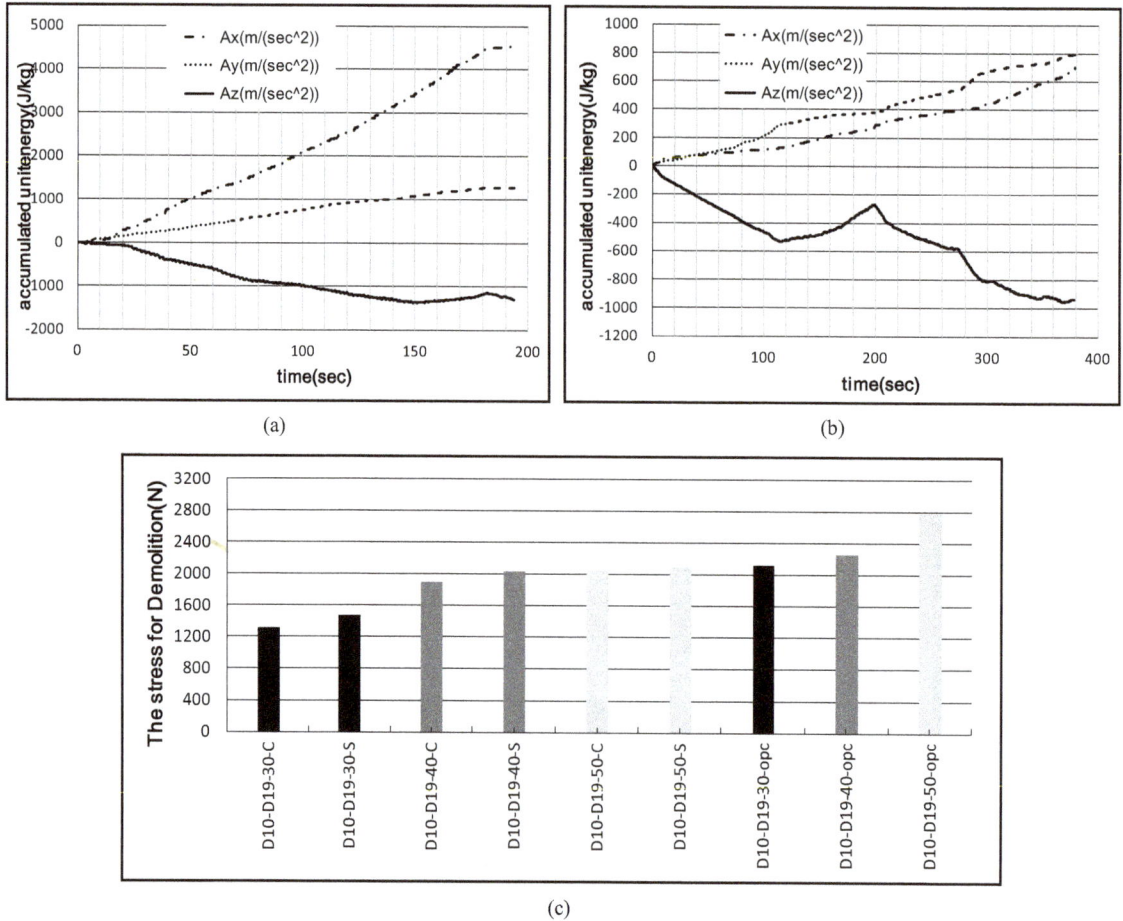

Figure 15. Experiment result of destructing the cross-reinforced concrete member after high frequency induction heating. (a) Specimen before heating (D19-10-30 mm) the center; (b) Specimen after heating (D19-10-30 mm) the center; (c) Resolved stress of reinforced concrete member.

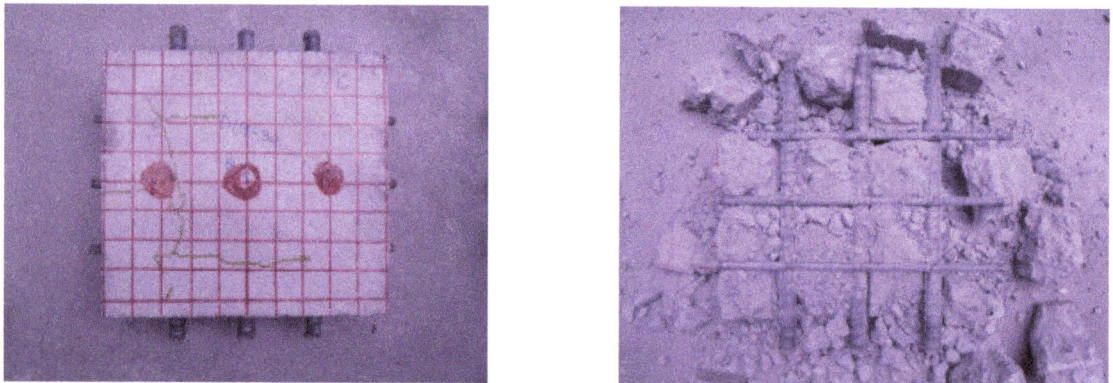

Figure 16. Experiment result of destructing the cross-reinforced concrete member after heating.

2) Using the high-frequency induction heating method, the surface of steel reinforcement can be heated up to 300˚C, at which the concrete-weakening temperature is the most effective. As the heating range is closely related to the diameter of the heating coil, selective heating up to 30 mm in the horizontal direction is possible based on the heating coil. With cross-arranged steel reinforcement, a covering thickness of 40 mm is possible for an effective heating, and a covering thickness of 50 mm or above is possible for an effective heating by thermal conduction through the upper steel reinforcement.

3) When high-frequency induction heating is applied to the structure in the field, steel reinforcement can be cut by exposing the steel reinforcement using the hitting method after weakening the concrete by the occurrence of cracks caused by the heated expansion pressure and thermal conduction. Thus, effective work is possible owing to weakening caused by rapid heating of the exposed steel reinforcement.

The high-frequency induction heating technology proposed in this study may be applicable to the field of dismantling existing high-rise buildings, fields where workers cannot use heavy equipment, and fields that are sensitive to pollutants such as noise and dust. Particularly in the field of dismantling single structures with localized dismantling, for example, repair or reinforcement, highly efficient dismantling is expected to be possible.

Acknowledgements

This work was supported by the GRRC program of Gyeonggi province. [(GRRC HANKYONG 2011-B05), Carbon Neutral Wall and Floor Elements for Smart Distribution Center].

References

[1]   World Economic Forum (2014) The Global Energy Architecture Performance Index Report 2014. Cologny.

[2]   Robert, W. and Messler, J.R. (2004) Joining of Materials and Structures: From Pragmatic Process to Enabling Technology. Elsevier Inc., Amsterdam.

[3]   Gatta, D. (2003) Generation and Management of Construction and Demolition Waste in Greece—An Existing Challenge. *Resources, Conservation and Recycling*, **40**, 81-91. http://dx.doi.org/10.1016/S0921-3449(03)00035-1

[4]   Mizutani, R. and Yoshikai, S. (2011) A New Demolition Method for Tall Building. Kajima Cut & Take down Method. *CTBUH Journal*, **5**.

[5]   Sealey, B.J., Phillips, P.S. and Hill, G.J. (2011) Waste Management Issues for the UK Ready-Mixed Concrete Industry. *Resources. Conservation and Recycling*, **32**, 321-324. http://dx.doi.org/10.1016/S0921-3449(01)00069-6

[6]   Uttam, K. and Balfors, B. (2014) 9—Green Public Procurement (GPP) of Construction and Building Materials. Eco-Efficient Construction and Building Materials, Life Cycle Assessment (LCA), Eco-Labelling and Case Studies, 166-195. http://dx.doi.org/10.1533/9780857097729.1.166

[7]   Sthiannopkao, S. and Wong, M.H. (2013) Handling E-Waste in Developed and Developing Countries: Initiatives, Practices, and Consequences. *Science of The Total Environment*, **463-464**, 1147-1153. http://dx.doi.org/10.1016/j.scitotenv.2012.06.088

[8]   Lin, F. (2012) The Development Path of Japanese Green Architecture under Energy Policy—Taking Misawa Home as an Example. *Energy Procedia*, **14**, 1305-1310. http://dx.doi.org/10.1016/j.egypro.2011.12.1093

[9]   Li, C.E., Burke, N., Gerdes, K. and Patel, J. (2013) The Undiluted, Non-Catalytic Partial Oxidation of Methane in a Flow Tube Reactor—An Experimental Study Using Indirect Induction Heating. *Fuel*, **109**, 409-416. http://dx.doi.org/10.1016/j.fuel.2013.02.055

[10] Panão, M.R.O., Correia, A.M. and Moreira, A.L.N. (2012) High-Power Electronics Thermal Management with Intermittent Multijet Sprays. *Applied Thermal Engineering*, **37**, 293-301. http://dx.doi.org/10.1016/j.applthermaleng.2011.11.031

[11] Wrzuszczak, M. and Wrzuszczak, J. (2005) Eddy Current Flaw Detection with Neural Network Applications. *Measurement*, **38**, 132-136. http://dx.doi.org/10.1016/j.measurement.2005.04.004

[12] Myrhaug, D. and Holmedal, L.E. (2014) Wave-Induced Current for Long-Crested and Short-Crested Random Waves. *Ocean Engineering*, **81**, 105-110. http://dx.doi.org/10.1016/j.oceaneng.2014.02.017

[13] Xiao, C.-L. and Huang, H.-X. (2014) Optimal Design of Heating System for Rapid Thermal Cycling Mold Using Particle Swarm Optimization and Finite Element Method. *Applied Thermal Engineering*, **64**, 462-470. http://dx.doi.org/10.1016/j.applthermaleng.2013.12.062

[14] Postacchini, M. and Brocchini, M. (2014) A Wave-by-Wave Analysis for the Evaluation of the Breaking-Wave Celerity. *Applied Ocean Research*, **46**, 15-27. http://dx.doi.org/10.1016/j.apor.2014.01.005

[15] http://www.ndt-ed.org/EducationResources/CommunityCollege/MagParticle/Physics/HysteresisLoop.htm

[16] Hukushima, Y. (2009) Experimental High Frequency Induction Heating Collision Instruction Manual (IMC-ADH-502). Imecs Corporation, Japan.

[17] Sadeghipour, K., Dopkin, J.A. and Li, K. (1996) A Computer-Aided Finite Element/Experimental Analysis of Induction Heating Process of Steel. *Computers in Industry*, **28**, 195-205. http://dx.doi.org/10.1016/0166-3615(95)00072-0

[18] Lee, K.S. and Hwang, B. (2014) An Approach to Triangular Induction Heating in Final Precision Forming of Thick

Steel Plates. *Journal of Materials Processing Technology*, **214**, 1008-1017. http://dx.doi.org/10.1016/j.jmatprotec.2013.11.002

[19] Kima, W., Suha, C.-Y., Roha, K.-M. and Choa, S.-W. (2013) Mechanical Properties of (W, Ti)C and (W, Ti)C–NiAl3 Cermet Consolidated by the High-Frequency Induction-Heating Method. *Journal of Alloys and Compounds*, **568**, 73-77. http://dx.doi.org/10.1016/j.jallcom.2013.02.187

[20] Cho, K.-H. (2012) Coupled Electro-Magneto-Thermal Model for Induction Heating Process of a Moving Billet. *International Journal of Thermal Sciences*, **60**, 195-204. http://dx.doi.org/10.1016/j.ijthermalsci.2012.05.003

8

Modelling Recycling Targets: Achieving a 50% Recycling Rate for Household Waste in Denmark

Amanda Louise Hill[1]*, Ole Leinikka Dall[2], Frits Møller Andersen[3]

[1]Department of Development and Planning, Aalborg University, Aalborg, Denmark
[2]Institute of Chemical Engineering, Biotechnology and Environmental Technology, University of Southern Denmark, Odense, Denmark
[3]DTU Management Engineering, Technical University of Denmark, Roskilde, Denmark
Email: *alhill@plan.aau.dk

Abstract

Within the European Union (EU) a paradigm shift is currently occurring in the waste sector, where EU waste directives and national waste strategies are placing emphasis on resource efficiency and recycling targets. The most recent Danish resource strategy calculates a national recycling rate of 22% for household waste, and sets an ambitious goal of a 50% recycling rate by 2020. This study integrates the recycling target into the FRIDA model to project how much waste and from which streams should be diverted from incineration to recycling in order to achieve the target. Furthermore, it discusses how the existing technological, organizational and legislative frameworks may affect recycling activities. The results of the analysis show that with current best practice recycling rates, the 50% recycling rate cannot be reached without recycling of household biowaste. It also shows that all Danish municipalities will need to make efforts to recover all recyclable fractions, and that the increased recycling efforts of only selected municipalities will not be sufficient to reach the target.

Keywords

Recycling Rate, Recycling Potential, Household Waste, Econometric Modeling, Waste Management

---
*Corresponding author.

1. Introduction

1.1. Background

Within the European Union (EU) a paradigm shift is currently occurring in the waste sector, where waste policies and strategies are moving away from a focus on public health, and toward an agenda of resource efficiency. This shift in paradigm is apparent in the so-called Waste Framework Directive (WFD) [1] and the EU Commission's Roadmap to a Resource Efficient Europe [2] and in the national waste strategies of the Member States; In Denmark, the newest national waste strategy has been published instead as a resources strategy [3]. The WFD sets a target of 50% recycling of municipal solid waste to be fulfilled by 2020.

In March of 2013, the European Environment Agency released a report detailing the recycling efforts in all the EEA countries which found that the majority of countries were not on track to reaching the target set in the WFD. The report looked at the annual rate of increase in recycling over the period 2001 to 2010, and considered if this rate of increase were continued to 2020, would the 50% recycling target be met. 5 countries have already achieved the target, and a further 6 are on track, while Denmark is amongst the remaining 21 countries that will need to increase its efforts if it is to achieve the recycling rate [4].

Despite the fact that the EEA considered Denmark would need to increase efforts in order to achieve the target set out in the WFD, the new national resource strategy (DRS) released in October of 2013 sets an even more ambitious national target [3]. While on the surface the target appears to be the same; 50% recycling rate of household waste by 2020, a difference in the calculation method means that certain fractions of waste that are included in the recycling rate as reported to the EU are not included in the recycling rate as calculated for the DRS. Denmark reports its recycling rate to the EU based on the fourth calculation option given in connection with the WFD, where all waste that is collected under the municipal waste ordinance is considered [5]. According to this definition Denmark had a household recycling rate of 42% in 2010. The DRS defines the recycling rate to be calculated from waste that is collected from households, and specifically; biowaste, paper, cardboard, glass, plastic, wood and metal. Electronic waste, hazardous waste, and other wastes that are collected according to separate ordinances are not included, giving a recycling rate of only 22% in 2013 [3]. The term biowaste here refers to the "wet" portion of household waste which is primarily composed of food waste, although different municipalities may have differing definitions of biowaste due to technical or hygiene considerations.

In order to increase the recycling rate, one of two actions is necessary; either the amount of non-recyclable waste needs to be decreased through prevention strategies which gives a greater proportion of waste going to recycling, or recyclable waste from the mixed waste stream needs to be diverted away from incineration and toward material recovery activities. This study assumes that all efforts to increase the recycling rate will be made through diversion of mixed household waste from incineration to recycling. Therefore this analysis of how to achieve the recycling target focuses on the composition and generation of waste in the mixed household waste stream.

While the DRS sets a very ambitious goal for municipalities to achieve, it purposefully does not address how the goal should be achieved, with the intention of allowing each individual municipality to find the solution that best fits their needs. This open approach is taken to ensure that no municipality is locked into a technological option that does not suit its needs: some municipalities are characterized as rural areas, with widely spread, single-family dwellings, while other municipalities are densely populated, inner-city areas with a high proportion of multi-family dwellings, while still others are a mixture of rural and urban. Given this diversity, it is unlikely that one technological option would be optimal for all municipalities. However, the relatively small amounts of household waste that each municipality is responsible for, render many technologies unfeasible on the municipal scale.

1.2. State of the Art—Current Methods for Planning Recycling Systems

Since the 1970's Denmark's waste management planning and investment in infrastructure in the waste sector has focused on the expansion of incineration and shifting waste from landfill to incineration. In the previous national waste strategy, which covered the years 2009 to 2012, the goal was to recycle at least 65% of all waste, with no specific targets for household waste other than for the paper fraction [6]. In that strategy, the econometric model called FRIDA was used to project the amount of combustible waste that would be produced in comparison to the incineration capacity. However, this approach neglects to include the information of how much combustible waste will be diverted from incineration to recycling if recycling targets are to be reached. While the flow of waste from one stream to another through differences in collection will not impact the total generation of waste [7], it has the potential to affect how much waste is available for both recycling and incineration.

In order to develop an efficient strategy for achieving the recycling target, it is essential to know what the future generation of household waste will be, both in terms of amount and composition [7] [8]. To determine this information, the Danish authorities currently rely on the use of econometric models, such as the FRIDA model, which extrapolates economic and waste generation data from the past to predict the future generation of waste. However, these models do not currently take into account the political goals that are set. This investigation is an attempt to incorporate information of policies and targets with the output of the FRIDA model. For a detailed description of the model, see the article by Andersen and Larsen [9].

In a literature review of the use of mathematical models for predicting waste generation, it was found that the majority of studies in this field are carried out as part of an optimization and planning exercise [10]. This study is novel in that it applies the waste generation model for the purpose of investigating the feasibility of a political goal. Reference [11] also discusses the need to develop existing models to incorporate more aspects into an analysis.

2. Research Approach

2.1. Determining the Generation and Composition of Waste Expected in 2020

The FRIDA model was used to determine the projected generation of household waste in 2020. The model creates a projection for the future generation of waste based on historical data of how much waste has been produced and economic projections prepared by the Danish Ministry of Finance. The historical data for waste generation is sourced from the Danish Environmental Protection Agency's database which underwent major changes in the way data was collected from waste handlers in 2010. The data in the years since the change have been unreliable as the sector learns how to use the new system. Therefore, the figures used in this projection are based on the 2009 data. The data represents total household waste and includes all waste fractions that originate in households, as well as down-stream wastes such as the slag from incineration. The design of the Danish data collection system reflects the EU definition of municipal waste, which gives the national recycling rate that is reported according to the so-called Waste Framework Directive [1] [5]. Many of the fractions included in this data as household waste are not relevant according the DRS target as it relates explicitly to household waste, and not to municipal waste. Further to this, there are certain waste fractions that originate in households that are not included in the DRS definition of waste as they are legislated under different ordinances, such as in the case of batteries and WEEE. Categories of municipal and household waste that are explicitly not included in the DRS definition of household waste were not included in the projection.

Surveys to determine the composition of mixed household waste are costly and time consuming, as it requires that waste be analysed on a large scale to eliminate systematic errors such as those due to regional and seasonal differences. Perhaps for this reason, the number of studies detailing the composition of Danish household waste have been limited; only three studies have been published determining the relative amounts of different fractions of waste [12]-[14]. Two of the studies [12] [13] both interpret a single set of data, while the third study [14] builds on the data from the two previous studies and also collects a new data set. The new data collected in the 2013 study is however not representative of the entire Danish population as participating households were located in a single geographical area, and also the participating households were aware that their waste was being monitored which could compromise the validity of the results. Despite the difference in collection methods and the time difference better the samples, all three studies give a similar picture of the composition of mixed household waste. In this study, the composition of household waste is based on the data reported in [12]. It is acknowledged that the quality of this data is not desirable as the composition of the waste may have changed over this time, and further uncertainty is added considering that the composition of mixed waste is likely to continue to change in the future. In order to determine if the data was sufficiently accurate for this study, it was compared to more recent Swedish data for mixed household waste composition [15]. Sweden is geographically close and culturally similar to Denmark, and it can therefore be assumed that waste composition would be quite similar in the two countries. The comparison showed that the compositions of Danish waste from 2001 and Swedish waste from 2011 were quite similar for most waste fractions. The greatest difference was seen in the fractions of biowaste, where the ratio in Swedish data was lower than that seen in Danish household waste, and glass where the ration in the Swedish waste was higher than the Danish. Importantly, both sets of data show that biowaste is the dominant fraction of mixed household waste by weight. While it is acknowledged that the data relied upon for the composition of mixed household waste is, due to its age, not as accurate as desired, it is considered to be the best representation available, and of sufficient quality to determine which waste fractions should be targeted for recycling.

2.2. Determining the Recycling Potential of Household Waste Fractions

The information of the composition of the mixed household waste was integrated with the FRIDA projection to determine what potential each fraction of the mixed household waste has for contributing to the 50% recycling target. In order to identify how much of that potential could be recovered for recycling, in other words, how much of the potential could be utilized, the projected potentials were compared to best practice recycling rates. A recent report looked into the potential recycling rate that could be achieved in Denmark for different fractions of waste based on both Danish experiences and experiences of Sweden and Germany [14]. The report considered 2 different rates for recycling for single family households, and a lower rate for multi-family dwellings. The two different rates were applied in this study with the higher recycling rate being used as a best case scenario, and the lower rate being applied as a lower recycling rate. The assumed recycling rates can be seen in **Table 1**. In the case of metals, the lower assumed recycling rate is lower than the projected recycling rate for 2020 considering no changes. In this case, the projected recycling rate was used as a lower value.

The data was then represented in 5 different scenarios. Scenario 1 represents a "best case", where all municipalities recycle all fractions optimally. Scenario 2 assumes the high recycling efficiency can be obtained for all dry recyclable fractions, but that biowaste recycling is not adopted by any municipalities not currently collecting biowaste for recycling. Scenarios 3 and 4 assume the lower rate of recycling efficiency, with all municipalities collecting biowaste in Scenario 3, while in scenario 4biowaste is not expanded past current efforts. Scenario 5 assumes that there is no change in recycling activities through to 2020 and is included as a baseline. The scenarios are designed to test the importance of biowaste recycling, as it has been recognized by other authors as being necessary to recycle in order to achieve high recycling rates [14] [15] [17].

3. Results

As shown in the **Figure 1** and **Figure 2**, the baseline projection shows that the total amount of waste expected in the future will continue to increase. While both the combustible and recycling streams are expected to increase, the combustible stream is projected to experience a greater increase than the recycling stream. Consequently without taking any action either to increase recycling rates or to prevent the generation of combustible waste, the recycling rate in 2020 is lower than it is today (although it is important to note that based on 2009 figures the current recycling rate as calculated for this study is slightly higher than that published in the DRS). In the data supplied from the Danish waste database we can see that the majority of the waste in the combustible category is collected through the daily mixed waste collection. The projected increase in waste generation can be seen in **Figure 1**.

Table 1. Recycling rates applied to mixed household waste. Based on data from [16].

Waste fractions	Applied recycling rates		
	Upper recycling rate	Lower recycling rate	Recycling rate in 2009
Biowaste	0.7	0.5	0.06[a]
Paper and Carsboard	0.9	0.7	0.49
Plastic	0.6	0.4	0.09
Glass	0.85	0.85	0.58
Metal	0.8	0.6	0.76

[a]2009 recycling rate for biowaste applied in scenarios 1 and 3.

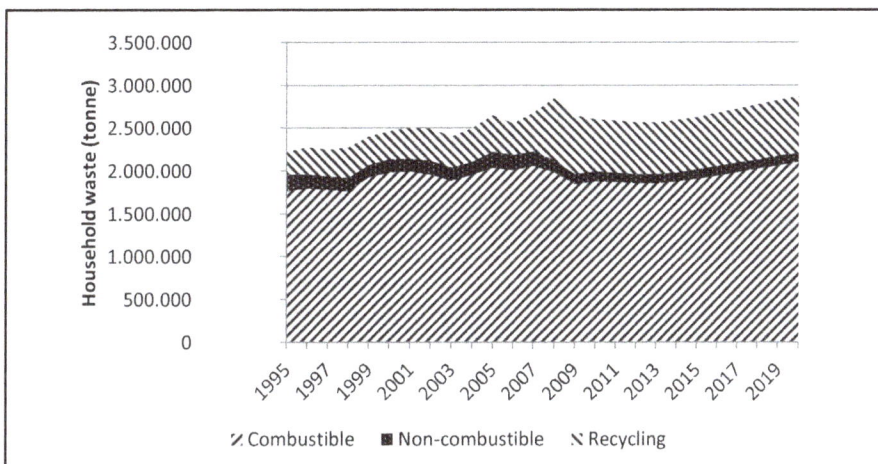

Figure 1. Baseline projection of waste generation and handling in tonne per year.

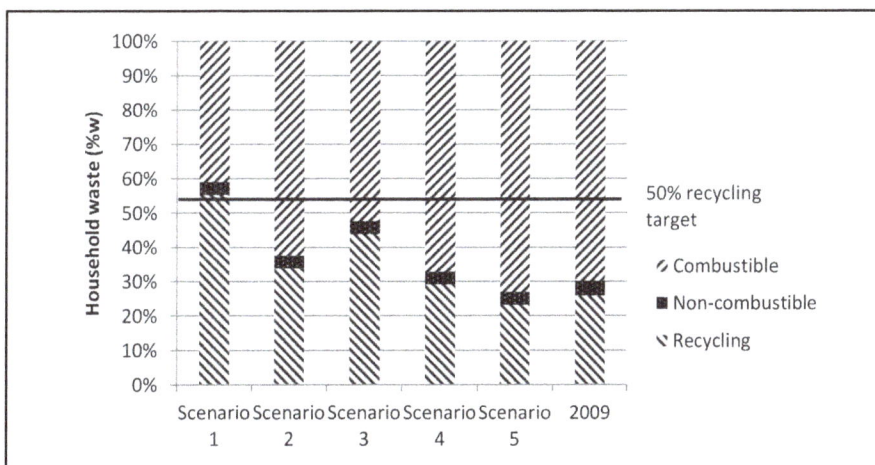

Figure 2. Handling of waste in alternative scenarios. The recycling rate reaches the 50% target only in scenario 1 which assumes a high recycling rate for all fractions.

While it was possible to project in tonnes how much waste will be diverted from incineration to recycling, it was not possible to determine exactly what effect this will have on total amounts of waste going to Danish incineration plants. This is due to the fact that generation of domestic industrial waste is not included. However, we can see that in the most optimistic scenario, the amount of waste diverted away from incineration represents 22% of the total 4.2 million tonnes of installed incineration capacity in Denmark [18].

Scenario 5, the base scenario, shows that without any efforts to increase the recycling we can only expect a recycling rate of 23%. The best case scenario, scenario 1, exceeds the target with a recycling rate of 55%. Scenarios 2, 3 and 4 all project recycling rates lower than the 50% target. The flows of waste in each of the scenarios can be seen in **Figure 2**. In **Figure 3**, the total amount of each fraction recycled in each scenario is represented.

4. Discussion

4.1. Potential for Recycling

The biowaste fraction is, by weight, the largest fraction of the mixed household waste and is also one of the least recycled fractions. Biowaste therefore represents the greatest potential for reaching the target, as can be seen in **Figure 3**. Only a handful of the 98 Danish municipalities currently have collection and recycling systems for household biowaste, and based on this current effort the projected recycling rate for household biowaste is only

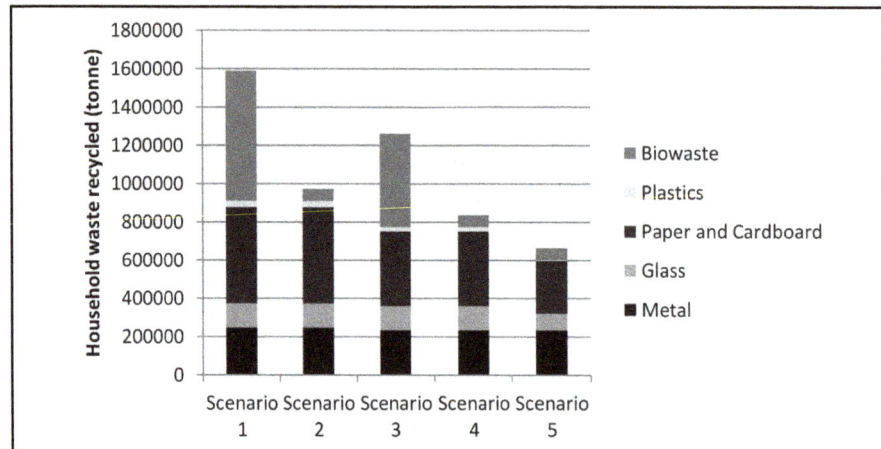

Figure 3. The amount in tonne of each fraction recycled in the different scenarios. This figure shows the amount in tonne recycled of each fraction in the different scenarios. The greatest variation and potential to contribute to the recycling rate is in the biowaste fraction.

expected to be 0.06% in 2020, which is far short of the theoretical recycling rate that could be achieved of 50% - 70%.

In terms of the dry recyclables, the plastics fraction represents the largest potential, as it represents a large percentage by weight of the mixed waste. Also plastics are currently not widely collected from households for recycling, with only small amounts being collected at recycling stations through bring systems. While paper and glass are already collected either with kerbside or residential collection points in most municipalities, some potential still exists for further recycling of these fractions. Metals represent only a small fraction, and a system already exists to recycle many metals from the ashes after incineration.

The 5 scenarios show, the potential of all the dry recyclable streams is not sufficient to reach the 50% recycling target. In scenario 2, where recycling of biowaste is not expanded beyond today's efforts, the projected recycling rate is only projected to be 34%. This indicates that the efforts required to reach the recycling target of 50% by 2020 will require a significant effort from all municipalities. However, this analysis assumes that all efforts to increase the recycling rate should come from the mixed household waste stream. Other waste streams may also have a good potential to contribute to the recycling rate. Two possible waste streams that could be investigated are the "bulky waste" that is collected from residential areas, and the "small and large combustible" waste stream collected at the recycling stations. The compositions of these two streams are currently unknown, and they are likely to differ between municipalities. Alternatively, a large potential may exist in increasing garden waste collection. The projection indicates that this alone could increase the Danish recycling rate to a level which would achieve the WFD target, however garden waste is not included as one of the household waste streams contributing to the DRS target.

4.2. Technological Considerations

While the biowaste fraction shows the greatest potential for contributing to the recycling target, it is possibly the fraction with the most technological hurdles. The biowaste fraction is also sometimes referred to as the "wet fraction" as it has a high water component and is biodegradable. These characteristics of the waste have implications for how it should be treated in respect to the other fractions. Source separation prevents contamination of organic material with potentially hazardous substances in the mixed stream, and inversely contamination of dry recyclables with organic matter. Unlike dry recyclables which are stable and can be stored for longer periods of time, biowaste should be collected regularly from residents to reduce smells and attracting pests and rodents.

A few Danish municipalities already collect biowaste for treatment at anaerobic digestion facilities [19], and some experiences also exist from biowaste collection trials in Denmark and neighbouring Sweden [12] [20] [21]. The methods of collection differ in different municipalities, and even more technological solutions exist in other EU countries. The municipality of Vejle in central Denmark has one of the most efficient systems which is based on optical technology, where green coloured bags containing biowaste are mechanically separated from black bags containing mixed waste.

Dry recyclables represent their own share of difficulties. Recycling collection schemes can in general be sorted into different categories: central sorting or source separation. In most instances the quality of recyclables achieved through source separation schemes is higher than that achieved through central sorting. The quality of the recyclable materials affects what purposes it can be used for and also the price for which it can be traded. However, source separation can be a politically unpopular option as it occupies more time and space for the citizen. The infrastructure requirements at the street level can also be a major cost. For example, in the Danish municipality of Aarhus, recycling bins for several different dry fractions have been installed below street level. While this is considered a low-cost option once the infrastructure has been established, the cost of expanding such a system to include all recyclable streams would likely be too costly. However, the question of central sorting versus source separation is not an either-or question, as some level of source separation is required for central sorting systems, and sorting of recyclables is also necessary in source separated waste. This is particularly important for the recycling of plastics, as different types of plastic should be separated for recycling.

In the case of metals, there is currently a high rate of recycling due to a system where they are harvested from the ashes after incineration. However, it must be considered how the calorific value of the mixed waste will be changed if biowaste and dry recyclables are removed. Socio-economic analyses would have to be conducted to determine if the increased quality of metals recovered through a collection scheme would justify the cost of separating the metals before incineration.

A third technological option, which is currently being trialed in Copenhagen, is a central sorting of mixed household waste, where both organic matter and dry recyclables are recovered for recycling. This process, called ReneScience, uses enzymes to break-down organic matter into a high energy "soup", which can be stored and used to produce biofuels. The other materials in the mixed waste stream; metals, glass, plastic, are not affected by this enzymatic process and are available for recycling after a mechanical sorting process. However, one implication of this technology for established recycling rates is that paper handled in this system is recycled together with other organic components, and is therefore not available for traditional paper recycling. Also, as the technology is still in a test phase, it is likely not feasible that it can be implemented on a nation-wide scale in the short time-span before the 2020 deadline.

In Denmark one of the major technological challenges is the low quantities of waste. Given the high costs associated with establishing a mechanical sorting plant for recyclables, it is likely that most municipalities will need to co-invest in such facilities. Another technological challenge for Denmark is the lack of domestic recycling operations which could set a specific demand for the quality of recyclables. This means that dry recyclables recovered by municipalities are largely destined to be sold on the global materials markets.

4.3. Legislative Considerations

The organizational structure of the waste sector is such that each municipality is responsible for the collection and treatment of all household waste produced within their boundary. Until 2010, this municipal responsibility also included industrial waste produced within the municipal boundary, but a recent liberalization of industrial waste diverts recyclable industrial waste away from municipal facilities, and potentially also combustible industrial waste. Each municipality has recycling stations where citizens can bring their pre-sorted recyclables, from where they are sold on to recycling facilities in other countries, but collection of recyclables from residential areas is in general limited to only paper and glass. This organizational structure and the technological focus on incineration has meant that recycling efforts have been decentralized and investments in advanced sorting or treatment technologies have not taken place.

The decentralized nature of waste handling in Denmark can be considered a barrier to the establishment of central sorting facilities. In order to ensure a large enough waste stream for economic viability, it would require several municipalities to collectively invest in a central sorting facility. There is a good precedent for this kind of collaboration between Danish municipalities in the field of waste management, such as in the collectively owned waste incineration facilities. The majority of Denmark's 27 dedicated waste incineration facilities are owned by a collective of municipalities [18]. This indicates that collaboration between municipalities for the purpose of economically efficient waste management is an institutionalized aspect of the DK waste sector. Therefore it would be reasonable to assume that Danish municipalities could collaborate for the purpose of investing in central sorting facilities. However, it is worthy to note that the federal government supported and encouraged the building of incineration facilities, through the establishment of economic tools, and requiring municipalities to ensure that there is available incineration capacity to treat waste that cannot be recycled. If we are to see a similar scale of investment in central sorting facilities as was seen for incineration facilities in the 70's and 80's, it may be necessary for the Danish government to support the investment with similar legislative tools.

The goal of 50% recycling of household waste by 2020 is not the only ambitious goal Denmark has set itself on environmental grounds. Some of these environmental goals have the potential to form a synergy with the recycling target. While anaerobic digestion of household organic waste has previously shown to give no socio-economic benefit in Denmark [12], targets to utilize at least 50% of animal slurry for energy production to be fossil-fuel free mean that there is potential to create a market where organic waste is in demand. In order to make the anaerobic digestion of animal slurry economically viable, AD facilities demand a high quality waste stream, such as household organic waste for co-digestion [22] [23]. Therefore, if both these targets are to be met, there is likely to be an increasing demand for source-separated household organic waste. While some technologies exist to separate organic waste from mixed waste, such as the ReneScience technology, the resulting organic waste fraction is not as of high a quality than that seen in source-separated waste, as it tends to have more contaminants such as small pieces of plastic [24]. These contaminants remain in the sludge after the digestion process, and their presence can affect the usability of the sludge for purposes such as spreading on farmland.

While Danish regulation currently allows for the use of organic household waste as a soil amendment, there exist tight restrictions for its use, the legislation currently only includes the use of organic household waste that is source-separated [25]. It is expected that the regulation of waste for agricultural purposes is likely to become more restrictive in the future, rather than more encompassing.

4.4. Integrating Political Goals into the FRIDA Model

Through integrating the recycling target into the FRIDA projection, we get both a better idea of what fractions should be recycled in order to achieve the target, as well as being able to determine how much waste will be left over for recycling. If information such as this is integrated into econometric projections performed at the level of individual incineration plants, it would be possible for plant owners to have a better idea of how waste flows for incineration will change in relation to the installed incineration capacity. As it is economically desirable to operate an incineration plant at full capacity, many incineration plants will choose to import industrial waste from abroad or invest in biomass resources. Since 2008, quantities of waste going to incineration have been falling in Denmark. Currently, many incineration plants have been utilizing stored waste to make up the difference. However this is a limited supply and with the large amounts of waste that will be diverted due to increased recycling, new sources of waste will have to be found. Some Danish incineration plants are already importing waste, with imports representing up to 40% of waste at some plants. Being able to project how quantities of combustible household waste are likely to fall due to increased recycling can help plant owners to create long-term plans for importing of waste.

5. Conclusions and Future Perspectives

The investigation into the composition of the mixed household waste reveals that the potential for the dry recyclable fractions to contribute to the recycling rate of household waste is not sufficient to reach the goal. Only in a scenario where all recyclable fractions, including organic waste, are recycled at an optimal rate will these efforts lead to a recycling rate over 50% for all household waste. Neither organic waste recycling nor dry materials recycling alone has the potential to reach the household recycling rate of 50% and a coordinated effort incorporating all recyclable streams will be necessary. Furthermore, it will be necessary for all Danish municipalities to recycle all fractions: The target will not be reached if each municipality only targets 1 or 2 fractions for increased recycling. As the investigation shows that significant efforts will need to be made by all municipalities in order to achieve the goal, it is recommended that other household waste streams be further investigated to determine what potential lies there. For example, mixed streams such as those collected as "bulky waste" or as "small and large combustible" at recycling stations could similarly be diverted to the recycling stream. The first step would be to establish what the composition of these waste fractions is.

The analysis shows that if the recycling target is reached, the amount of waste diverted from incineration represents around 20% of the current installed incineration capacity. While it is not possible to draw any detailed conclusions about the effect of this, without investigating the individual waste flows and economies of individual incineration plants, it can be concluded that in 2020 there will either be closures of less-economical waste incineration plants, or an increased import of combustible industrial waste.

While it has been possible to integrate the political goal into the FRIDA projection, thereby modelling the effect of increased recycling on the future waste streams, this modelling that has been undertaken in this study is somewhat underdeveloped, and is not capable of feeding information back into the FRIDA model. If political goals such as recycling targets could be integrated into projection models such as FRIDA, then the models could also give a much more detailed view at the strategic level of what goals are realistic to set, and could ensure that the government strategy is in harmony with the organizational and technological structure of the waste sector.

Acknowledgements

This work was supported by the Danish Strategic Research Council as part of the TOPWASTE project.

References

[1]   European Parliament and Council (2008) Waste Framework Directive.

[2]   European Commission (2011) Roadmap to a Resource Efficient Europe. OPOCE, Brussels.

[3]   Danish Ministry of the Environment (2013) Danmarkudenaffald: Genanvend mere—Forbrændmindre. Copenhagen.

[4]   Fischer, C., Gentil, E., Ryberg, M. and Reichel, A. (2013) Managing Municipal Solid Waste—A Review of Achievements in 32 European Countries—European Environment Agency (EEA). Luxembourg.

[5]   European Commission (2011) 2011/753/EU: Commission Decision of 18 November 2011 Establishing Rules and Calculation Methods for Verifying Compliance with the Targets Set in Article 11(2) of Directive 2008/98/EC. http://eur-lex.europa.eu/legal-content/EN/TXT/?uri=CELEX:32011D0753

[6]   Danish Ministry of the Environment (2009) Regeringens Affaldsstrategi 2009-12 1. Delstrategi. Copenhagen.

[7]   Beigl, P., Lebersorger, S. and Salhofer, S. (2008) Modelling Municipal Solid Waste Generation: A Review. *Waste Management*, **28**, 200-214. http://dx.doi.org/10.1016/j.wasman.2006.12.011

[8]   Chen, H.W. and Chang, N.-B. (2000) Prediction Analysis of Solid Waste Generation Based on Grey Fuzzy Dynamic Modeling. *Resources, Conservation and Recycling*, **29**, 1-18. http://dx.doi.org/10.1016/S0921-3449(99)00052-X

[9]   Andersen, F.M. and Larsen, H.V. (2012) FRIDA: A Model for the Generation and Handling of Solid Waste in Denmark. *Resources, Conservation and Recycling*, **65**, 47-56. http://dx.doi.org/10.1016/j.resconrec.2012.04.004

[10]   Abou Najm, M., El-Fadel, M., Ayoub, G., El-Taha, M. and Al-Awar, F. (2002) An Optimisation Model for Regional Integrated Solid Waste Management I. Model Formulation. *Waste Management & Research*, **20**, 37-45. http://dx.doi.org/10.1177/0734242X0202000105

[11]   Pires, A., Martinho, G. and Chang, N.-B. (2011) Solid Waste Management in European Countries: A Review of Systems Analysis Techniques. *Journal of Environmental Management*, **92**, 1033-1350. http://dx.doi.org/10.1016/j.jenvman.2010.11.024

[12]   Petersen, C. and Domela, I. (2003) Sammensætningafdagrenovationogordninger for hjemmekompostering (Composition of Household Waste and Schemes for Home Composting, in Danish). Danish Ministry of the Environment, Copenhagen.

[13]   Riber, C., Petersen, C. and Christensen, T.H. (2009) Chemical Composition of Material Fractions in Danish Household Waste. *Waste Management*, **29**, 1251-1257. http://dx.doi.org/10.1016/j.wasman.2008.09.013

[14]   Møller, J. and Jensen, M.B. (2013) Miljø-ogsamfundsøkonomiskvurderingafmuligheder for øgetgenanvendelseafpapir,pap, plast, metal ogorganiskaffald fradagrenovation (Environmental and Socio-Economic Analysis of Possibilities for Increased Recycling of Paper, Cardboard, Metal and Organic Waste). Danish Ministry of the Environment, Copenhagen.

[15]   Bernstad, A., La Cour Jansen, J. and Aspegren, H. (2012) Local Strategies for Efficient Management of Solid Household Waste—The Full-Scale Augustenborg Experiment. *Waste Management & Research*, **30**, 200-212. http://dx.doi.org/10.1177/0734242X11410113

[16]   Bagge, L., Münter, C.F., Justesen, J.S., Carlsbæk, M., Heidemann, R. and Møller, M.S. (2010) Idékatalogtiløgetgenanvendelseafdagrenovation—Sortering i to ellerflerefraktioner (Idea Catalogue for Increased Recycling of Municipal Waste, in Danish). Danish Ministry of the Environment, Copenhagen.

[17]   La Cour Jansen, J., Spliid, H., Hansen, T.L., Svärd, A. and Christensen, T.H. (2004) Assessment of Sampling and Chemical Analysis of Source-Separated Organic Household Waste. *Waste Management*, **24**, 541-549.

http://dx.doi.org/10.1016/j.wasman.2004.02.013

[18] Affaldsforening, D., Industri, D. and Energi, D. (2012) BEATE Benchmarking afaffaldssektoren 2012 (BEATE Benchmarking of the Waste Sector 2012, in Danish).
http://mst.dk/media/mst/Attachments/Forbrndingbenchmarkingfor2011.pdf

[19] Hansen, E. (1992) Vejlesaffaldssystem, Miljøprojekt nr. 208 (The Waste System of Vejle, in Danish). Danish Ministry of the Environment, Copenhagen.

[20] Bernstad, A. and La Cour Jansen, J. (2012) Separate Collection of Household Food Waste for Anaerobic Degradation—Comparison of Different Techniques from a Systems Perspective. *Waste Management*, **32**, 806-815.
http://dx.doi.org/10.1016/j.wasman.2012.01.008

[21] Hansen, T.L., La Cour Jansen, J., Davidsson, A. and Christensen, T.H. (2007) Effects of Pre-Treatment Technologies on Quantity and Quality of Source-Sorted Municipal Organic Waste for Biogas Recovery. *Waste Management*, **27**, 398-405. http://dx.doi.org/10.1016/j.wasman.2006.02.014

[22] Lindboe, H.H. and Energistyrelsen, D. (1995) Progress Report on the Economy of Centralized Biogas Plants. Danish Energy Agency, Copenhagen.

[23] Nielsen, L.H., Hjort-Gregersen, K., Thygesen, P. and Christensen, J. (2002) Samfundsøkonomiskeanalyserafbiogasfællesanlæg (Socio-Economic Analysis of Centralised Biogas Plants, in Danish). Fødevareøkonomisk Institut, Frederiksberg, Denmark.

[24] Braber, K. (1995) Anaerobic Digestion of Municipal Solid Waste: A Modern Waste Disposal Option on the Verge of Breakthrough. *Biomass and Bioenergy*, **9**, 365-376.

[25] Danish EPA (2010) Anvendelseafaffaldtiljordbrugsformål (Application of Waste for Agricultural Purposes, in Danish). Copenhagen.

9

Composite Materials Damage Modeling Based on Dielectric Properties

Rassel Raihan[1*], Fazle Rabbi[2], Vamsee Vadlamudi[2], Kenneth Reifsnider[1]

[1]University of Texas at Arlington, Arlington, USA
[2]University of South Carolina, Columbia, USA
Email: *mdrassel.raihan@uta.edu

Abstract

Composite materials, by nature, are universally dielectric. The distribution of the phases, including voids and cracks, has a major influence on the dielectric properties of the composite materials. The dielectric relaxation behavior measured by Broadband Dielectric Spectroscopy (BbDS) is often caused by interfacial polarization, which is known as Maxwell-Wagner-Sillars polarization that develops because of the heterogeneity of the composite materials. A prominent mechanism in the low frequency range is driven by charge accumulation at the interphases between different constituent phases. In our previous work, we observed *in-situ* changes in dielectric behavior during static tensile testing, and also studied the effects of applied mechanical and ambient environments on composite material damage states based on the evaluation of dielectric spectral analysis parameters. In the present work, a two dimensional conformal computational model was developed using a COMSOL™ multi-physics module to interpret the effective dielectric behavior of the resulting composite as a function of applied frequency spectra, especially the effects of volume fraction, the distribution of the defects inside of the material volume, and the influence of the permittivity and Ohmic conductivity of the host materials and defects.

Keywords

Polymer Matrix Composite Materials, Dielectric Properties, Degradation of Composite Materials, Broadband Dielectric Spectroscopy (BbDS)

1. Introduction

The applications of composite materials are now widespread because of their various advantages over conven-

---

tional isotropic materials. These heterogeneous material system's properties can be tailored based on the needs of the application and design. Aerospace and automotive industries are using composite materials to reduce weight to increase fuel efficiency, and also for energy storage and structural stability. The automobile Company Volvo has developed structural composite materials which can store and discharge electrical energy while also being used as a car body structure, fabricated from carbon fibers and a polymer resin [1] [2].

It is necessary to understand the material state changes caused by applied mechanical, thermal, and electrical fields to design and synthesize an effective material system. These complex material systems degrade progressively under combined applied field conditions. To evaluate such material state changes there are many computational tools and methods but most of them do not give a direct and quantitative assessment of the damage state. Numerous experimental techniques and methods have also been developed to measure such material state changes but most of them do not give a direct and quantitative assessment of the damage state.

During the service life of composite materials many degradation processes occur and generally this degradation initiates and evolves by microdamage development, especially matrix microcracking and crack growth, delamination, fiber fracture, fiber-matrix debonding, and microbuckling [1]. In our previous work, **Figure 1**, we have shown that the analysis of the dielectric data gives us information about the types of material state changes throughout the mechanical life of composite materials [2] [3].

Broadband Dielectric Spectroscopy (BbDS) measures the interaction of EMF with a material system over a wide range of frequencies which is shown in **Figure 2**. The response to that broad frequency range typically contains information about molecular and collective dipolar fluctuations, and charge transport and polarization effects that occur at inner and outer boundaries as they affect the form of different dielectric properties of the composite material under study.

Various researchers have used finite element methods (FEM) to model the effective dielectric properties of periodic and random composites containing inclusions of various shapes [4] [5]. In this present work, we used

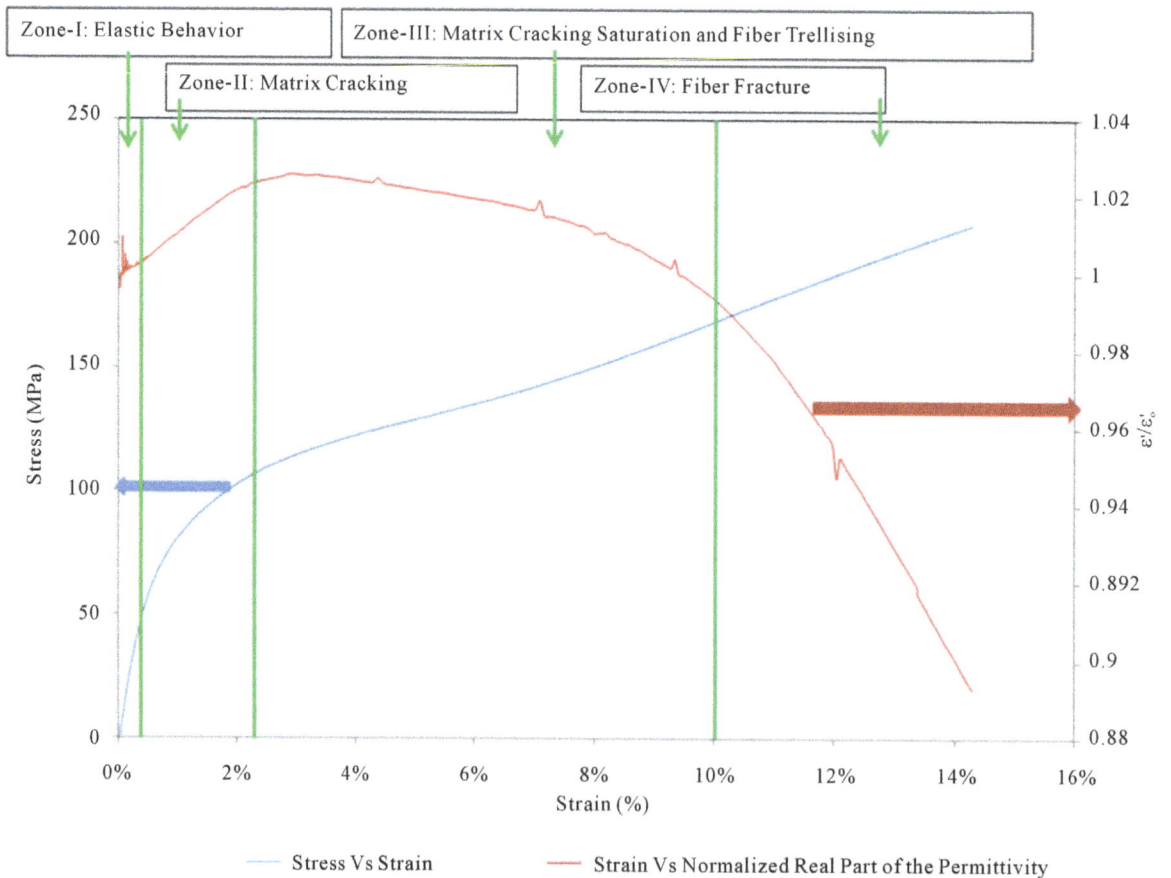

Figure 1. Response of the dielectric property in different zones of damage progression [2].

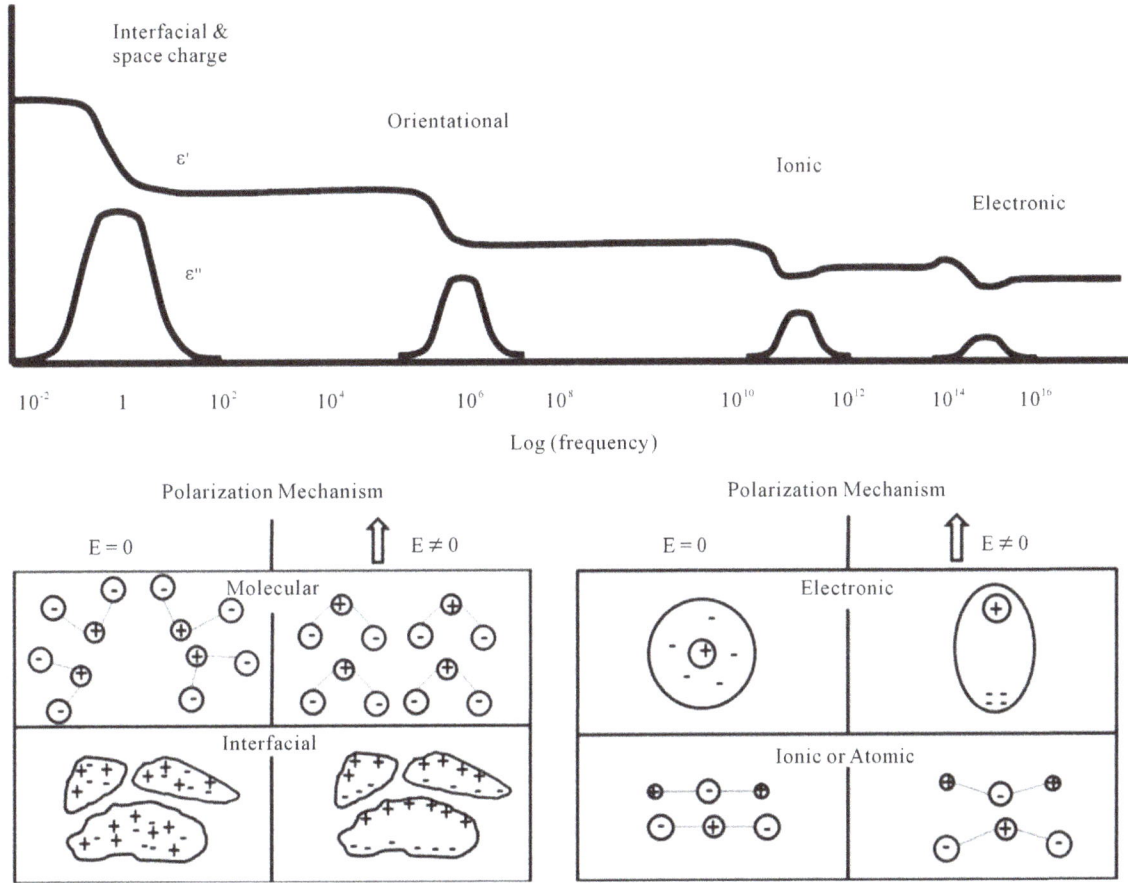

Figure 2. Dielectric responses of material constituents at broad band frequency ranges and different polarization mechanism.

COMSOL Multiphysics^{TM} for conformal modeling, and to reduce the complicacy of the model we only considered the interfacial polarization which is caused by the permittivity and conductivity difference between two constituents. We assumed that the composite materials were homogeneous, and represented defects/cracks as inclusions as shown in **Figure 3**.

2. Basic Equations

In classical dielectrics the relation between the applied electric field E and the dielectric displacement D is linear and can be expressed as [6],

$$D = \varepsilon_o \varepsilon_r E \tag{1}$$

where, ε_o is the permittivity of the free space and ε_r is the relative permittivity of the dielectric material.

If ρ is the charge density, from Maxwell's equations we know that the dielectric displacement follows the following relationship

$$\nabla \cdot D = \rho \tag{2}$$

For current density J we can state the following from the continuity equation

$$\nabla \cdot J = -\frac{d\rho}{dt} \tag{3}$$

From also Ohm's law we know

$$J = \sigma E \tag{4}$$

Here σ is the conductivity of the material.

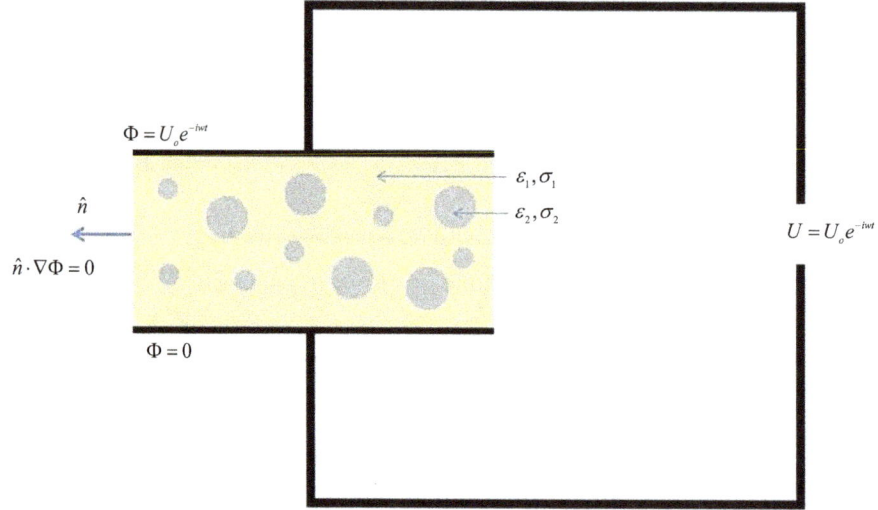

Figure 3. Schematic of the calculation of effective dielectric properties of composites.

So from Equations (2) and (3) we obtain

$$\nabla \cdot \left(\boldsymbol{J} + \frac{\mathrm{d}\boldsymbol{D}}{\mathrm{d}t} \right) = 0 \tag{5}$$

Now using 1, 4 and 5 we can write the following

$$\nabla \cdot \left(\sigma \boldsymbol{E} + \frac{\mathrm{d}\left(\varepsilon_o \varepsilon_r \boldsymbol{E} \right)}{\mathrm{d}t} \right) = 0 \tag{6}$$

In case of a sinusoidal applied electric field $\boldsymbol{E}$ of angular frequency ω

$$\nabla \cdot \left(\sigma + i\omega\varepsilon_o\varepsilon_r \right) \boldsymbol{E} = 0 \tag{7}$$

We know

$$\boldsymbol{E} = -\nabla\Phi \tag{8}$$

From Equation (7) and (8) we get

$$\nabla \cdot \left[\left(\sigma + i\omega\varepsilon_o\varepsilon_r \right) \nabla\Phi \right] = 0 \tag{9}$$

From Equation (9), we can tell that in a heterogeneous material the product of the physical properties (some form of the conductivity and permittivity) and the slope of the potential must be a constant as we cross material boundaries. For the quasi-static case with harmonic input fields, the gradient of that product vanishes. The interacting field is a result of the charge difference at the interface, and unless the conductivity and permittivity of adjacent material phases are identical, there is a disruption of charge transfer at the material boundary which results in internal polarization.

To solve Equation (9), we set the potential on the top electrode to be,

$$\Phi = U = U_o \mathrm{e}^{-i\omega t} \tag{10}$$

And on the bottom electrode,

$$\Phi = 0 \tag{11}$$

Boundary conditions on the interfaces are,

$$\Phi_1 = \Phi_2 \tag{12}$$

$$\varepsilon_1 \hat{\boldsymbol{n}} \cdot \nabla\Phi_1 = \varepsilon_2 \hat{\boldsymbol{n}} \cdot \nabla\Phi_2 \tag{13}$$

where $\hat{\boldsymbol{n}}$ is the unit vector normal to the interface surface. To eliminate fringe effects on the side planes, we set

$$\hat{n} \cdot \nabla \Phi = 0 \qquad (14)$$

here $\hat{n}$ is the unit vector normal to the side plane.

3. Results and Discussion

3.1. Two Phase Model

For this model, an undamaged composite material is considered to be a homogeneous material and the cracks (here as circular inclusions) are considered to be the second phase inside of that homogeneous material system. Permittivity and ohmic conductivity of the host material were taken to be $\varepsilon_1 = 5$ and $\sigma_1 = 10^{-13}$ S/m and for the inclusion permittivity and ohmic conductivity, $\varepsilon_2 = 2$ and $\sigma_2 = 10^{-15}$ S/m, were chosen which are values close to those of the ambient air permittivity and conductivity [7]. Because of the difference in the permittivities and conductivities of the phases, the accumulation of charge at the interphase boundaries causes an undulation of the space distribution of the potential which is shown in the **Figure 4** and **Figure 5**. **Figure 4** shows the potential distribution around the inclusion and **Figure 5** shows the potential distribution along the horizontal center line. It can be seen that around the boundary of the phases there is a potential nonlinearity (in the figure the nonlinearity is shown inside two ellipses) that is caused by the charge accumulation at the interface between the host material and inclusion. **Figure 6** shows that the space charge accumulation is higher, which is caused by the dissimilarity of the material properties around the inclusion boundary in the presence of the applied electric field.

Computer simulations were performed for different volume fractions of the inclusions. **Figure 7** shows that the space charge density increases with an increase of the inclusion volume fraction. In the frequency range above 1 Hz the space charge density is constant but below 1 Hz a nonlinear increase is observed in the space charge density around the inclusion interface as shown in **Figure 8**.

Figure 9 shows the change of the real and imaginary parts of the global permittivities with the increase of volume fraction of the inclusion as a function of frequency. At a high frequency the period of potential oscillations is not sufficient for charge accumulation but at low frequency the charge has enough time to accumulate around the interface which leads to interfacial polarization (Maxwell-Wagner-Sillar polarization); that is why there is an increase in the real part of the permittivity (shown in **Figure 10**) and dielectric loss at the lower frequencies.

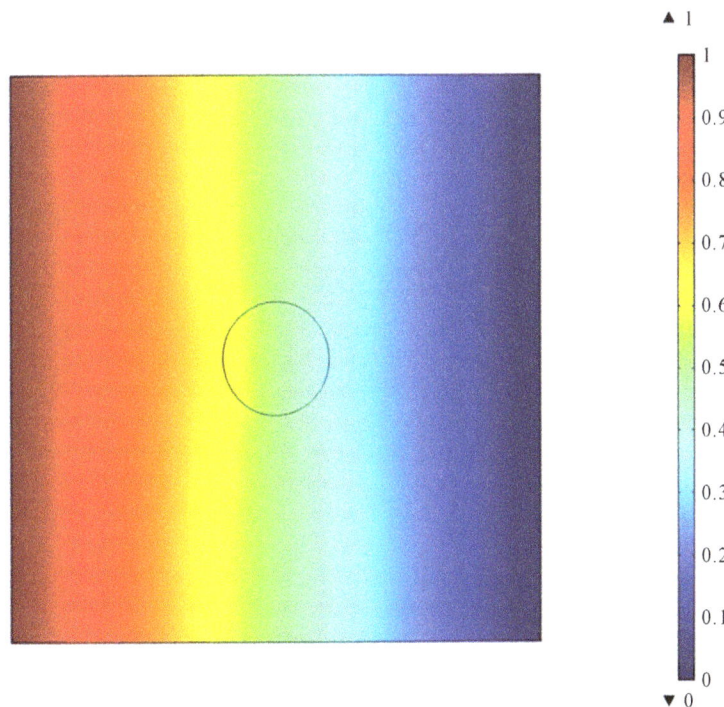

Figure 4. Potential distributions around the inclusion.

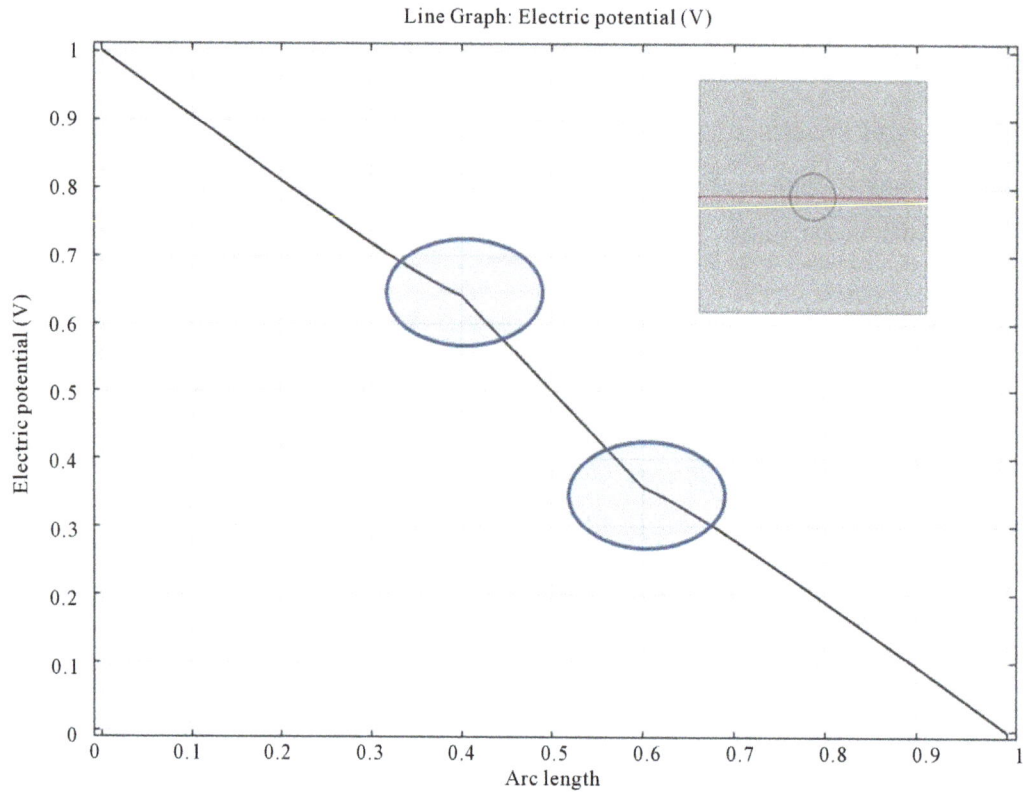

Figure 5. Potential distributions along the line.

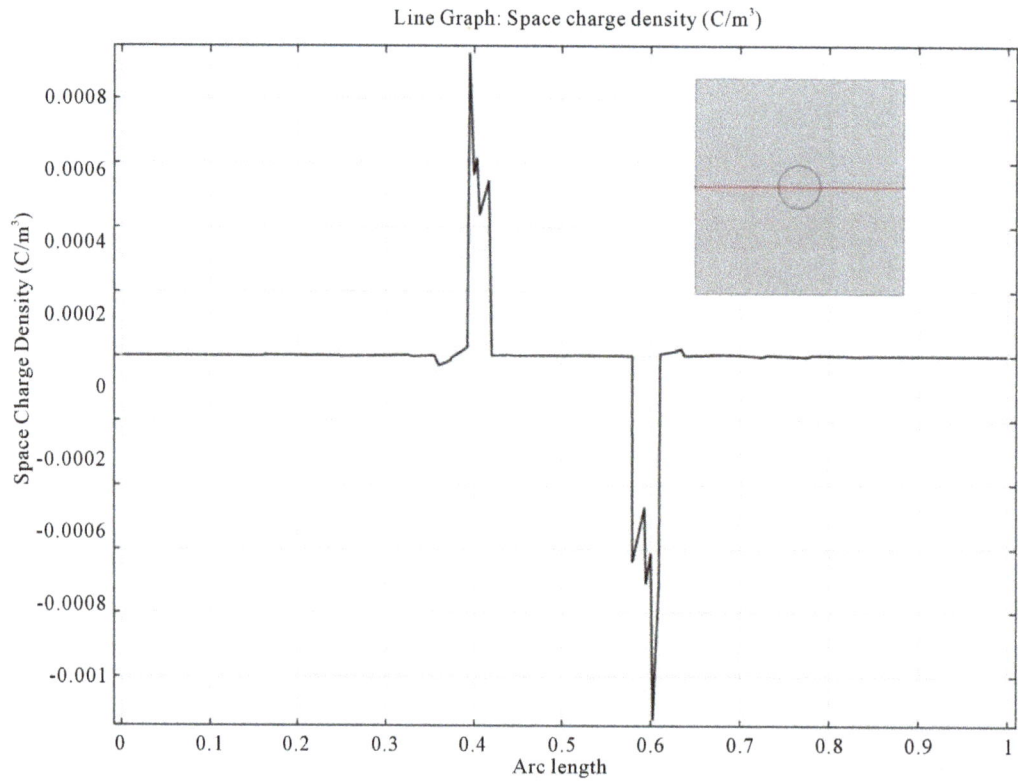

Figure 6. Space charge densities along the line.

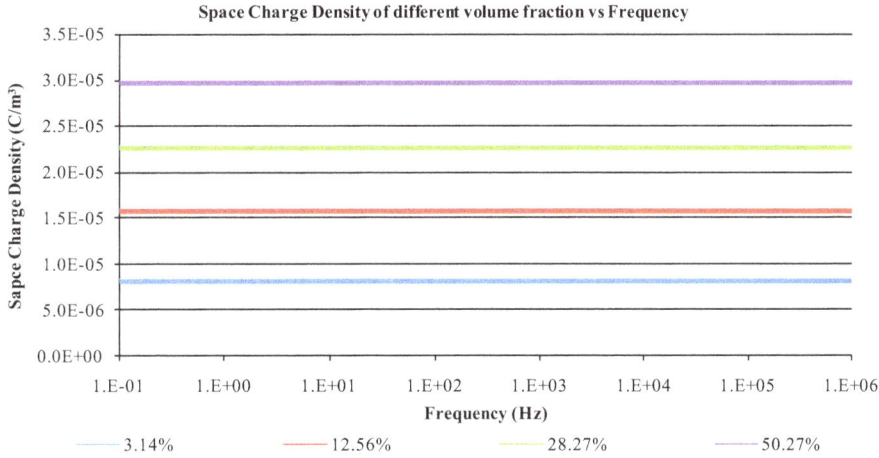

Figure 7. Space charge densities around the inclusion interface of different volume fraction in different frequency.

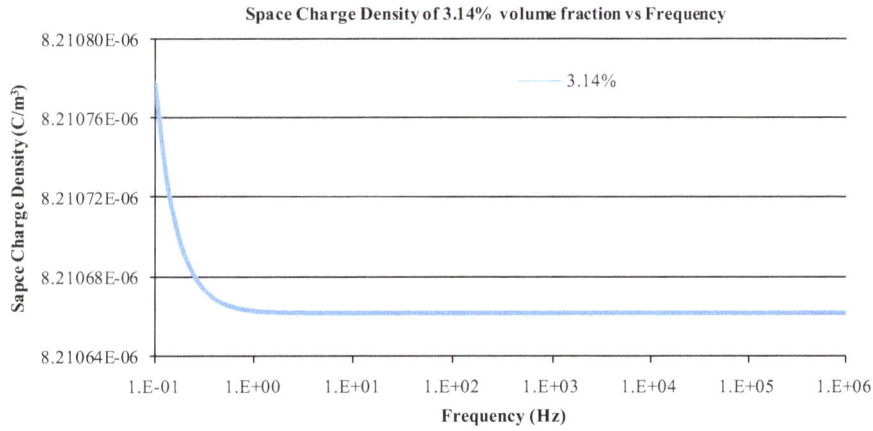

Figure 8. Space charge density change of 3.14% volume fraction of inclusion with frequency.

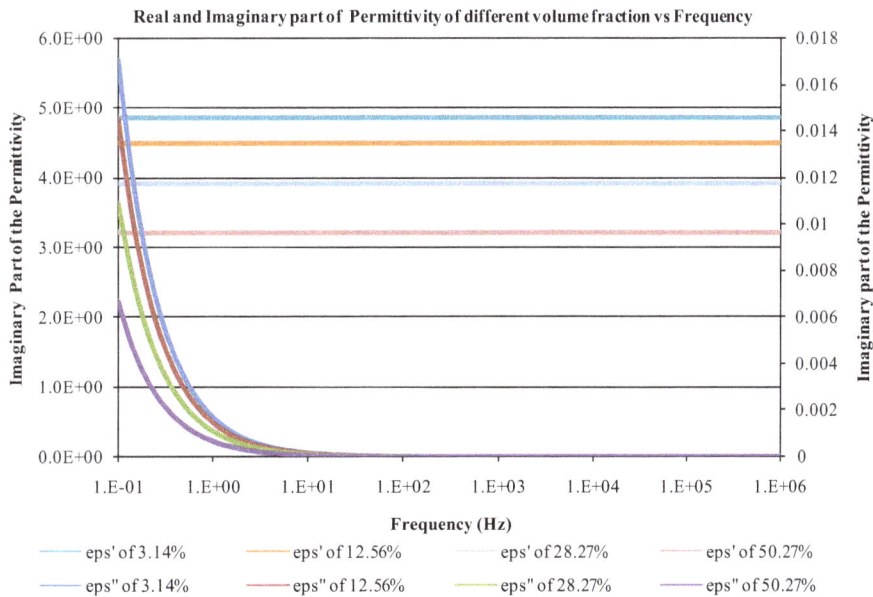

Figure 9. Real and imaginary part of the permittivity with different volume fraction of the inclusion.

For different volume fractions of inclusion, the real part of the permittivity was calculated from the computer simulation for a frequency of 10 Hz. **Figure 11** is the comparison of real part of the permittivity change with increasing volume fraction of the inclusion phase. The relation between the real part of the permittivity and volume fraction is almost linear, but when the real part of the permittivity is plotted with surface area fraction (surface area fraction is the ratio of inclusion surface to the material surface) of the inclusion there is clearly a nonlinear relationship predicted. For low volume fractions the effect of surface area fraction of the inclusion is more dominant than the volume fraction, but for the higher volume fractions it is opposite.

Figure 12 illustrates the comparison between computer simulation results with increasing inclusion volume/surface-area fraction and experimental results. **Figure 12(a)** and **Figure 12(b)** show slight increases in the real part of the impedance for low volume/surface-area fractions. **Figure 12(c)** shows the experimental results for the real part of the impedance change with the strain. The real part of the impedance increases below the low strain (below 5%) for the off axis sample where matrix microcracking is dominant and distributed throughout the material system.

3.2. Three Phase Model

Composite materials are filled with various additive materials to achieve the desired mechanical, thermal and electrical properties. Typical filler materials used for the present modeling are carbon or glass fibers. The use of these fibers as filler materials introduces a water sensitive component into the polymer composites. Glass fibers are well known for their water affinity on their surfaces. Currently, epoxies are widely used matrix materials in

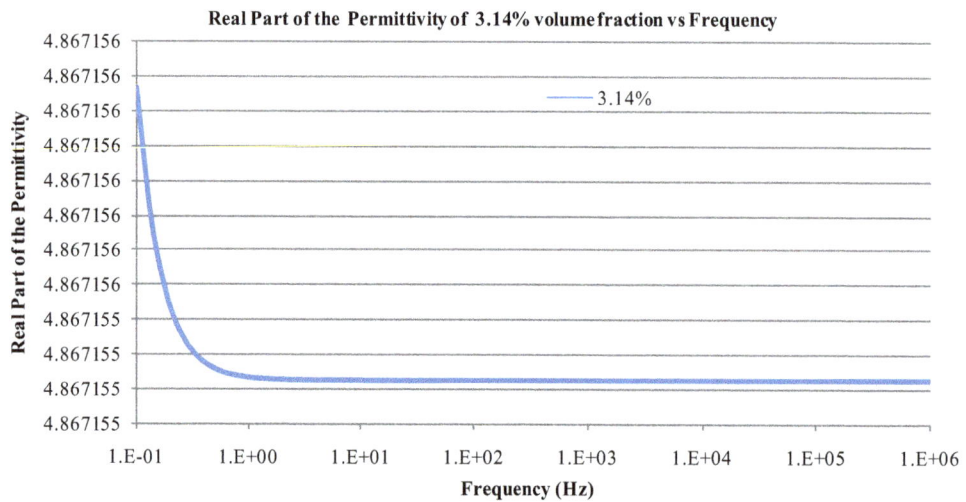

Figure 10. Real part of the permittivity of the material with 3.14% volume fraction of inclusion.

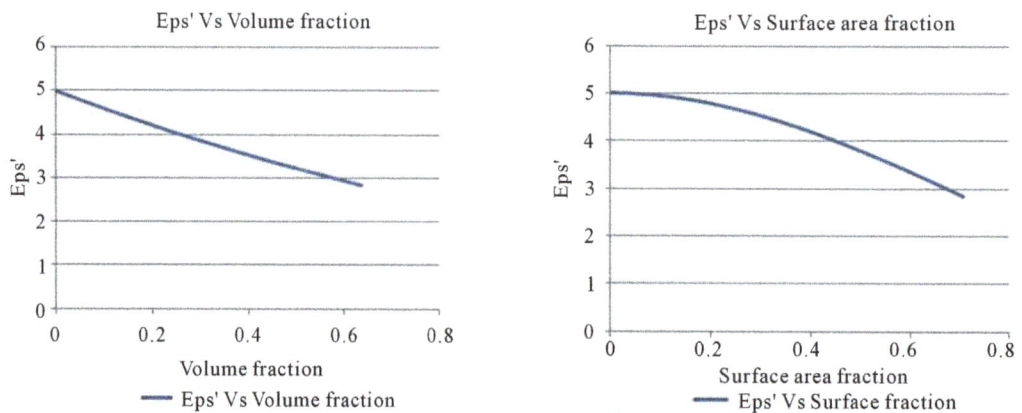

Figure 11. Eps' comparison with volume fraction and surfacearea fraction of the inclusion.

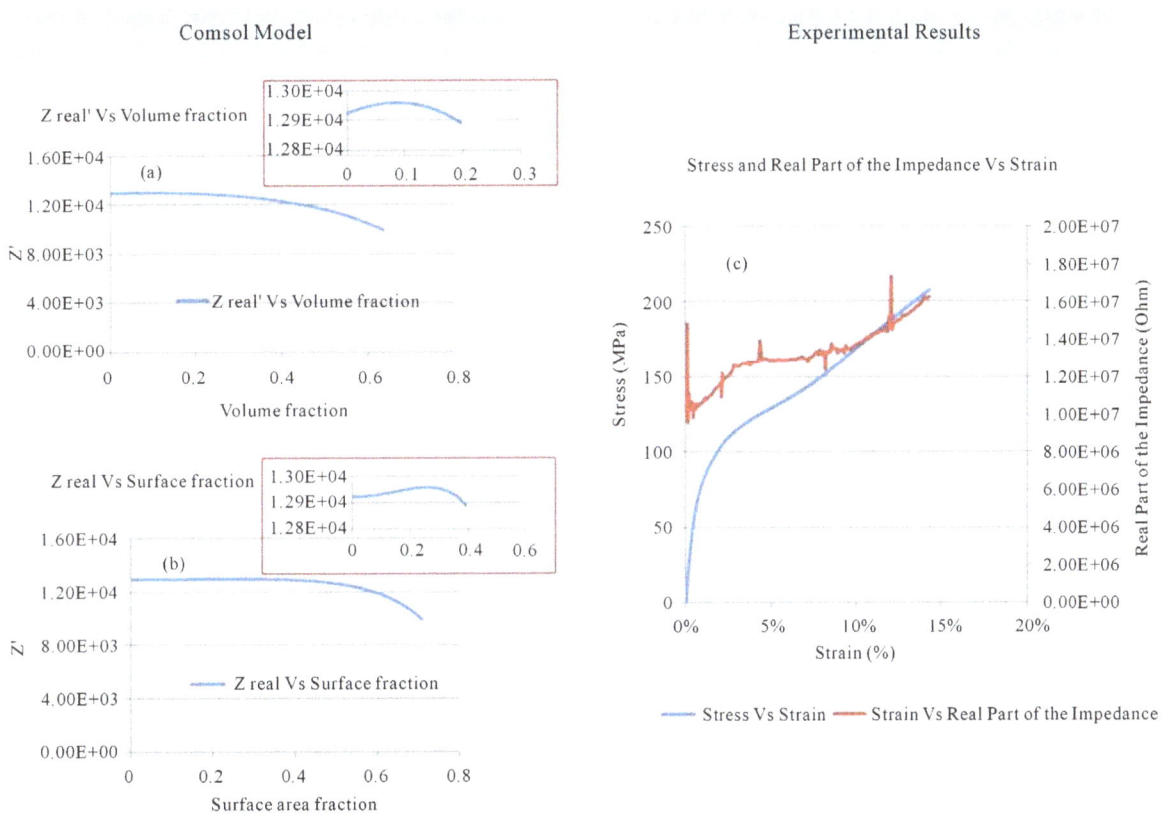

Figure 12. Comparison of Comsol™ simulation result and experimental results.

composite industries, which also have the potential of being sensitive to moist conditions or humid environments. Soles and Yee [8] found that a network of nanopores that is inherent in the epoxy structure helps free water to traverse the epoxy; they found the average size of nanopore diameters to vary from 5 to 6.1 Å and account for 3% - 7% of the total volume of the epoxy material. The approximate diameter of a kinetic water molecule is just 3.0 Å, so via the nanopores network the moisture can easily traverse into the epoxy. They also found that the volume fraction of nanopores does not affect the diffusion coefficient of water and argued that polar groups coincident with the nanopores are the rate-limiting factor in the diffusion process, which could explain why the diffusion coefficient is essentially independent of the nanopore content. In their **Figure 13** they explain how the water transport happens in epoxy networks.

There are many theories about the state of water molecules in polymers. Adamson [9] suggested that moisture can transfer in epoxy resins in the form of either liquid or vapor. It is proposed by Tencer [10] that it is also possible that vapor water molecules undergo a phase transformation and condense to the liquid phase. This condensed moisture was stated to be either in the form of discrete droplets on the surface or in the form of a uniform monolayer [11].

Water has a higher dielectric permittivity and conductivity than the glass fiber and matrix, so it has strong effects on the dielectric properties, *i.e.* relative permittivity and dielectric loss, of the material system. In the literature it is well established that water absorption increases the dielectric constant of the dielectric material [12]-[16]. This dielectric loss is observed in the low frequency range. Water diffuses through the interface and also weakens the interfacial strength of filler and matrix.

When composite materials go through degradation processes, microcracks typically form and these microcracks can also be filled with moist air, and condensed or adsorbed water layers can form on the surface of those defects. In our two phase model we saw an interfacial polarization (Maxwell-Wagner-Sillars polarization) that is present in the low frequency region of the frequency spectra. If a water layer is present on the surface of the defect it will become electrically conductive. Since the host material and defect have low electrical conductivity and permittivity is not significantly high, this will give rise to interfacial polarization.

Figure 14 illustrates the tri-layer computational model used for the study where yellow, gray and blue parts represent respectively the host material, defect, and a conductive layer. The permittivity and ohmic conductivity of the host material were taken to be $\varepsilon_1 = 5$ and $\sigma_1 = 10^{-13}$ S/m and for the inclusion the permittivity and ohmic conductivity had values of $\varepsilon_2 = 2$ and $\sigma_2 = 10^{-15}$ S/m. For the conductive layer a different permittivity ε_3 and conductivity σ_3 (higher than host and defect properties) was used to see the effect on the effective dielectric properties of the material system.

Figure 15 shows the potential distributions around the inclusion of the tri-layer model which is different than the potential difference shown in **Figure 4** for the two phase inclusion model. Because of the conductive layer around the inclusion there is a large undulation of the space distribution of the electric potential.

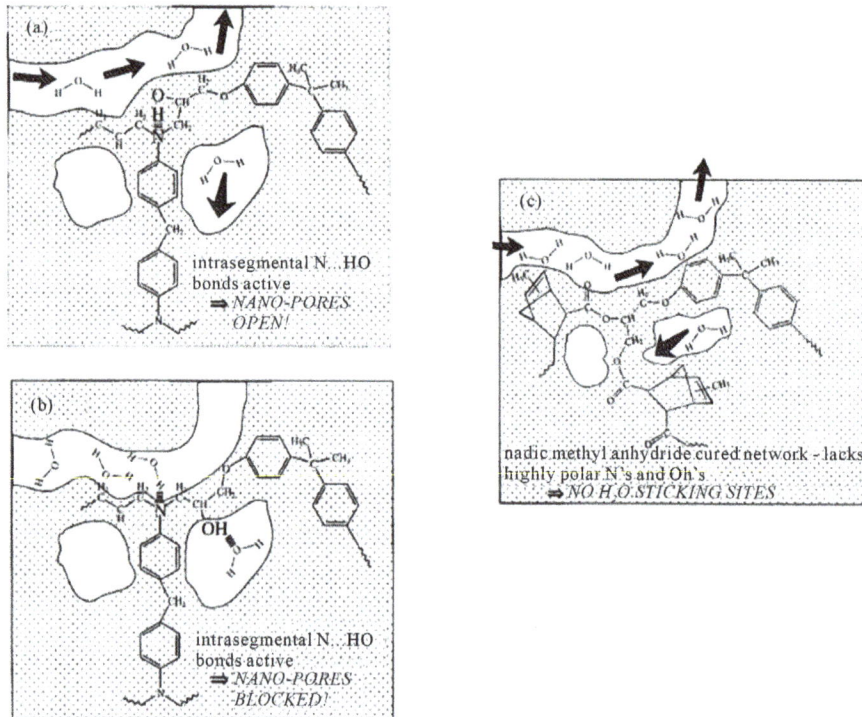

Figure 13. A plausible picture of moisture diffusion through the nanopores of an amine-containing epoxy (Figure from reference [7]).

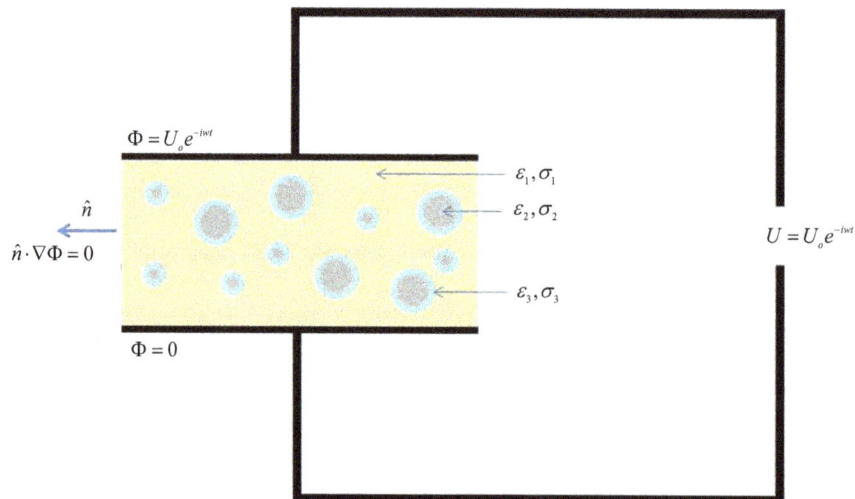

Figure 14. Tri-layer model.

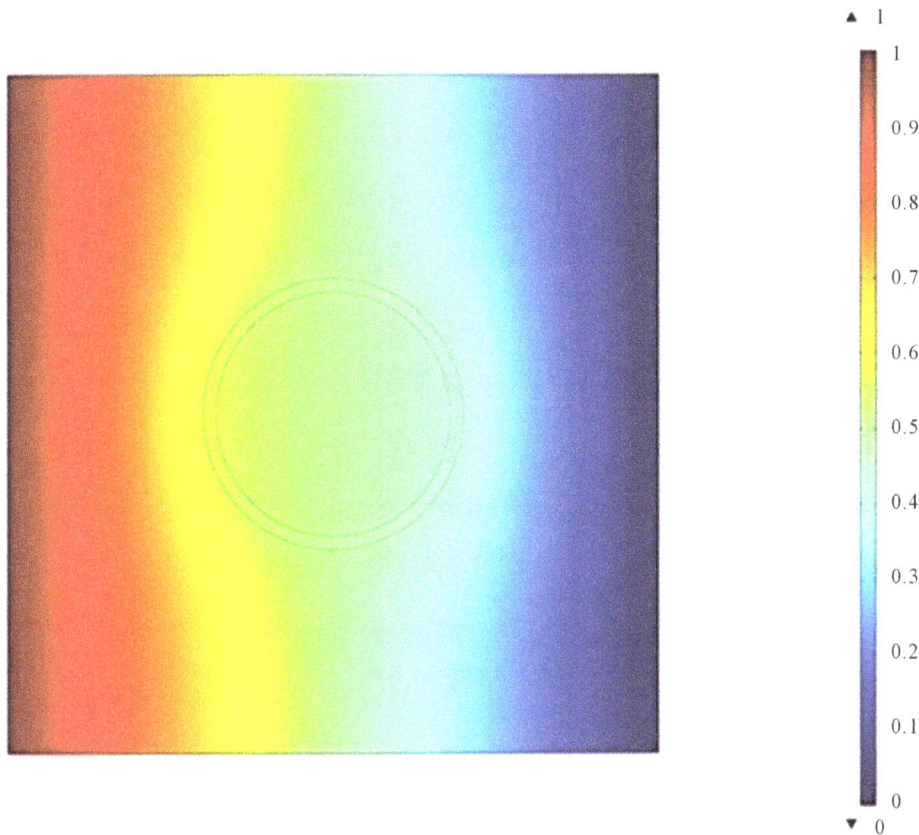

Figure 15. Potential distributions around the inclusion with a conductive layer.

For the tri-layer model, the total volume fraction is the sum of the volume fraction of the defect and the volume fraction of the conductive layer. For all of the cases of tri-layer modelling, the conductive layer thickness was specified as 0.5 micro meter. We observed that for two phase models, the real part of the permittivity was almost linear but in **Figure 16** we can see that for the three phase case there is an increase in real part of the permittivity for lower volume fractions and then a decreasing trend. As there is a conductive layer in between the defect and the host matrix, the interfacial polarization plays a vital role for this type of behavior. The subsequent decrease of the real part of the permittivity for higher volume fractions is caused by the dominance of the volume of the defects which is higher than the interfacial polarization contributed by the conductive layer.

It is also clear from **Figure 17** that for the same conductivity and permittivity values of the conductive layer, for low frequency the real part of the permittivity of the material system is higher than the value for higher frequency.

Figure 18 shows the dependence of dielectric constant on the conductivity of the conductive layer. The real part of the permittivity at different volume fractions for the same permittivity and frequency behave differently for different surface layer conductivities.

As shown in **Figure 19**, for the same volume fraction of the inclusion but variable frequency of the input field, there is a step-like increase of the real part of the permittivity over a narrow frequency range, and the dielectric loss also has the peak in that region, where the Maxell-Wagner-Sillars polarization dominates.

Figure 20 shows the relation of the real part of the permittivity with the frequency for all volume fractions of the inclusion. That simulation was also done for just the matrix material (this is the host material, as we considered it homogeneous). Since there were no other phases presents for that case, there was no chance of charge accumulation and there is no predicted change of the dielectric constant for the matrix-only case. We can see for higher volume fractions the dielectric relaxation strength (difference between the real parts of the permittivity at low frequency and high frequency) also increases. At higher frequency the real part of the permittivity drops as a function of volume fraction because charge accumulation does not occur at the interface at those frequencies.

Real Part of the Permittivity vs Total Volume Fraction of the Inclusion

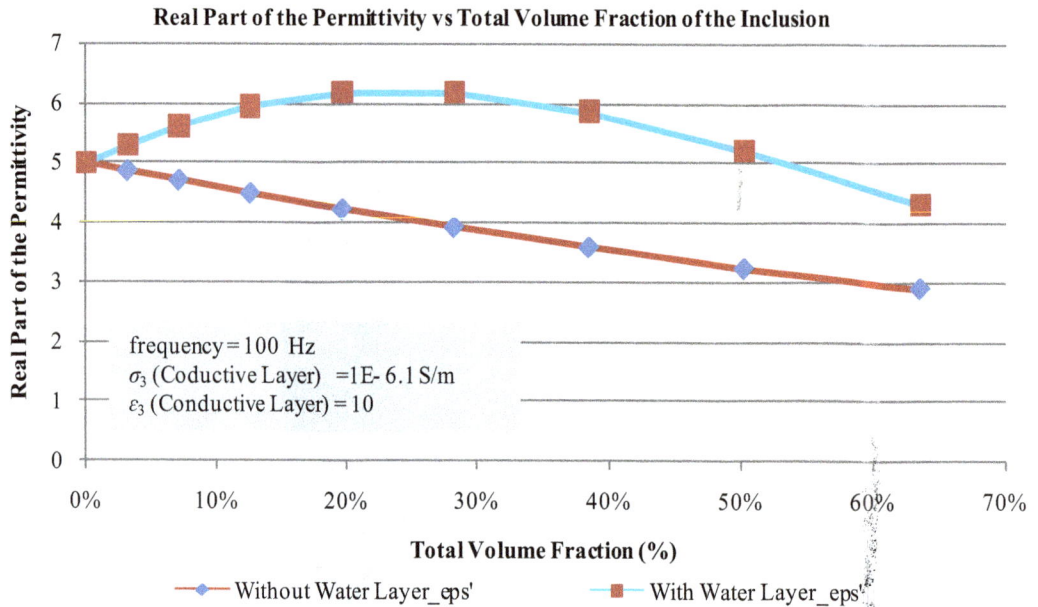

frequency = 100 Hz
σ_3 (Coductive Layer) = 1E- 6.1 S/m
ε_3 (Conductive Layer) = 10

Total Volume Fraction (%)

Without Water Layer_eps'   With Water Layer_eps'

Figure 16. Variation of real part of the permittivity with increasing volume fraction. For the tri-layer model, total volumefraction is the sum of the volume fraction of defect.

(a) Real Part of the Permittivity vs Total Volume Fraction of the Inclusion

Frequency = 10.8 Hz
σ_3 (Coductive Layer) = 1E-6.1 S/m
ε_3 (Coductive Layer) = 10

(b) Real Part of the Permittivity vs Total Volume Fraction of the Inclusion

Frequency = 100 Hz
σ_3 (Coductive Layer) = 1E-6.1 S/m
ε_3 (Coductive Layer) = 10

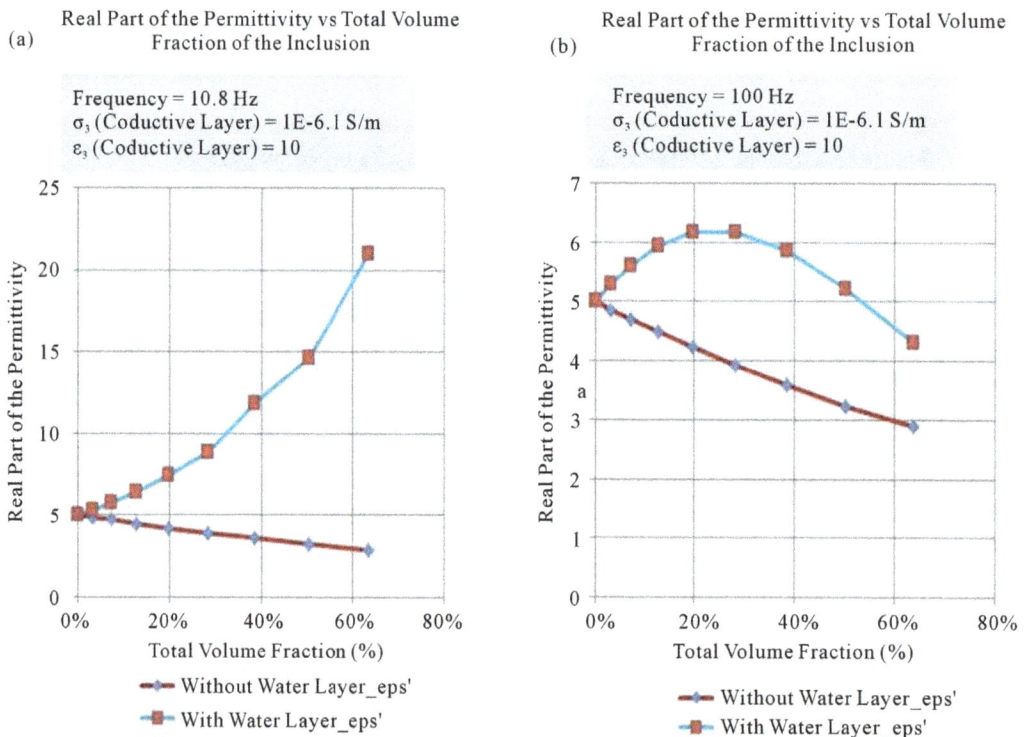

Total Volume Fraction (%)

Without Water Layer_eps'
With Water Layer_eps'

Without Water Layer_eps'
With Water Layer_eps'

Figure 17. Frequency dependency of the real part of permittivity for different volume fraction.

Dielectric loss (the imaginary part of the permittivity) also varies with volume fraction and it is illustrated in **Figure 21**. For high volume fraction dielectrics, the loss changes somewhat and the peak of the loss also increases.

The corresponding Cole-Cole plot, **Figure 22**, also shows the shift in relaxation for different volume fractions.

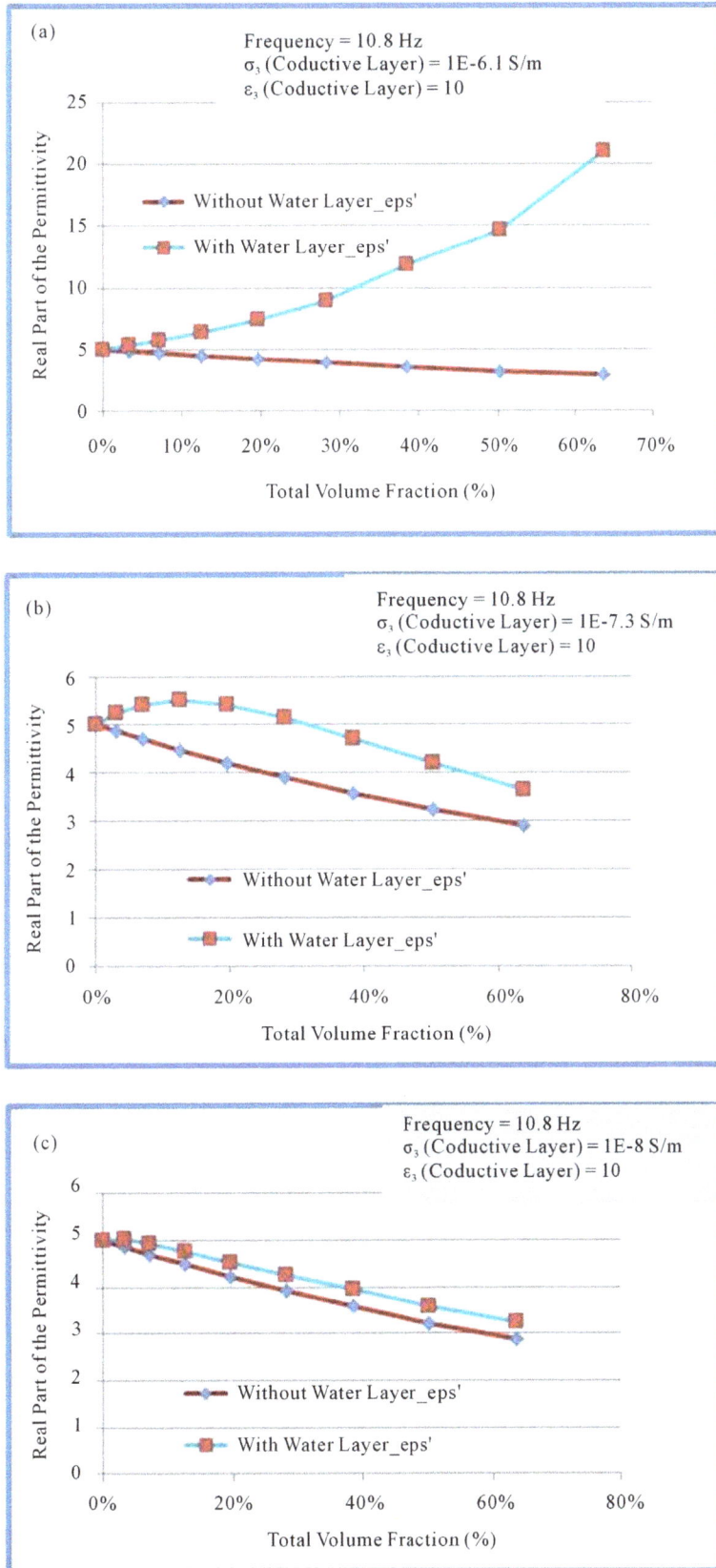

Figure 18. Dielectric properties dependence on the conductivity.

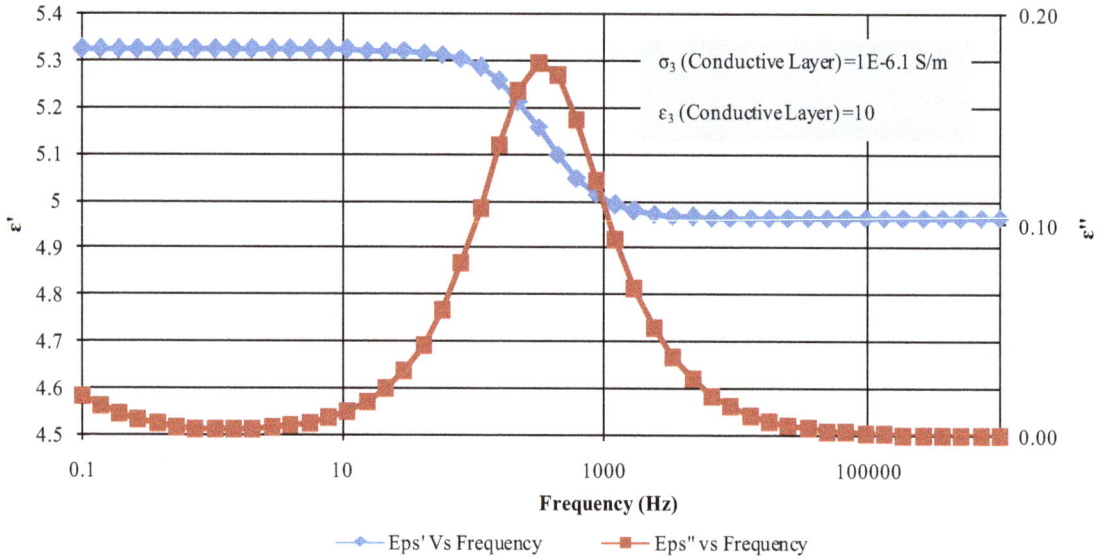

Figure 19. Real and imaginary part of the permittivity Vs frequency for 3.14% total volume fraction of the inclusion.

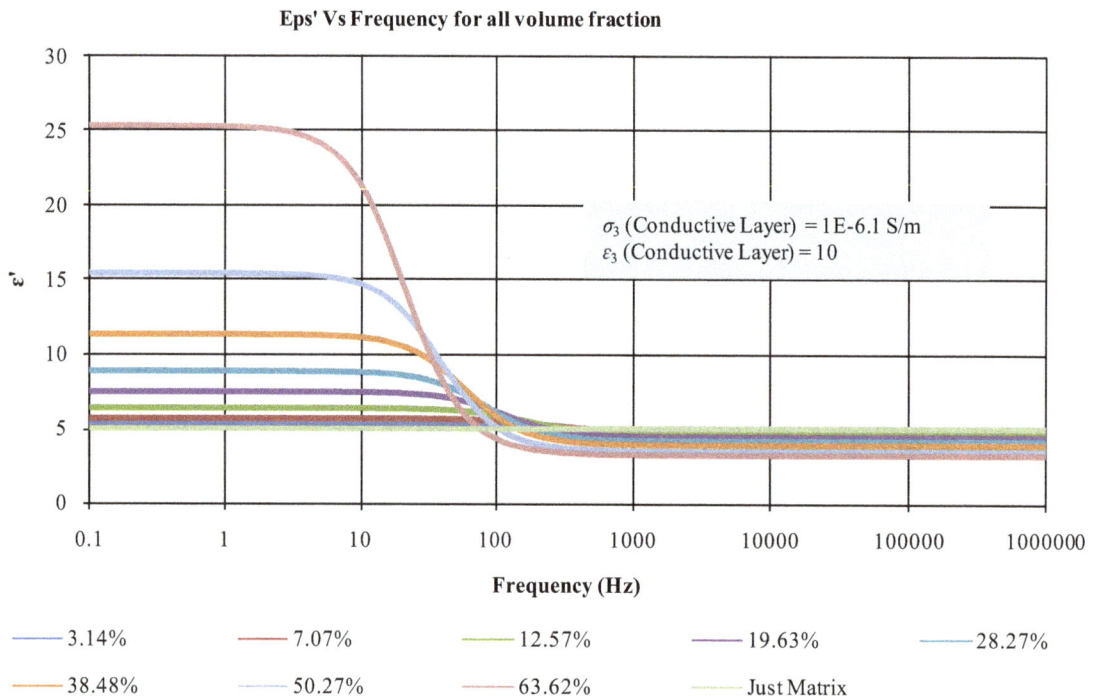

Figure 20. Real part of the permittivity in frequency spectra of all volume fractions.

3.3. Distributed Damage Model

A distributed damage model was created to see the effect of the distribution of the damage. A dielectric study was performed for a certain volume fraction of inclusion, and then that inclusion was divided into 10 inclusions while keeping the total volume fraction the same.

Figures 23-26 show the change of dielectric properties of a single damage volume and distributed damage volumes with the same amount of volume fraction without any conductive layer around the defects. The dielectric loss increased for the distributed damage because of the presence of more interfacial polarization.

Eps'' Vs Frequency for all volume fraction

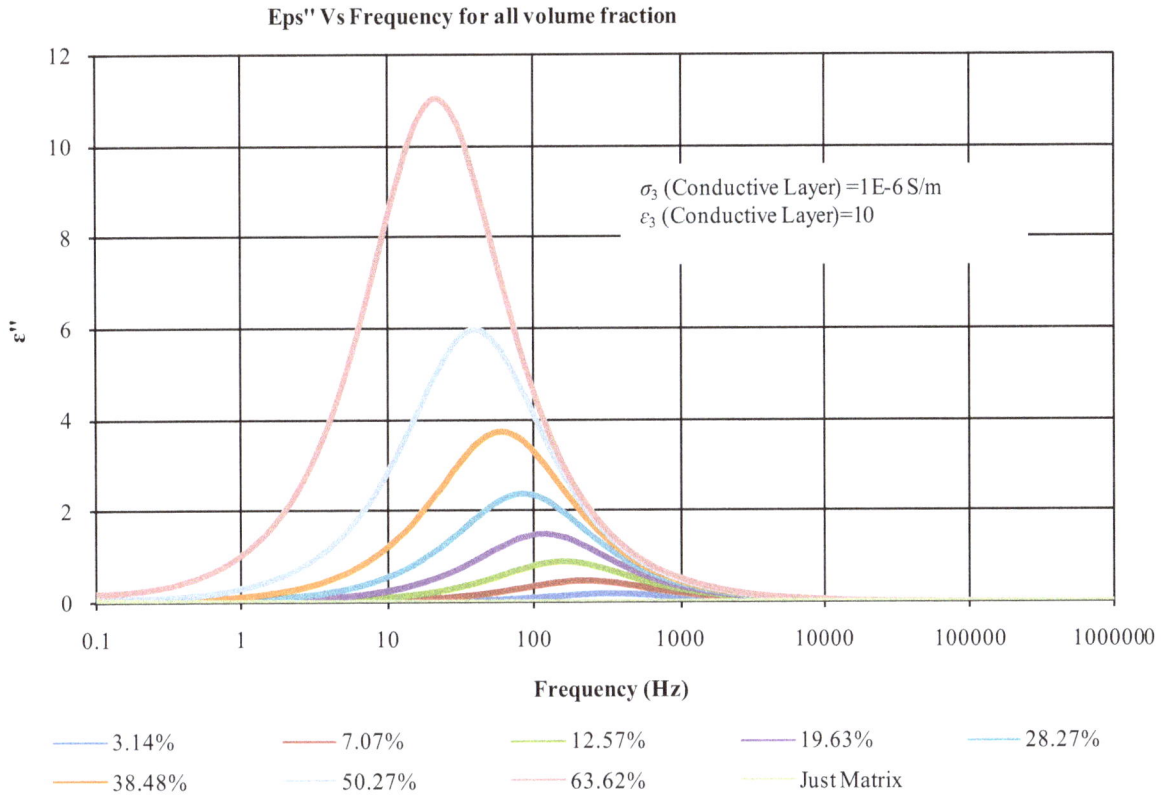

σ_3 (Conductive Layer) =1E-6 S/m
ε_3 (Conductive Layer)=10

Frequency (Hz)

| ——— 3.14% | ——— 7.07% | ——— 12.57% | ——— 19.63% | ——— 28.27% |
| ——— 38.48% | ······· 50.27% | ——— 63.62% | Just Matrix | |

Figure 21. Imaginary part of the permittivity in frequency spectra of all volume fractions.

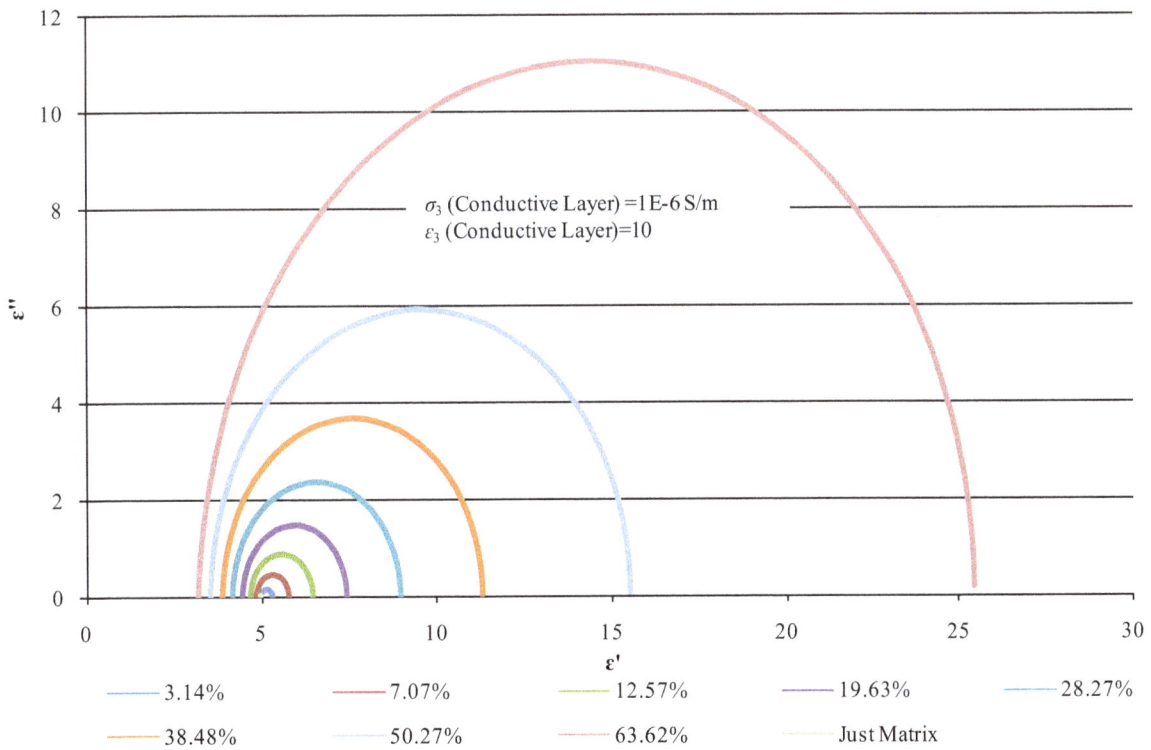

σ_3 (Conductive Layer) =1E-6 S/m
ε_3 (Conductive Layer)=10

ε'

| ——— 3.14% | ——— 7.07% | ——— 12.57% | ——— 19.63% | ——— 28.27% |
| ——— 38.48% | ——— 50.27% | ——— 63.62% | Just Matrix | |

Figure 22. Cole-Cole plot of different volume fraction.

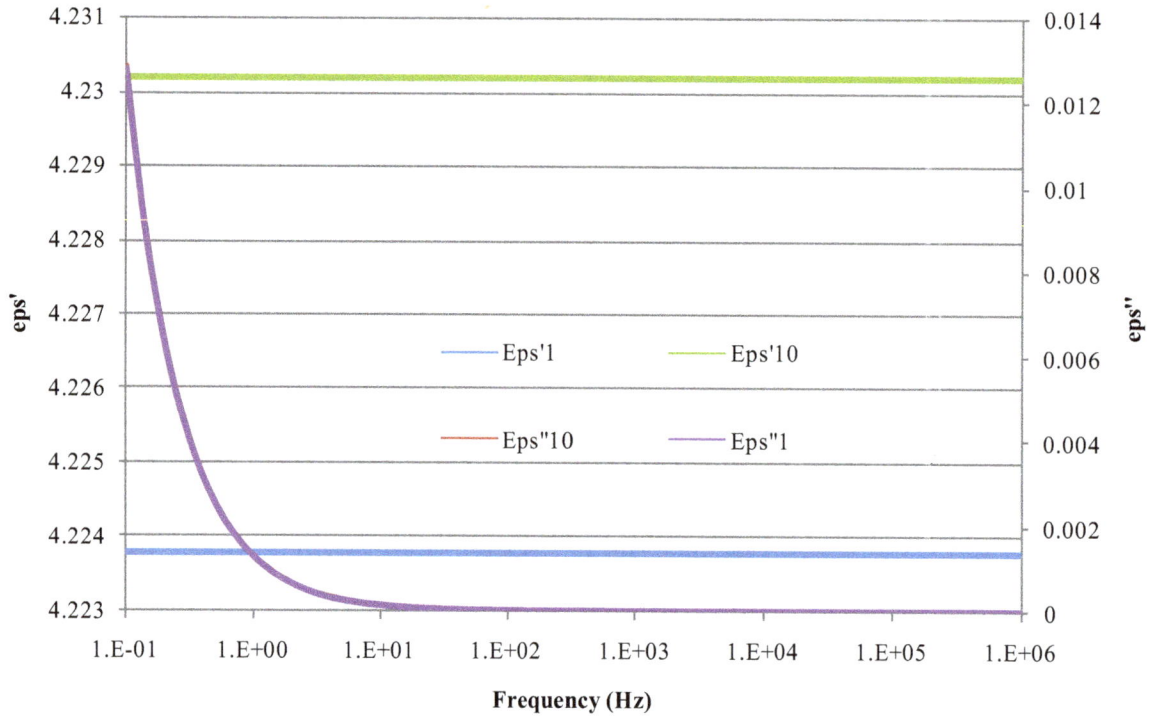

Figure 23. Dielectric Properties without conductive layer for same volume fraction but different number of inclusion.

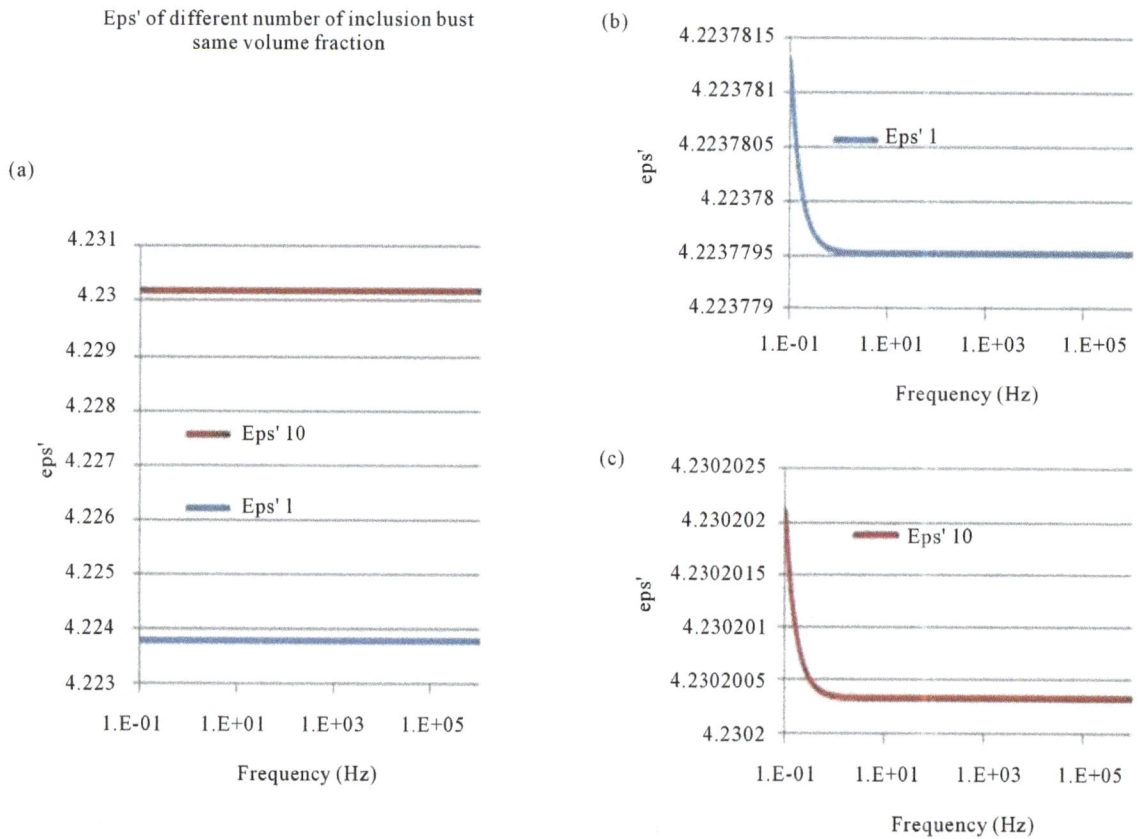

Figure 24. Real Part of the permittivity for different number of inclusion but same volume fraction.

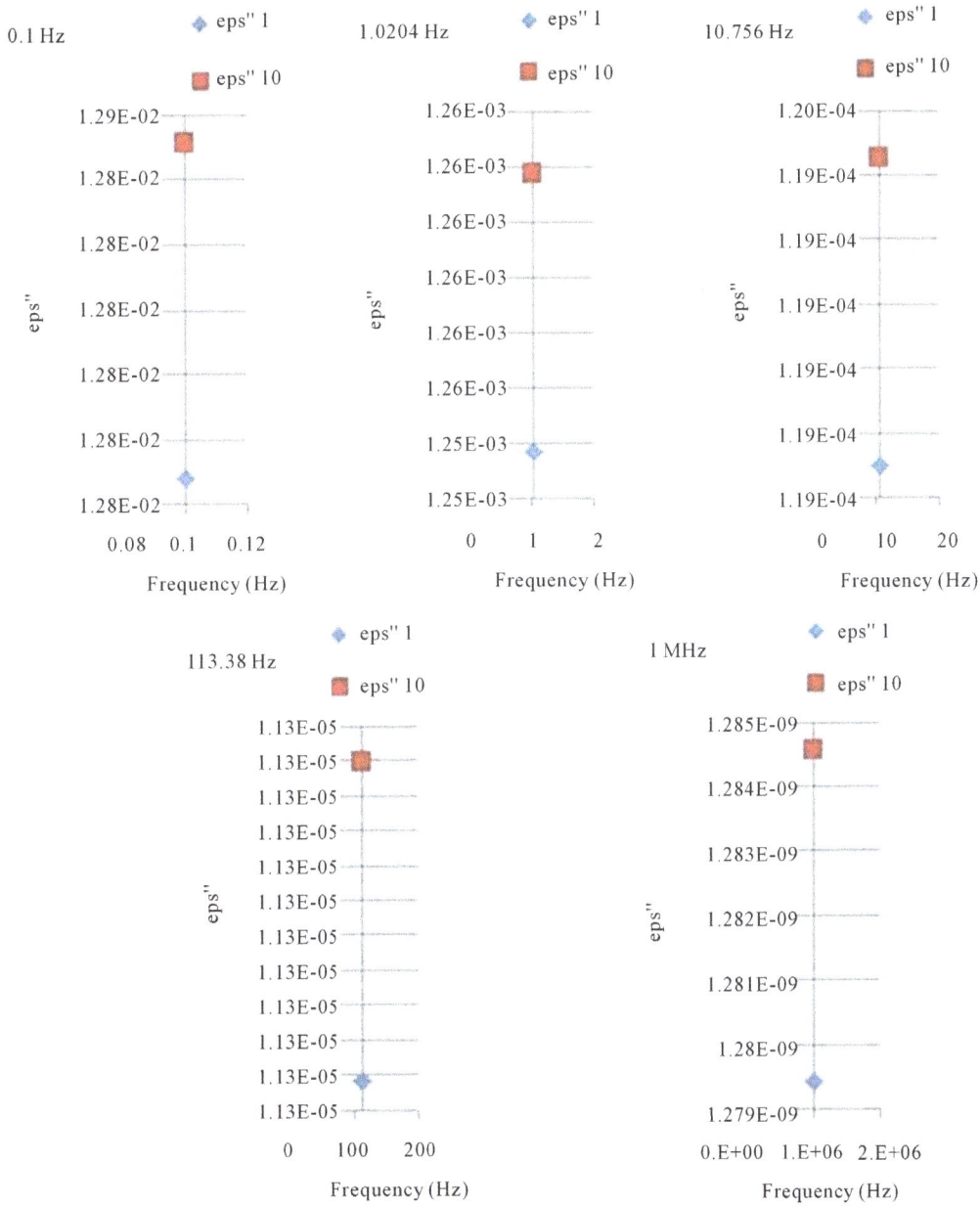

Figure 25. Dielectric losses at different frequency for different number of inclusion but same volume fraction without any conductive layer.

Figures 27-29 show the change of dielectric properties of a single damage phase and distributed damage with the same amount of volume fraction with a conductive layer around the defect. The dielectric loss increased for the distributed damage because of more interfacial polarization and it is more evident than the prior case because the conductive layer around the defect leads to increased interfacial polarization.

The difference between the static permittivity and the limiting high frequency dielectric permittivity is called the Dielectric relaxation strength (DRS), $\Delta\varepsilon$, as given in Equation (15).

$$\Delta\varepsilon = \varepsilon_s - \varepsilon_\infty \tag{15}$$

where, ε_s is the static permittivity and ε_∞ is the limiting high frequency dielectric constant. To calculate DRS from the experimental data we subtract the value of real part of the permittivity at 1 MHz from the value of real part of the permittivity at 0.1 Hz.

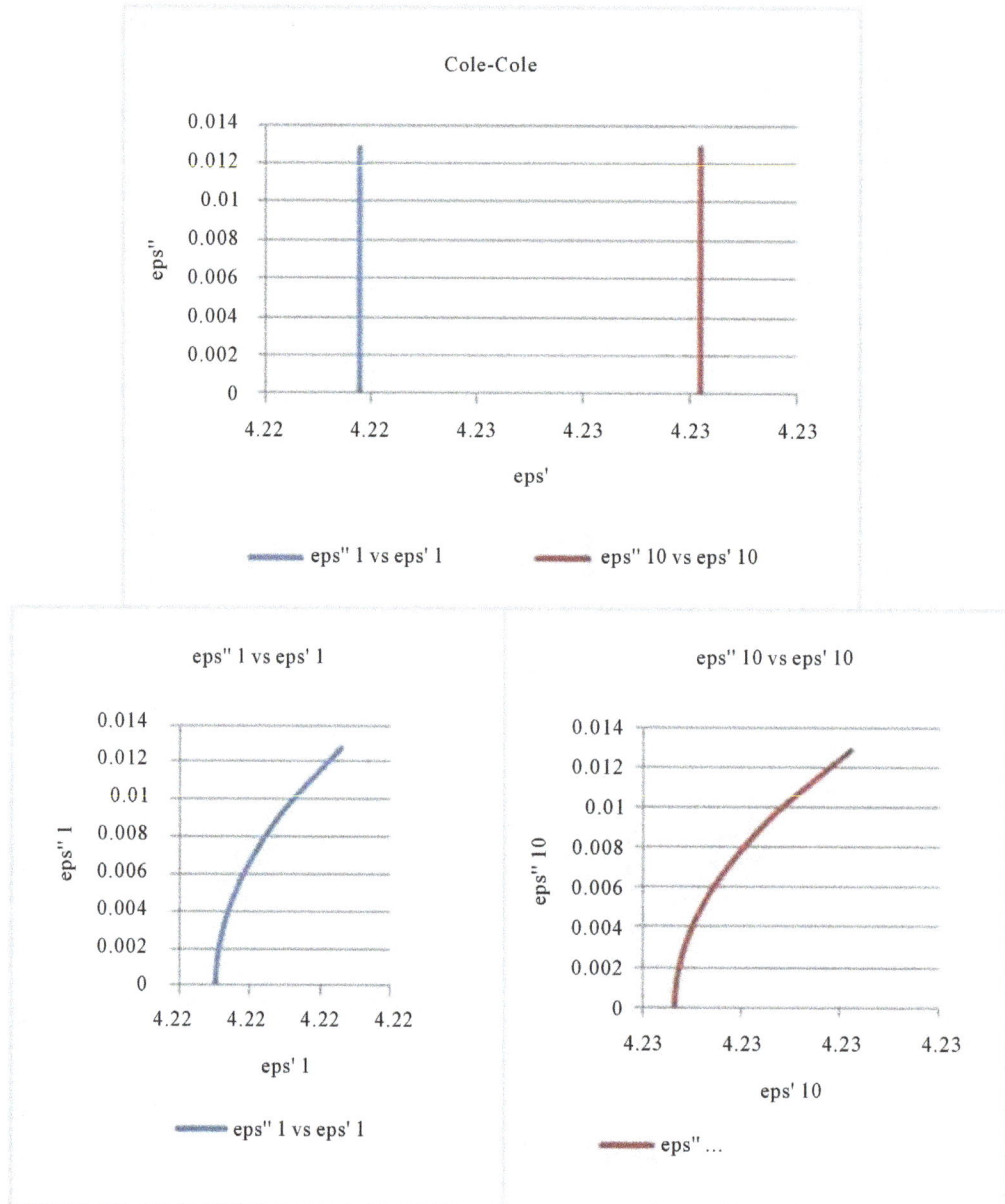

Figure 26. Cole-Cole plot of different number of inclusion without conductive layer.

Figure 30(a) shows the increase of DRS with the increase of damage state defects which is in agreement with what we saw in the computational data (**Figure 20** and **Figure 27**). **Figure 30(b)** shows the increase in dielectric loss with the increase of damage, and this loss increased more in the lower frequency region when the defects had conductive solution layers on their surface, which in also in agreement with the computational model (**Figure 21** and **Figure 28**).

4. Conclusions

In this paper, we have demonstrated a computational model to predict global dielectric property changes caused by increasing defects inside of a materials system. We show that the dielectric character of the defects, their volume, and the morphology of the defect surfaces play an important role in the overall dielectric properties of the materials system during degradation. The data presented here also demonstrate the possibility of using dielectric properties to model and interpret the progressive damage of heterogonous materials systems.

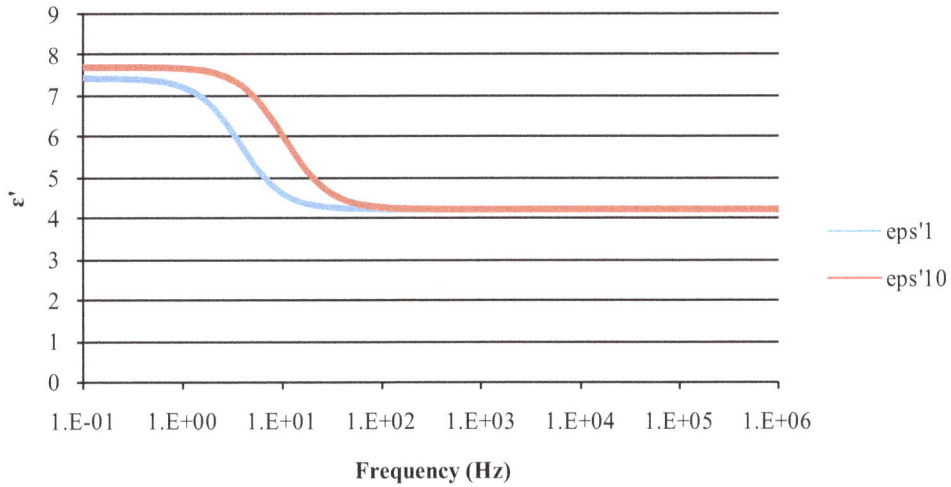

Figure 27. Real part of the Permittivity of different number inclusion but same volume fraction with conductive layer.

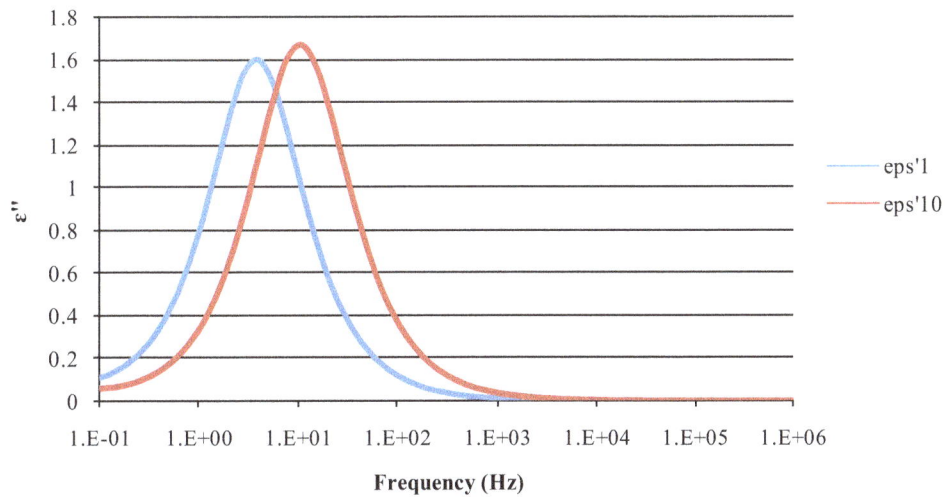

Figure 28. Imaginary part of the Permittivity of different number inclusion but same volume fraction with conductive layer.

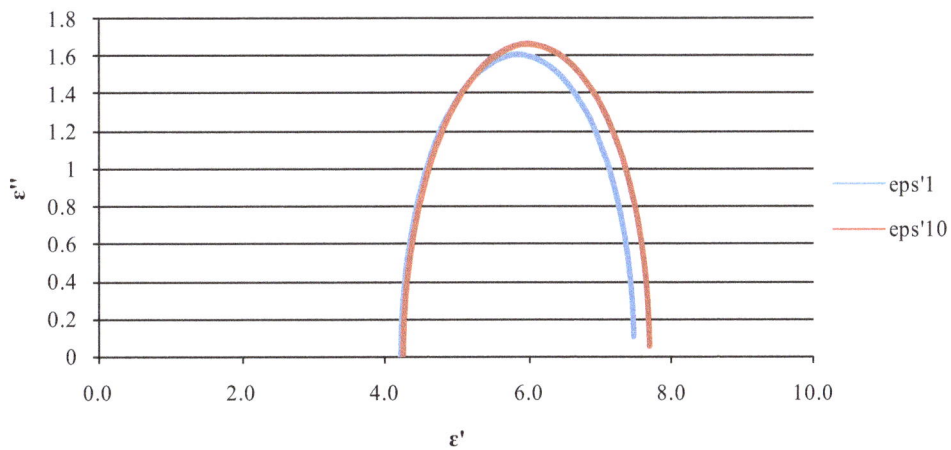

Figure 29. Cole-Cole plot of different number of inclusion with conductive layer.

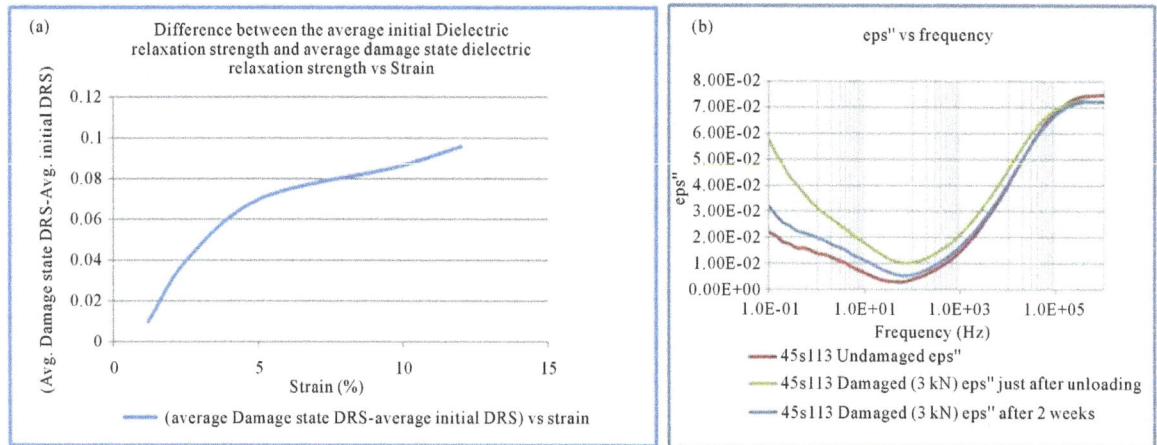

Figure 30. Experimental results of dielectric properties change of damaged composite materials.

In general, we have shown that the dielectric properties of heterogeneous systems are influenced by various physical factors: electrical and structural interactions between particles, heterogeneity of morphological and electrical properties of the constituent phases, frequency dependence of electrical phase parameters, intra-particle structure, particle shape, size, orientation and, volume and surface fraction of the constituent phases. This dependence complicates the determination of the electrical parameters of heterogeneous materials from the observed global dielectric relaxation spectra, but also presents us with an opportunity to recover important information not only about the electrical and structural properties of constituents but also about the interactions between constituents, including the parent materials and damage phases. Further theoretical and experimental investigation is required to fully understand the changes in dielectric spectra associated with many of the specific damage accumulation events and local details in heterogeneous material systems.

From the results presented in this paper, it can be concluded that analysis of the dielectric data gives us information about the type of material state changes throughout the mechanical life of a composite material. It should be emphasized that these changes in the dielectric properties are distinct and measurable changes in material state, and that they are caused by a non-conservative, non-equilibrium material response to the applied fields. Opportunities for further understanding include the identification of the material and physical limitations of this method of characterization, e.g., specimen size, material property ranges, and specimen shapes that are most and least suited to the approach. A robust study of the interpretation of dielectric data associated with specific damage modes and details is also needed.

References

[1]   Reifsnider, K.L. and Case, S.W. (2002) Damage Tolerance and Durability of Material Systems. John Wiley and Sons, New York.

[2]   Raihan, R., Adkins, J.M., Baker, J., Rabbi, F. and Reifsnider, K. (2014) Relationship of Dielectric Property Change to Composite Material State Degradation. *Composites Science and Technology*, **105**, 160-165. http://dx.doi.org/10.1016/j.compscitech.2014.09.017

[3]   Raihan, R., Reifsnider, K., Cacuci, D. and Liu, Q. (2015) Dielectric Signatures and Interpretive Analysis for Changes of State in Composite Materials. *ZAMM-Journal of Applied Mathematics and Mechanics/Zeitschrift für Angewandte Mathematik und Mechanik.* http://dx.doi.org/10.1002/zamm.201400226

[4]   Tuncer, E., Serdyuk, Y.V. and Gubanski, S.M. (2001) Dielectric Mixtures—Electrical Properties and Modeling.

[5]   Brosseau, C., Beroual, A. and Boudida, A. (2000) How Do Shape Anisotropy and Spatial Orientation of the Constituents Affect the Permittivity of Dielectrich Eterostructures? *Journal of Applied Physics*, **88**, 7278-7288. http://dx.doi.org/10.1063/1.1321779

[6]   Baker, J., Adkins, J.M., Rabbi, F., Liu, Q., Reifsnider, K. and Raihan, R. (2014) Meso-Design of Heterogeneous Dielectric Material Systems: Structure Property Relationships. *Journal of Advanced Dielectrics*, **4**, 1450008. http://dx.doi.org/10.1142/S2010135X14500088

[7]   Pawar, S.D., Murugavel, P. and Lal, D.M. (2009) Effect of Relative Humidity and Sea Level Pressure on Electrical Conductivity of Air over Indian Ocean. *Journal of Geophysical Research*: *Atmospheres*, 114(D2).

[8]   Soles, C. and Yee, A. (2000) A Discussion of the Molecular Mechanisms of Moisture Transport in Epoxy Resins. *Journal of Polymer Science, Part B*: *Polymer Physics*, **38**, 792-802. http://dx.doi.org/10.1002/(SICI)1099-0488(20000301)38:5<792::AID-POLB16>3.0.CO;2-H

[9]   Adamson, M.J. (1980) Thermal Expansion and Swelling of Cured Epoxy Resin Used in Graphite/Epoxy Composite Materials. *Journal of Material Science*, **15**, 1736-1745. http://dx.doi.org/10.1007/bf00550593

[10] Tencer, M. (1994) Moisture Ingress into Nonhermetic Enclosures and Packages—A Quasisteady State Model for Diffusion and Attenuation of Ambient Humidity Variations. *IEEE 44th Electronic Components Technology Conference*, Washington DC.

[11] Shirangi, M.H. and Michel, B. (2010) Mechanism of Moisture Diffusion, Hygroscopic Swelling, and Adhesion Degradation in Epoxy Molding Compounds. *Moisture Sensitivity of Plastic Packages of IC Devices*, Springer, 29-69. http://dx.doi.org/10.1007/978-1-4419-5719-1_2

[12] Banhegyi, G. and Karasz, F.E. (1986) The Effect of Adsorbed Water on the Dielectric Properties of $CaCO_3$ Filled Polyethylene Composites. *Journal of Polymer Science Part B*: *Polymer Physics*, **24**, 209-228. http://dx.doi.org/10.1002/polb.1986.090240201

[13] Banhegyi, G., Hedvig, P. and Karasz, F.E. (1988) DC Dielectric Study of Polyethylene/$CaCO_3$ Composites. *Colloid and Polymer Science*, **266**, 701-715. http://dx.doi.org/10.1007/BF01410279

[14] Cotinaud, M., Bonniau, P. and Bunsell, A.R. (1982) The Effect of Water Absorption on the Electrical Properties of Glass-Fibre Reinforced Epoxy Composites. *Journal of Materials Science*, **17**, 867-877. http://dx.doi.org/10.1007/bf00540386

[15] Reid, J.D., Lawrence, W.H. and Buck, R.P. (1986) Dielectric Properties of an Epoxy Resin and Its Composite I. Moisture Effects on Dipole Relaxation. *Journal of Applied Polymer Science*, **31**, 1771-1784. http://dx.doi.org/10.1002/app.1986.070310622

[16] Paquin, L., St-Onge, H. and Wertheimer, M.R. (1982) The Complex Permittivity of Polyethylene/Mica Composites. *IEEE Transactions on Electrical Insulation*, **5**, 399-404. http://dx.doi.org/10.1109/TEI.1982.298482

10

Characterization of Solid Wastes from Aluminum Tertiary Sector: The Current State of Spanish Industry

Roberto Galindo[1], Isabel Padilla[1], Olga Rodríguez[1], Ruth Sánchez-Hernández[1], Sol López-Andrés[2], Aurora López-Delgado[1]

[1]National Centre for Metallurgical Research, CENIM-CSIC, Madrid, Spain
[2]Department of Crystallography and Mineralogy, Faculty of Geology, UCM, Madrid, Spain
Email: alopezdelgado@cenim.csic.es

Abstract

Aluminum recycling is an important activity that allows returning this metal to the market saving energy and resources. This activity generates slag and dross, both hazardous materials, which are recovered by other industries (tertiary sector). In that process, new wastes are produced, but most of them are disposed in security storage facilities because of their hazardousness and scarce marketable value. In Spain, the statistical data analysis on waste reveals that this sector is increasing every year. This study aims to characterize the wastes generated by the tertiary aluminum industries in Spain. Samples were collected in different aluminum recycling industries and characterized by chemical analyses, X-ray fluorescence, X-ray diffraction and particle size determination. Wastes rich in aluminum oxide and alkaline elements also comprise metallic aluminum and aluminum nitride. Such components are the main responsible for the waste hazardousness since they generate toxic gases in the presence of water. Besides, their fine granulometry ($x_{50} < 30$ μm) also contributes highly to the hazardousness.

Keywords

Hazardous Aluminum Waste, Characterization, Recycling, Slag/Dross

1. Introduction

Aluminum is a product with growing consumption in most areas of everyday life such as transport, engineering, construction and packaging, due to its remarkable properties. The worldwide implementation of aluminum has been successful due to the cost reduction in manufacturing process and the huge possibilities of recovery. The Life Cycle Assessment (LCA) of aluminum is composed of several activities (industries) which form a loop that allows minor costs in market. The LCA of aluminum in 2010 revealed that around 50% of total ingot production was produced from secondary aluminum industry generated in Spain, and around 20% from secondary aluminum imported [1].

The primary production is focused on obtaining the aluminum from bauxite, an expensive activity due to its high energy consumption. This metal is sold to smelters that transform these ingots into sheets, foil or profiles (semi-production). The secondary aluminum industry is based on end-of-life products (scrap), slag and skimming from primary industry and semi-production. This industry smelts wastes and inserts them again into the semi-production chain. And finally, wastes from previous businesses (especially slag or dross) are recycled by the tertiary aluminum industry, which uses different processes to recover the residual aluminum present in this kind of waste materials.

The aluminum recycling process has been already detailed by Tsakiridis [2], who also described the characterization and utilization of aluminum salt slag. The conventional process used in this industry consists of grinding the slag or dross and sieving it to recover the metal [3]. Several fractions are separated according to their granulometry. The coarsest fractions are sold to other industries because of their high aluminum content. However, the finest fraction, captured by suction systems and treated in sleeve filters, is not a marketable product due to its lower aluminum content. Accordingly, it is the main waste generated by the tertiary industry [4].

After the grinding and sieving treatment, a water leaching process is also commonly used to separate aluminum from the non metallic material as well as to dissolve the salts present in the dross [3]. The reaction of some components present in dross with water, such as aluminum nitride, aluminum carbide and metallic aluminum, leads to release toxic gases, namely ammonia, methane and hydrogen, and thus causing new compounds in the solid waste as aluminum hydroxides [5]-[7]. As a consequence of this recycling process, a low aluminum recovered/waste treated rate is obtained.

Due to the huge and growing demand for aluminum around the world, significant quantities of wastes are still generated despite the optimization of recycling, and specific control measurements are required, especially for the hazardous wastes.

There is no precise information about the benefit of using dry or wet techniques in the processes used by the tertiary industry for slag recycling. Neither are available exact data on the residues generated by tertiary industries. Accordingly, the characterization of these materials is an important tool to develop their best treatment and disposal pathway. In this sense, there are few examples of characterization of these powders from tertiary industry which can content several components in such heterogeneous mixture, such as corundum, spinel, aluminum nitride, metallic aluminum, quartz, calcite, iron oxide, etc. [8].

So, the aim of this work is to characterize samples of fine solid waste generated by tertiary industry in Spain, to acquire a better understanding about this particular hazardous waste, by means of its composition (chemical and mineralogical) and particle size.

2. Current State

2.1. The Aluminum Recycling Industry in Spain

Spain has a large industrial aluminum sector, but very fragmented in many medium and small companies. Industries include primary production (being the third largest producer in Europe with 17% of total production in 2012) [9] as well as secondary production (more than 35% of total aluminum production), and waste recycling, where the best available techniques (BAT) for the non-ferrous metal industries are applied [10] [11]. The main industrial areas that perform the recovery and disposal of hazardous wastes from aluminum industry are mainly located in the north and centre of Spain, which are traditional areas of the metallurgical activities. **Figure 1** indicates the main industrial sites where recovering and disposal activities are performed for hazardous wastes from aluminum industry in Spain (wastes are identified by the European Waste Catalogue (EWC) code, **Table 1**).

2.2. Legal Framework for Hazardous Wastes in the Aluminum Industry

The Spanish legal framework for hazardous wastes is based on the transposition of the Directive 2008/98/EC of the European Parliament on waste which indicates the following waste hierarchy: prevention, preparing for re-

Figure 1. Main Spanish industrial sites for aluminum production where hazardous waste treatment is performed (wastes with EWC code: 100304 = ■; 100308 = ■; 100309 = ■; 100321 = ■).

Table 1. European Waste Catalogue (EWC) for hazardous in the aluminum thermal metallurgy.

EWC	Hazardous waste
100304	Primary production slag
100308	Salt slag from secondary production
100309	Black dross from secondary production
100315	Skimming that are flammable or emit, upon contact with water, flammable gases in dangerous quantities
100317	Tar-containing wastes from anode manufacture
100319	Flue-gas dust containing dangerous substances
100321	Other particulates and dust (including ball-mill dust) containing dangerous substances
100323	Solid wastes from gas treatment containing dangerous substances
100325	Sludge and filter cakes from gas treatment containing dangerous substances
100327	Wastes from cooling-water treatment containing oil
100329	Waste from treatment of salt slag and black dross containing dangerous substances

use, recycling, other recovery and disposal. Furthermore, these wastes are classified according to the European Waste Catalogue (EWC) (**Table 1**) as the Spanish regulation, MAM/304/2002, indicates [12].

The Spanish Register of Emissions and Pollutant Source [13] provides information of hazardous wastes transfer for both recovery and disposal activities, being a useful source of the quantity of wastes generated by aluminum industries, due to the strong fragmentation and the lack of information provided by other sources or references. According to this register, the majority of hazardous wastes, in terms of transfer tons, are represented by the primary smelting slag (EWC 100304), salt slag from secondary smelting (EWC 100308), black dross from secondary smelting (EWC 100309) and other particulates and dust (including ball mill dust) containing hazardous substances (EWC 100321).

The evolution of total waste transfer (disposal and recovery) in tons per year during the period 2007-2012 in Spain has varied according to aluminum industry demand (**Figure 2**).

As can be seen, primary smelting slag and salt slag from secondary production amounted to more than 100,000 tons per year. In general, from 2007 to 2009 there was a drop in the primary slag and salt slag transfer,

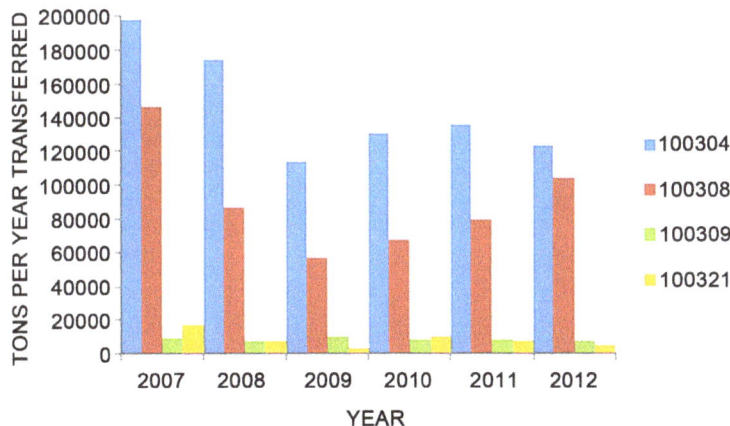

Figure 2. Total tons per year transferred by the main hazardous wastes in the aluminum industry in Spain. 100304: primary smelting slag; 100308: secondary smelting salt slag; 100309: secondary smelting black dross; 100321: particulates and dust, including ball mill dust (Official statistical data compiled from [13]).

while from 2010 onwards a slight increase occurred in this transfer as well as in the aluminum manufacturing, showing the corresponding rise in the waste transfer. In 2012, the secondary industry suffered a strong growth crisis in comparison to the primary industry. This fact was clearly reflected by the similar quantities of waste transfer obtained. On the other hand, black dross from secondary smelting and other particulates (dust containing hazardous substances) were transferred less than 20,000 tons per year from secondary smelting.

Figure 3 shows the quantity of hazardous waste per tons and year transferred in Spain during the same period, 2007-2012. As shown, primary production slag (100304) presented low recovery rates and its disposal into secure landfill was intense especially since 2009. On the other hand, salt slag smelting (100308) and black dross (100309) from secondary aluminum industry achieved high recovery rates over the period. In the case of other particulates (100321) recovery percentages were around 50% referred to the total transfer.

Regarding the fine powder coming from the bag filter suction systems used in the aluminum slag milling, it is considered as hazardous waste included in the EWC 100321 group. Nevertheless, this waste generated by the Spanish tertiary industry is not recovered. So, its disposal into secure landfill is the current option selected for its control, as a profitable recovery is not possible.

3. Experimental

Dried fine powder samples of a bulk of hazardous wastes from different Spanish tertiary aluminum industries were collected and kept into polyethylene bags. Samples come from the bag filter suction system of the aluminum slag milling process. Powders were labeled as Sx (x: 1 - 11). Next, bulk samples (2 kg) were divided into eight representative samples by a Laborette 27, rotary cone sample divider. Representative samples were characterized as follows: Determination of particle size distribution was carried out using a Sympatec Helos 12LA by ultrasonic dispersion in isopropyl alcohol. Sample composition analysis was performed by X-ray fluorescence (XRF), using a wavelength dispersive X-ray fluorescence spectrometer (Bruker, S8 Tiger), with rhodium anode and 4 kW excitation power. Measurements were done on fusion disks prepared with lithium tetraborate and lithium metaborate at 60 kV and 170 mA tube setting and using crystals LiF (220), LiF (200), PET, and XS-55 as analyzer crystals. X-ray diffraction (XRD) was performed in order to determine the main crystal phases present in samples by means of a Bruker D8 Advance Diffractometer with $CuK\alpha$ radiation, with 2θ from 5° to 85°, at a scan rate 2θ of 0.02°, 5 s per step. Tube setting was 40 kV and 30 mA. Aluminum nitride content in hazardous wastes was analyzed by Kjeldhal method using an automatic steam distilling unit UDK 130 A by Velp Scientifica, and subsequent titration with 1 M HCl. The content of metallic aluminum was determined by treating samples with a 10% HCl solution to solve aluminum acid soluble compounds. Filtrate was analyzed by atomic absorption spectrometry (AAS) in a Varian Spectra model AA-220FS equipment.

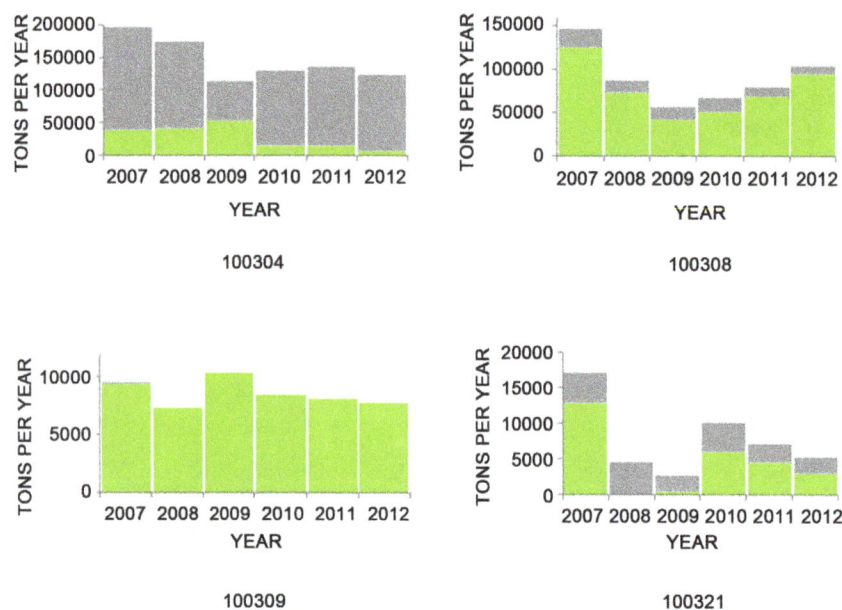

Figure 3. Amount of hazardous waste (100304: primary smelting slag; 100308: secondary smelting salt slag; 100309: secondary smelting black dross; 100321: particulates and dust, including ball mill dust) transferred in the Spanish aluminum industry (statistical data compiled from [13]).

4. Results and Discussion

The particle size distribution for all investigated samples is shown in **Table 2**. As can be seen, samples present a high percentage of fine particles. Cumulative size distributions calculated at 50% (x_{50}) indicate particle diameters less than 30 µm for most of the samples, except for S5. These small size samples allow that fraction of particles can be inhaled by humans or penetrate into airways is relatively high as the cumulative distribution indicates at 10% (x_{10}) [14]. Thus, it is observed that diameter sizes are around 2.5 µm in most cases, except for S5 sample which contains larger particle size not only for x_{10} but also for all fractions. It is noticeable that the sample S11 exhibits the highest x_{90} value but numbers for x_{10} and x_{50} are very low. Furthermore, no significant differences have been found when different plans locations are compared. In general, samples consisted of very fine grain size which contributes to their hazardousness but also to their workability and handling properties. The distribution of particle size, determined by laser diffraction, showed for the most of wastes a trimodal tendency. Several examples of the distribution density curves are collected in **Figure 4**. Nevertheless, the sample S11 showed a quite different curve with a multimodal profile where peaks corresponding to small, medium and large size are appeared within a small range of distribution density, moreover, as above commented, this sample is the responsible for the grain size maximum value (164.0 µm) of **Table 2**.

Concerning chemical composition, results from FRX are collected in **Table 3** along with their statistical values: average value ($\bar{x}$) and standard deviation (σ). The major component ranging between 53 wt% - 81 wt% corresponds to aluminum compounds which are expressed as Al_2O_3. Percentages lower than 10 wt% are obtained for other components, except for S5 where the highest contents of silica and sodium are found. Its high sodium content is attributable to the employ of high amounts of salt in the melting process to get a higher aluminum recovery; indeed, the Al_2O_3 content is the lowest one in this sample. The high percentage of TiO_2 (>5 wt%) in samples S4 and S6 is noticeable and it may be associated with the different kinds of scrap used in secondary industry, and similarly, other metals (Fe, Cr, etc.). The more efficiency achieved in the scrap classification prior to melting process the lower the content of those metals in the waste. Minor metals such as copper, zirconium, manganese, lead, and chromium are grouped in column "others". The high values obtained for the standard deviation of the average composition reveal the high variability of the chemical composition of samples. The main environmental problem related to these wastes, come from the contents of aluminum nitride and metallic aluminum (also aluminum carbide). Their high reactivity with water or even humidity in air, leads to formation of toxic and explosive gases such as ammonia, hydrogen and methane [2]. **Table 4** shows the contents of alumi-

Table 2. Particle size distribution of aluminum solid wastes from the bag filter suction systems used in the aluminum slag milling and statistical values ($\bar{x}$, σ).

Sample	Maximum particle diameter (μm)		
	x_{10}	x_{50}	x_{90}
S1	2.4	29.0	81.1
S2	2.1	17.8	81.9
S3	0.9	5.1	40.4
S4	1.0	6.0	43.7
S5	7.1	47.4	154.7
S6	2.2	28.0	119.0
S7	3.2	27.4	80.7
S8	2.4	25.9	99.7
S9	2.3	27.9	89.9
S10	2.3	17.3	65.8
S11	1.5	18.6	164.0
$\bar{x}$	2.5	22.8	92.8
σ	1.7	11.8	39.9

Table 3. Waste composition (main components) obtained by XRF and expressed as oxides (wt%).

Sample	XRF									
	Al_2O_3	MgO	Fe_2O_3	TiO_2	SiO_2	CaO	K_2O	Na_2O	Cl	Others
S1	79.3	4.7	0.6	2.7	2.9	2.2	1.0	2.5	2.0	2.3
S2	79.6	4.3	0.7	3.5	2.8	2.2	0.2	1.8	0.7	4.1
S3	80.9	3.4	0.7	3.4	2.7	2.1	0.3	1.7	0.9	4.0
S4	71.6	4.5	1.9	5.6	6.2	6.0	0.3	-	-	4.1
S5	53.4	6.8	1.2	0.5	17.8	2.9	1.1	10.5	3.5	2.4
S6	80.5	1.3	0.9	8.3	2.9	1.6	3.0	0.43	-	1.0
S7	77.9	3.8	1.3	0.4	7.6	2.5	1.1	2.81	0.9	1.7
S8	76.7	4.3	3.4	2.0	5.2	3.5	0.7	1.0	0.8	2.5
S9	76.9	4.1	0.4	3.2	2.6	2.0	0.6	3.2	3.4	3.7
S10	81.0	4.2	0.7	1.1	3.9	3.8	0.4	0.3	-	4.6
S11	71.2	4.3	1.8	1.5	10.6	2.9	0.6	1.7	1.5	4.1
$\bar{x}$	75.0	4.0	1.4	2.8	6.3	2.8	0.7	2.2	1.2	3.3
σ	8.0	1.2	1.0	2.3	4.7	1.1	0.7	2.7	1.2	1.3

Table 4. Contents of aluminum nitride (AlN) (wt%) and metallic aluminum (Al_M) (wt%).

Sample	S1	S2	S3	S4	S5	S6	S7	S8	S9	S10	S11	$\bar{x}$	σ
AlN	11.8	15.3	13.7	13.7	1.40	23.6	12.6	12.7	16.0	19.7	9.8	13.3	5.4
Al_M	15.6	7.7	17.6	5.4	17.4	7.4	11.9	7.4	8.7	1.2	29.6	11.8	7.5

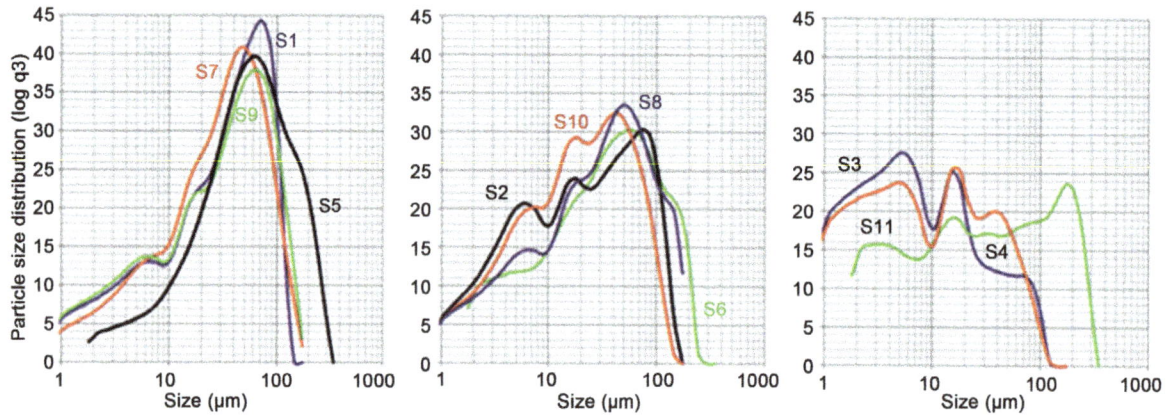

Figure 4. Particle size distribution curves of aluminum wastes.

num nitride (calculated from the results of the determination of ammonia by Kjendhal method) and metallic aluminum in the wastes. A high variability in results is observed as showed by the high value of standard deviation for both parameters. Concerning AlN, the lowest value (1.4 wt%) is observed for S5 which is the sample with the highest content of salts and silica; and the highest value (23.6 wt%) corresponds to S6. Gil in his study about Spanish slags in the secondary industry pointed contents in Aluminum Nitride lower than 5% [15]. In this study, authors report higher contents of this compound in samples collected from tertiary industries in which slag is milled and the finest fraction is enriched in aluminum nitride and salts. Related to metallic aluminum contents, the highest value is obtained for S11. This sample also exhibits the highest x_{90} value which could be attributable to its highest metallic aluminum content, because of its spherical particles hardly crushed in the milling process in comparison with oxidized phases and salts.

From the aluminum nitride contents, the volume of ammonia gas, which can be generated by the wastes in presence of humidity, can be calculated by the reaction (1) [16]. Values range between 8 - 129 Nm^3 per ton of waste as can be seen in the **Figure 5**.

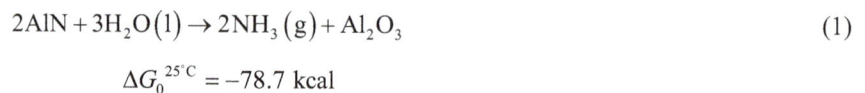

$$2AlN + 3H_2O(l) \rightarrow 2NH_3(g) + Al_2O_3 \tag{1}$$

$$\Delta G_0^{25°C} = -78.7 \text{ kcal}$$

From the metallic aluminum contents, the volume of hydrogen evolved per ton of waste range between 14 - 368 Nm^3 as shown in the reaction (2) [15] (**Figure 5**).

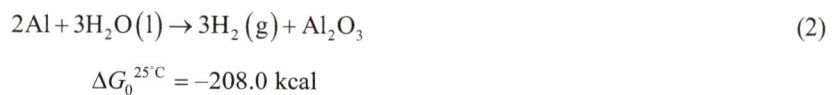

$$2Al + 3H_2O(l) \rightarrow 3H_2(g) + Al_2O_3 \tag{2}$$

$$\Delta G_0^{25°C} = -208.0 \text{ kcal}$$

From the mineralogical point of view, wastes comprise a mixture of both amorphous and crystalline phases as observed in the complex X-ray diffraction patterns (**Figure 6**). The crystalline phases mainly consist of not-well defined peaks with variable intensity. They are assigned to corundum, aluminum nitride, metallic aluminum and spinel (aluminum phases), along with other phases as quartz, calcium carbonate and Na/K chloride. The ratio of these phases in the different samples is variable as a result of different factors such as scrap composition and the thermal treatment of slag, resulting in a high heterogeneity of the chemical composition of samples [2] [17]. The presence of high contents of salts in slag promotes the formation of a great number of tiny disordered phases in wastes, being S3 and S4 samples, those which present the most formation of this non-crystalline component. In this sense, peaks of sodium chloride (halite) with strong intensity are clearly observed in particular for all samples, except for S6 sample. Crystalline phases are mainly composed of not well defined peaks with variable intensities assigned to aluminum nitride, corundum and spinel. Quartz and calcium carbonate show low intensities. The specific thermal treatments received by the slag and the salt content explain differences between samples. Thus, slag produced at higher temperatures present their corundum and spinel phases with a better growth, as seen in S1, S2, S5 and S11. Peaks of iron oxide could not be unambiguously assigned and for samples with high content of TiO$_2$, peaks corresponding to this compound were also poorly detected. The high background of all

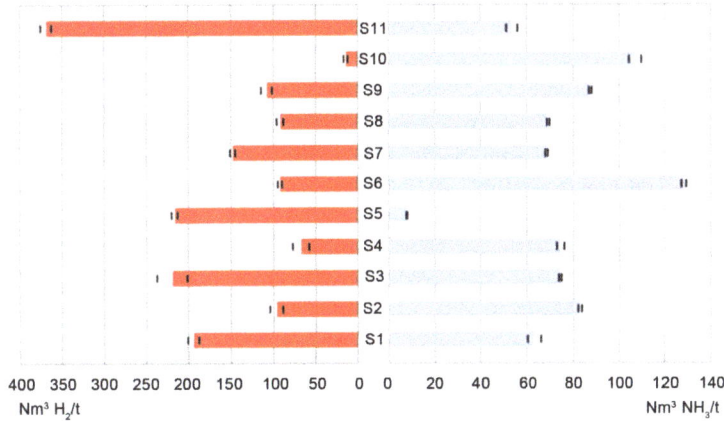

Figure 5. Ammonia and Hydrogen released by samples in presence of humidity, expressed in Nm^3 per ton of waste. (| |, indicates the standards deviations of measurements).

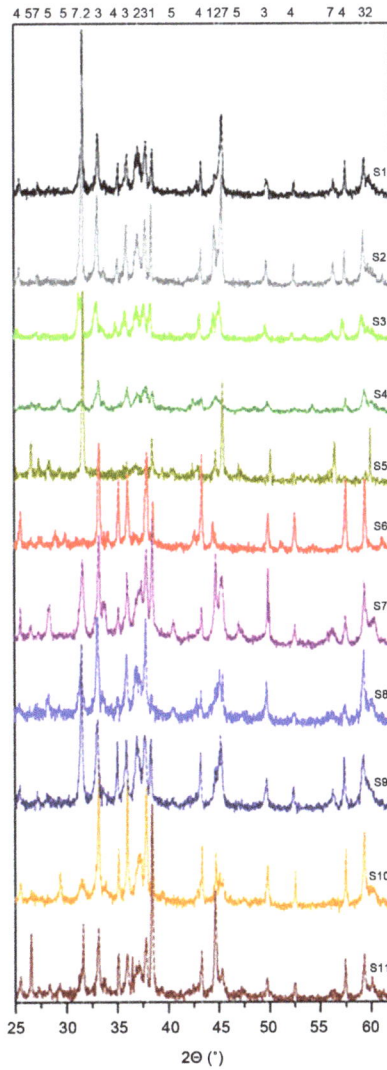

Figure 6. X-ray diffraction patterns of aluminum waste samples (1 = Al, aluminum; 2 = $MgAl_2O_4$, spinel; 3 = AlN, aluminium nitride; 4 = Al_2O_3, corundum; 5 = SiO_2, quartz; 6 = $CaCO_3$, calcium carbonate; 7 = NaCl, halite).

XRD patterns may inform about the existence of very low crystalline or amorphous compounds. On the other hand, no significant differences in composition have been observed when different locations were studied.

5. Conclusion

Hazardous wastes from Spanish tertiary aluminum industry were characterized. Wastes consisted of a heterogeneous blend of crystalline and non-crystalline phases, whose distribution depends on the thermal treatment of the slag, the saline content added during remelting process to recover the highest amount of aluminum, and the scrap composition from the secondary industry. In general, the content of aluminum (as oxide) represents more than 75% in weight, but the participation on it of aluminum nitride, metallic aluminum and corundum is very variable. The particle size of these wastes is relatively small, since 50% of grains are smaller than 30 μm of diameter. By their composition, these wastes could not be landfilled. Relatively high amounts of toxic gases (ammonia and hydrogen) can be evolved to atmosphere. Others processes that convert the nitride and aluminum metal content of these residues into value-added products are necessary for the profitability of this recycling sector.

Acknowledgements

Authors thank the MINECO (Spain) for financing the project CTM2012-34449 and aluminum tertiary companies for supplying the wastes. R. Sánchez-Hernández thanks MINECO for the grant BES-2013-066269.

References

[1]   Sevigné-Itoiz, E., Gasol, C.M., Rieradevall, J. and Gabarrell, X. (2014) Environmental Consequences of Recycling Aluminum Old Scrap in a Global Marked. *Resources Conservation and Recycling*, **89**, 94-103. http://dx.doi.org/10.1016/j.resconrec.2014.05.002

[2]   Tsakiridis, P.E. (2012) Aluminium Salt Slag Characterization and Utilization—A Review. *Journal of Hazardous Materials*, **217-218**, 1-10. http://dx.doi.org/10.1016/j.jhazmat.2012.03.052

[3]   Das, B.R., Dash, B., Tripathy, B.C., Bhattacharya, I.N. and Das, S.C. (2007) Production of η-Alumina Form Waste Aluminium Dross. *Minerals Engineering*, **20**, 252-258. http://dx.doi.org/10.1016/j.mineng.2006.09.002

[4]   López-Delgado, A. and Tayibi, H. (2012) Can Hazardous Waste Become a Raw Material? The Case Study of an Aluminum Residue: A Review. *Waste Management and Research*, **30**, 474-484. http://dx.doi.org/10.1177/0734242X11422931

[5]   Krnel, K., Drazic, G. and Kosmac, T. (2004) Degradation of AlN Powder in Aqueous Environments. *Journal of Materials Research*, **19**, 1157-1163. http://dx.doi.org/10.1557/JMR.2004.0150

[6]   López, F.A., Medina, J., Gutierrez, A., Tayibi, H., Peña, C. and López-Delgado, A. (2004) Treatment of Aluminium Dust by Aqueous Dissolution. *Revista de Metalurgia*, **40**, 389-394. http://dx.doi.org/10.3989/revmetalm.2004.v40.i5.294

[7]   Shinzato, M.C. and Hypolito, R. (2005) Solid Waste of Aluminium Recycling Process: Characterization and Reuse of Its Valuable Constituents. *Waste Management*, **25**, 37-46. http://dx.doi.org/10.1016/j.wasman.2004.08.005

[8]   Gonzalo-Delgado, L., López-Delgado, A., López, F.A., Alguacil, F.J. and López-Andrés, S. (2011) Recycling of Hazardous Waste from Tertiary Aluminium Industry in a Value-Added Material. *Waste Management and Research*, **29**, 127-134. http://dx.doi.org/10.1177/0734242X10378330

[9]   Renda, A., Pelkmans, J., Egenhofer, C., Marcu, A., Schrefler, L., Luchetta, G., Simonelli, F., Genoese, F., Valiante, D., Mustilli, F., Colantoni, L., Infelise, F., Stoefs, W., Teusch, J., Timini, J., Wieczorkiewicz, J., Zavatta, R., Giannotti, E. and Stecchi, G.M. (2013) Assessment of Cumulative Cost Impact for the Steel and the Aluminium Industry—Final Report Aluminium. Centre of European Policy Studies, Brussels.

[10] Observatorio Industrial del Metal (2010) El sector del reciclaje de metales en España. Ministry of Industry, Energy and Tourism, Madrid.

[11] Joint Research Center (2014) Best Available Techniques (BAT) Reference Document for the Non-Ferrous Metal Industries. Institute for Prospective Technological Studies Sustainable Production and Consumption Unit European IPPC Bureau. European Commission Final Draft, 496-505.

[12] ORDEN MAM/304/2002, de 8 de febrero, por la que se publican las operaciones de valorización y eliminación de residuos y la lista europea de residuos. In: BOE num. 43, de 19 de febrero de 2002, 6494-6515. http://www.boe.es/diario_boe/txt.php?id=BOE-A-2002-3285

[13] PRTR-España (2012) Ministry of Agriculture, Food and Environment, Madrid. http://www.en.prtr-es.es

[14] Marshall, J. (2013) PM 2.5. *Proceedings of the National Academy of Sciences of the United States of America, PNAS*, **110**, 8756. http://dx.doi.org/10.1073/pnas.1307735110

[15] Gil, A. (2005) Management of the Salt Cake from Secondary Aluminum Fusion Processes. *Industrial and Engineering Chemistry Research*, **44**, 8852-8857. http://dx.doi.org/10.1021/ie050835o

[16] López, F.A., Peña, M.C. and López-Delgado, A. (2001) Hydrolysis and Heat Treatment of Aluminum Dust. *Journal Air & Waste Management Association*, **51**, 903-912. http://dx.doi.org/10.1080/10473289.2001.10464314

[17] Huang, X.-L., El Badawy, A., Arambewela, M., Ford, R., Barlaz, M. and Tolaymat, T. (2014) Characterization of Salt Cake from Secondary Aluminum Product Ion. *Journal of Hazardous Materials*, **273**, 192-199. http://dx.doi.org/10.1016/j.jhazmat.2014.02.035

11

Improved High Performance Recycling of Polymers by Means of Bi-Exponential Analysis of Their Fluorescence Lifetimes

Heinz Langhals*, Dominik Zgela, Thorben Schlücker

Department of Chemistry, LMU University of Munich, Munich, Germany
Email: *Langhals@lrz.uni-muenchen.de

Abstract

Technical polymers could be identified by means of their remarkably strong auto fluorescence. The mono-exponentially obtained time constants of fluorescence decay were applied for a rough assignment of the polymeric materials whereas bi-exponential analysis allowed a fine classification such as for special batches and for preceding contaminations. Chemically similar materials such as LDPE (low-density polyethylene), HDPE (high-density polyethylene) and UHDPE (ultra-high-density polyethylene) could be as well identified as contaminations of mineral oil in PET (polyethylene terephthalate). Furthermore, the fluorescence spectra could be characterized by means of five Gaussian functions in the visible allowing a redundant assignment to the fluorescence lifetimes. Thus, efficient sorting of polymers was possible for high performance recycling.

Keywords

Recycling, Polymers, Fluorescence Decay, Fluorescence Spectroscopy, Gaussian

1. Introduction

There is an increasing need for the recycling of organic polymers (plastics) both for economy and avoiding environmental pollution. The pacific trash vortex is one of the most prominent and impressive examples for the latter and even gets the name Great Pacific Garbage Patch. The majority of technical polymers are thermoplasts and may be re-used by melt and moulding again. However, the immiscibility and the incompatibility of different organic polymers are therefore the main obstacle, because high performance materials require uniformity where

---

*Corresponding author.

a content of external material as low as 5% downgrades the value of polymers appreciably. As a consequence, pure polymers are required for an efficient recycling where a sorting becomes necessary for mixed collected materials. Sorting is also of interest for chemical processing of polymers because such processes operate most stable with uniform starting materials. The machine-based recognition of polymers is a prerequisite for such processes where methods using the density or electrostatic properties are described [1]-[3]. Optical methods [4] [5] are more attractive because of simple, stable and efficient technology where fluorescence is advantageous [6]-[10] because of unproblematic light path and detection. The doping of polymers with fluorescent markers [10] and their re-identification by the spectral resolution of their fluorescence in combination with a binary coding is described in preceding papers [11] [12], however, a secure pre-treatment of the polymers is a prerequisite for such methods. The use of the remarkably strong autofluorescence of polymers [13] is an attractive alternative because no pre-treatment is necessary and even the workup of deposited material will become possible.

2. Materials and Methods

Spectroscopy: Fluorescence spectra: Varian Cary Eclipse; fluorescence lifetimes: PicoQuant FluoTime 300; Pico Quant PicoHarp 300 (PC-405 laser; 403 nm).

The preferred wavelengths for the detection of the fluorescence decays were located by means of lifetime dependant fluorescence spectra. Starting at the maximum of the fluorescence intensity $I_{(t)}$ with $t = 0$ and a delay of 3 ns the fluorescence spectrum was recorded with sampling until 100 ns. The maximum λ_{max} of this fluorescence spectrum was used for the determination of the fluorescence lifetimes τ because of an amplification of slower component of decay (the spectrum of the shorter period between 0 and 3 ns was recorded for comparison). The fluorescence intensities were accumulated with a repetition rate of 20 MHz during 10 s; slightly increased noise is obtained for a shortened accumulation time of 1.0 ms. The fluorescence lifetimes were obtained by exponential fitting (exponential tail fit) of the respective sections according to the equation $I_{(t)} = A \cdot e^{-t/\tau}$ by means of the software FluoFit from PicoQuant. The shape of the laser pulse was neglected due to its small FWHM. The decay time τ_1 was calculated from the period between 0 and 3 ns and the decay time τ_2 between 3 and 43 ns. Thus, all values are relative and are not independent on the used analytical equipment.

Materials: The technical polymers Luran® (styrene, polyacrylonitile copolymer from BASF), Delrin® (polyoxymethylene from DuPont) and Ultramid® (polyamide with glass fibre from BASF) were applied without further treatment. Spectroscopic grade solvents were applied. Tectosil® from Wacker Chemie AG; PET (polyethylene terephthalate) flakes from Inter Recycling GmbH, applied without further purification.

Purification of PET: PET-Flakes were treated with stirring with a mixture of 3% aqueous NaOH (100 mL) and 15% aqueous sodiumdodecylsulphate solution (SDS, 50 mL) at 85°C for 2 h, washed with distilled water, dried in air at room temperature and then at 60°C for 16 h.

3. Results and Discussion

3.1. First Order Decay of Autofluorescence

The spectra of the autofluorescence of polymers are obtained with optical excitation in the UV, such as with common light sources at 365 nm or at short wavelengths in the visible region such as with well-available lasers at 403 nm. The majority of the fluorescence emission extends in the visible at short wavelengths. The fluorescence decay can be mainly described with a first-order exponential function and the lifetime τ can be precisely determined and is characteristic for the polymeric material [13]. The special advantage of first order is the independence of the time constant as well from the initial intensity as the starting time of the measurement. The thus obtained time constants are reported for typical technical polymers in **Figure 1**.

Here we see that there is a broad range of τ between 0.2 until 5 ns where the individual time constants are as characteristic as fingerprints. Remarkably, the chemically very similar and otherwise difficult to discriminate types of polyethylene, LDPE (low-density polyethylene), HDPE (high-density polyethylene) and UHDPE (ultra-high-density polyethylene) can be unambiguously distinguished; this may be a consequence of different microscopic rigidities of these materials. Moreover, different methods of manufacturing such as PET for bottles and for PET plates or different types of silicones can be well categorized; see **Figure 1**. The various types of silicone dehesives result in well distinguishable, however similar lifetimes τ. The availability of a further criterion for characterization would be of interest.

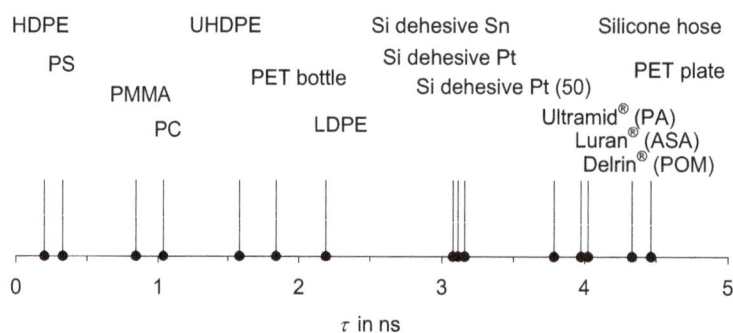

Figure 1. Fluorescence lifetimes τ in ns of technical polymers. Reproducibility is given within 1%.

3.2. Bi-Exponential Components of Fluorescence Decay

We found that the mono-exponential decay dominates resulting in the quite well acceptable first order kinetics. However, there is a bi-exponential component in the decay process. Thus, the lifetimes τ from **Figure 1** mean an averaging of both components depending on their relative contributions. We analyzed the fluorescence decay bi-exponentially and additionally obtained the two time constants τ_1 and τ_2. All three time constants τ, τ_1 and τ_2 are reported in **Table 1**.

The two time constants τ_1 and τ_2 of the bi-exponential data processing allow the two-dimensional characterization of the polymers as is shown in **Figure 2**; Luran® (No. 16) is omitted for clearness because recording the very high value of τ_1 would compress the abscissa.

The two-dimensional presentation of the fluorescence lifetimes τ_1 and τ_2 in **Figure 2** allows a clear distinguishing even for similar materials and supplements the τ values of **Figure 1**. It becomes simpler to distinguish as well between the widely applied polyethylenes LDPE, HDPE and UHDPE (filled circles) as differently processed materials such as the silicone elastomer Tectosil® (triangles) and PET (squares). The differently processed silicone dehesives (diamonds) become more clearly separated where the catalyst for cross-linking (Sn or Pt) seems to influence predominantly τ_2, whereas the addition of a hardener (Pt (50)) seems to increase τ_1. This corresponds to the very low values of τ_1 for the elastomer Tectosil®.

3.3. Test for Contaminated PET

PET is widely applied both for foodstuff such as bottles for water and soft drinks and for technical liquids such as the mineral oils diesel or engine oil. The latter lipophilic liquids can diffuse into the polymeric material such as a plasticizer and would be slowly released. As a consequence, PET for foodstuff must be carefully separated from the latter material. We simulated such contaminations by the contact with the mineral oils diesel and engine oil for one week and purified PET with alkaline detergents according to a technical standard procedure (see materials and methods).

Figure 3 clearly indicates that a pre-treatment of PET results in a lowering of the fluorescence lifetime τ. This is most pronounced for diesel with low molecular weight and low viscosity and slightly less pronounced for engine oil with higher molecular weight and viscosity; this corresponds to observations with silicone dehesives and Tectosil® (see **Figure 2**). A technical cleaning of the contaminated PET increases the fluorescence lifetime again, however, it does not reach the high value of the neat material by far. Even clean commercial recycling flakes exhibit a slightly diminished fluorescence lifetime.

The detection of contaminations becomes more extended by means of the two-dimensional presentation of the fluorescence lifetimes τ_1 and τ_2 in **Figure 4**. The recycling flake with τ close to the neat material given in **Figure 3** becomes much more separated by τ_1 and τ_2 in **Figure 4**. The alterations by means of purifying exhibit different effects for diesel and engine oil. As a consequence, neat material can be well separated.

3.4. Further Extensions

Additional optical information about the polymers could be redundantly used for independent characterization of polymers in order to further increase of the reliability or for fine-tuning of the detection concerning the identifi-

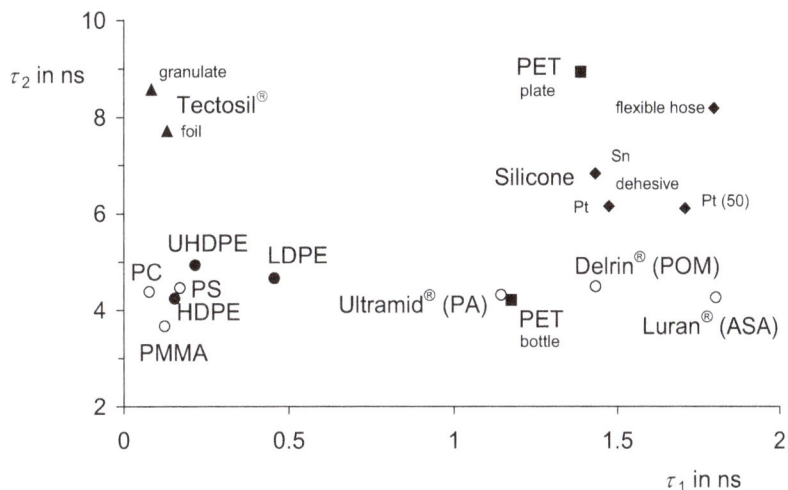

Figure 2. Two-dimensional characterization of polymers by means of their constants τ_1 and τ_2 of bi-exponential data processing of fluorescence decay. Filled circles: The polyethylenes LDPE, HDPE and UHDPE; squares: PET; diamonds: Silicones; triangles: The silicone elastomer Tectosil®.

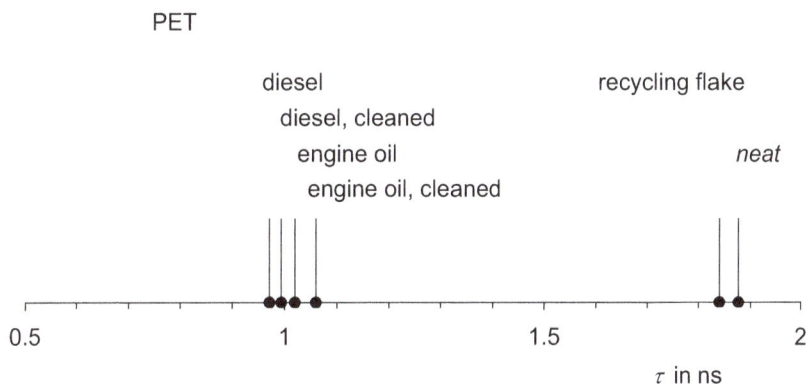

Figure 3. Fluorescence lifetimes τ of pre-treated PET; expanded range.

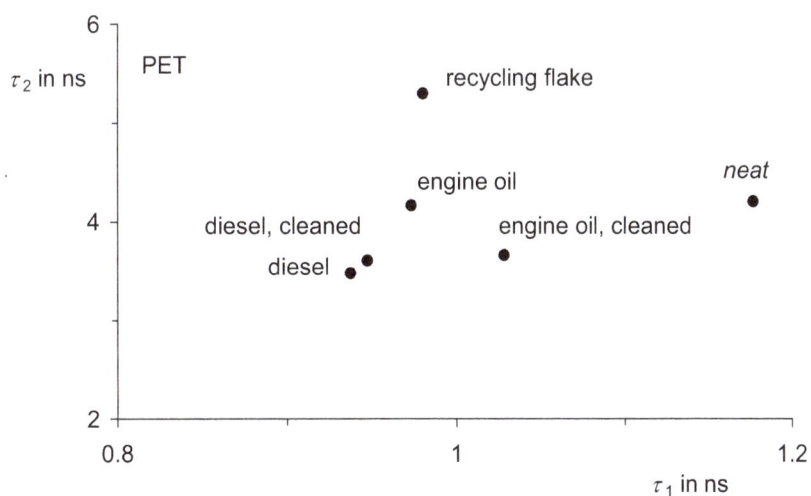

Figure 4. Two-dimensional characterization of PET with various pre-treatment by means of their constants τ_1 and τ_2 of bi-exponential data processing of fluorescence decay.

Table 1. Time constants for the autofluorescence decay of technical polymers. For mono-exponential data processing τ and for bi-exponential τ_1 and τ_2.

Polymeric material	τ	τ_1	τ_2	No.
Polymethylmethacrylate (PMMA)	0.841	0.124	3.669	1
Polystyrene (PS)	3.290	0.171	4.457	2
Polycarbonate (PC)	1.038	0.077	4.379	3
Polyethylenterephthalate (PET) bottle[a]	1.840	1.176	4.205	4
Polyethylenterephthalate (PET) plate	4.466	1.387	8.933	5
Polyethylene LDPE	2.19	0.456	4.655	6
Polyethylene HDPE	<0.2	0.155	4.238	7
Polyethylene UHDPE	1.58	0.217	4.932	8
Silicone Tectosil® granulate		0.132	7.709	9
Silicone Tectosil® foil		0.084	8.572	10
Silicone Dehesive Sn	3.078	1.432	6.825	11
Silicone Dehesive Pt	3.162	1.473	6.149	12
Silicone Dehesive Pt (50)	3.114	1.707	6.106	13
Silicone hose	4.333	1.793	8.180	14
Delrin® (POM)	4.024	1.433	4.487	15
Luran® (ASA)	3.976	1.199	4.259	16
Ultramid® (PA)	3.784	1.145	4.313	17

a) PET bottle for soft drinks.

cation of certain batches. The fluorescence spectra of polymers are comparably broad and seem to be similar and less characteristic [11] [12]. However, we found that the spectra can be precisely characterized by means of Gaussian functions [14] according to Equation (1) where $I_{(\lambda)}$ is the fluorescence intensity dependent on the wavelength λ.

$$I_{(\lambda)} = \sum_{i=0}^{n} I_i \cdot e^{-100 \frac{\left(\frac{1}{\lambda} - \frac{1}{\lambda_i}\right)^2}{2\sigma_i^2}} \tag{1}$$

I_i is the intensity of the entire Gaussian band i, λ_i the position and σ_i the widths. λ is in the denominator because of the reciprocal relation with the energy of the basic Gaussian band. The factor 100 in the exponent simplifies the interconversion between the wavelength in nm and the energy-linear kK (10,000 cm^{-1}).

We applied Gaussian analyse to the technical high-performance polymers Luran® (ASA), Ultramid® (PA) and Delrin® (POM) as typical examples and found that $n = 4$ to 5 Gaussian functions are sufficient for a precise description of the experimental spectra in the visible; see **Table 2**.

The Fluorescence spectrum of Luran® in the visible (>400 nm) is perfectly described by means of four Gaussian bands; see **Figure 5**. Thus, 12 parameters (see **Table 2**) are sufficient for the complete description of the broad spectrum and require minimal 12 values [15] of fluorescence intensity at different wavelengths.

Similar results are obtained for Ultramid® where also four Gaussian bands are sufficient for characterization; see **Figure 6**.

The behaviour of Delrin® is slightly different because of strong fluorescence in the UV close to the visible where five bands concern the visible and one or two bands in the UV should be considered because of tailing into the visible; see **Figure 7**.

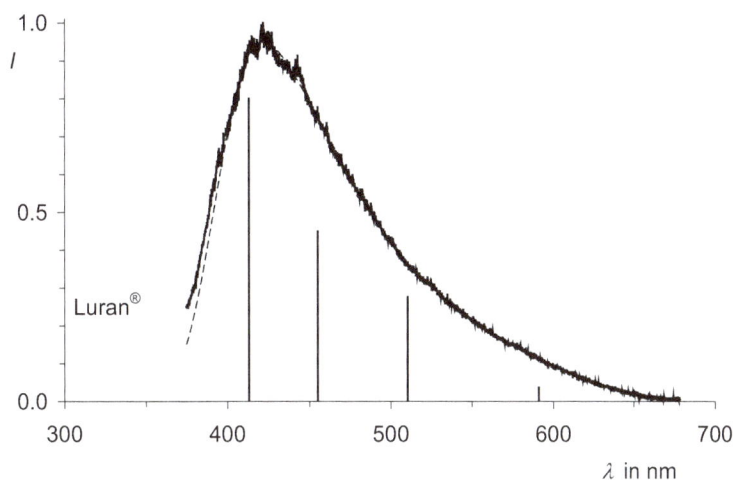

Figure 5. Fluorescence spectrum of Luran® (solid, thick curve) and simulated spectrum (thin, dashed curve, mainly covered by the experimental spectrum) composed of the bands of a Gaussian analysis (>400 nm) on the basis of the parameters of **Table 2**. Bars: Positions and intensities of the Gaussian bands.

Table 2. Parameter of the sum of Gaussian functions according to Equation (1) for the technical polymers Ultramid® (PA), Luran® (ASA) und Delrin® (POM). The numbers i of the bands are given in brackets, the wavelengths λ in nm and σ in kK (10,000 cm^{-1}). The intensities I are derived from the normalized experimental spectrum in the investigated interval.

Polymer	Ultramid®	Luran®	Delrin®
λ_{max} (1)	420.93	413.08	359.96
$2\,\sigma^2$ (1)	2.352	3.628	0.942
I_{max}(1)	0.796	0.799	2.758
λ_{max} (2)	459.09	454.96	385.82
$2\,\sigma^2$ (2)	2.180	3.501	1.291
I_{max} (2)	0.650	0.448	0.398
λ_{max} (3)	499.79	510.34	418.26
$2\,\sigma^2$ (3)	2.369	5.630	1.115
I_{max} (3)	0.161	0.276	0.242
λ_{max} (4)	540.03	590.78	443.58
$2\,\sigma^2$ (4)	3.179	1.643	0.495
I_{max} (4)	0.045	0.036	0.047
λ_{max} (5)			458.34
$2\,\sigma^2$ (5)			2.591
I_{max} (5)			0.140
λ_{max} (6)			484.67
$2\,\sigma^2$ (6)			54.301
I_{max} (6)			0.013
λ_{max} (7)			504.46
$2\,\sigma^2$ (7)			1.667
I_{max} (7)			0.044

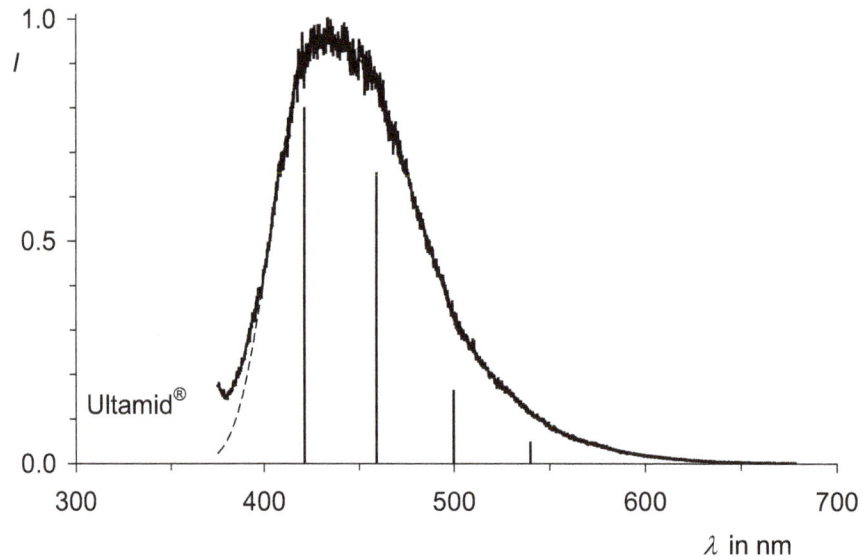

Figure 6. Fluorescence spectrum of Ultramid® (solid, thick curve) and simulated spectrum (thin, dashed curve, mainly covered by the experimental spectrum) composed of the bands of a Gaussian analysis (>400 nm) on the basis of the parameters of **Table 2**. Bars: Positions and intensities of the Gaussian bands.

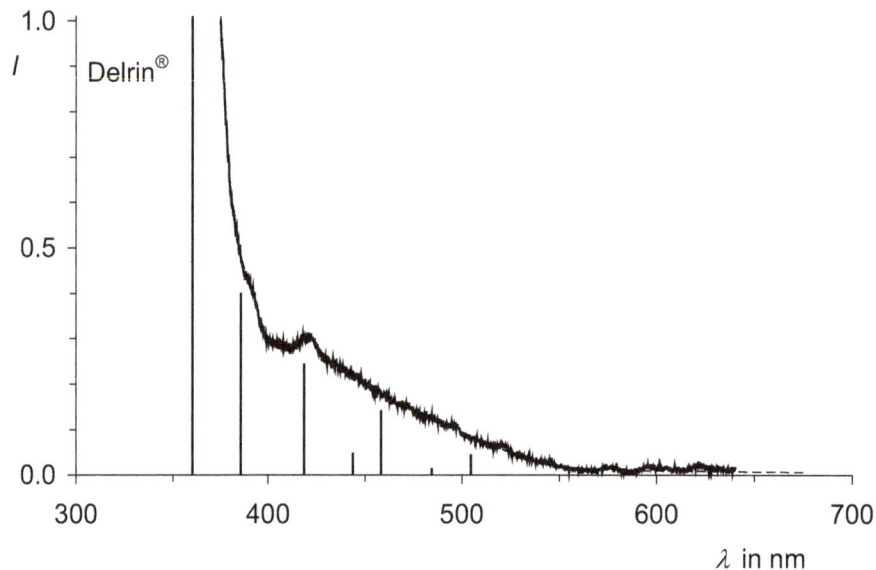

Figure 7. Fluorescence spectrum of Delrin® (solid, thick curve) and simulated spectrum (thin, dashed curve, mainly covered by the experimental spectrum) composed of the bands of a Gaussian analysis (>375 nm) on the basis of the parameters of **Table 2**. Bars: Positions and intensities of the Gaussian bands.

3.5. Practical Setup

For the determination of the first order time constant it is not necessary to record the whole exponential decay but the sampling of two fluorescence intensities [16]-[18], conveniently before and after the first half life is sufficient; for further details, see ref. [13]. This sampling needs not to be at definite times, but may be obtained by integration over a defined interval and thus, increases the signal to noise ratio. Moreover, an absolute determination of intensities is not necessary because of unproblematic calibration. Four points are necessary for bi-exponential decays and for some cases three would be enough.

As a consequence, a convenient setup is the application of a laser with ns or sub ns light pulses for optical excitation and phase shifted detections of the fluorescence intensities; two detections are necessary for mono-exponential decay and four for bi-exponential. A spectral resolution to 12 until 15 wavelengths allows the further identification by means of the data of the Gaussian analyses. The optical detection can be accumulated for the improvement of the signal to noise ratio where a frequency of repetition of 15 MHz may be conveniently reached. Thus, half a ton of material can be separated per hour by the application of convenient techniques.

4. Conclusion

Technical polymers can be identified by means of the mono-exponential lifetime of their autofluorescence. A bi-exponential recording allows an advanced classification useful for the discrimination of chemical similar materials such as LDPE, HDPE and UHDPE, different processing or the recognition of contaminates. The fluorescence spectra of polymers can be simplified by means of Gaussian analyses and applied as a source for redundant information of fine tuning for the recognition of certain batches. The optical method allows automation for sorting plastics in the dimensions of half a ton per hour.

Acknowledgements

This work was supported by The BMBF, CIPSM cluster in Munich, and the Fonds der Chemischen Industrie. We thank Dr. Moritz Ehrl for technical assistance. We thank Karin Haslböck and Dipl.-Ing. Dipl.-Wl-Ing. Thomas Böhme for material support.

References

[1]   Nemeth, E., Schubert, G., Albrecht, V. and Simon, F. (2005) The Triboelectric Charging of Mixtures of Plastics in the Technology of Recycling. *Aufbereitungs-Technik*, **46**, 35-46.

[2]   Nemeth, E., Simon, F., Albrecht, V. and Schubert, G. (2006) Separating a Mixture of Plastics Comprises Electrostatic Sorting after Plasma Treatment and Triboelectric Charging. *Ger. Patent*, DE 102004024754 B3 (May 12. 2004); *Chemical Abstracts*, **144**, Article ID: 392348.

[3]   Gohs, U., Albrecht, V., Husemann, K., Reinsch, E., Schuenemann, R. and Simon, F. (2009) Waste Plastic Mixture *i.e.* Completely Crushed Polyethylene- and Polypropylene-particles, Separation Method for e.g. Electrical Industry, Involves Subjecting Plastic Mixture to Ttriboelectric Charging, and Separating Pure Polyolefin Particles. *German Patent Application*, DE 102007055765 A1 (Dec. 11, 2007); *Chemical Abstracts*, **151**, Article ID: 57663.

[4]   Huth-Fehre, Th., Feldhoff, R., Kantimm, Th., Quick, L., Winter, F., Cammann, K., van den Broek, W., Wienke, D., Melssen, W. and Buydens L. (1995) NIR—Remote Sensing and Artificial Neural Networks for Rapid Identification of Post Consumer Plastics. *Journal of Molecular Structure*, **348**, 143-146.
http://dx.doi.org/10.1016/0022-2860(95)08609-Y

[5]   Michaeli, W., Plessmann, K.W., Andrassy, B., Breyer, K. and Laufens, P. (1998) Qualitative and Quantitative Characterisation of Mixed Polymers Using Near-Infrared-Spectroscopy (NIR). *Polymer Recycling*, **3**, 287-293; *Chemical Abstracts*, **132**, Article ID: 181307.

[6]   Corbet, E.C., Frey, J.G., Groce, R.I. and Hendra, P.J. (1994) An Investigation into the Applicability of Luminescent Tagging to Polymer Recovery. *Plastics, Rubber and Composite Processing and Application*, **21**, 5-11.

[7]   Rafi Ahmad, S. (2000) Marking of Products with Fluorescence Tracers in Binary Combinations for Automatic Identification and Sorting. *Assembly Automation*, **20**, 58-65. http://dx.doi.org/10.1108/01445150010311617

[8]   Alam, M.K., Stanton, S.L. and Hebner, G.A. (1994) Near Infrared Spectroscopy and Neural Network for Resin Identification. *Spectroscopy*, **9**, 31-39.

[9]   Scott, D.M. (1995) A Two-Colour Near-Infrared Sensor for Sorting Recycled Plastic Waste. *Measurement Science and Technology*, **6**, 156-159. http://dx.doi.org/10.1088/0957-0233/6/2/004

[10] General Electric Company (inv. Hubbard, S., Potyrailo, R., Schottland, P. and Thomas, V.) (2005) Tagging Materials for Polymers, Methods, and Articles Made Thereby. US-Patent, 2005/0095715.

[11] Langhals, H., Schmid, T., Herman, M., Zwiener, M. and Hofer, A. (2013) Binary Fluorescence Labeling for the Recovery of Polymeric Materials for Recycling. *International Journal of Science, Environment and Technology*, **7**, 124-132.

[12] Langhals, H., Schmid, T., Herman, M., Zwiener, M. and Hofer, A. (2012) Marking of Polymer Materials with Fluorescence Dyes for Their Clear Automatic Sorting. *German Patent Application*, DE 102012012772.3 (June 22, 2012);

Chemical Abstracts, **160**, Article ID: 63983.

[13] Langhals, H., Zgela, D. and Schlücker, T. (2014) High Performance Recycling of Polymers by Means of Their Fluorescence Lifetimes. *Green and Sustainable Chemistry*, **4**, 144-150. http://dx.doi.org/10.4236/gsc.2014.43019

[14] Langhals, H. (2002) The Rapid Identification of Organic Colorants by UV/Vis-Spectroscopy. *Analytical and Bioanalytical Chemistry*, **374**, 573-578. http://dx.doi.org/10.1007/s00216-002-1473-x

[15] Langhals, H. (1982) The Determination of the Composition of Binary Liquid Mixtures by Means of Fluorescence Measurements. *Fresenius' Zeitschrift für Analytische Chemie*, **310**, 427-428. http://dx.doi.org/10.1007/BF00483019

[16] Ballew, R.M. and Demas, J.N. (1989) An Error Analysis of the Rapid Lifetime Determination Method for the Evaluation of Single Exponential Decays. *Analytical Chemistry*, **61**, 30-33. http://dx.doi.org/10.1021/ac00176a007

[17] Woods, R.J., Scypinski, S., Cline Love, L.J. and Ashworth, H.A. (1984) Transient Digitizer for the Determination of Microsecond Luminescence Lifetimes. *Analytical Chemistry*, **56**, 1395-1400. http://dx.doi.org/10.1021/ac00272a043

[18] Meier, R.J., Fischer, L.H., Wolfbeis, O.S. and Schäferling, M. (2013) Referenced Luminescent Sensing and Imaging with Digital Color Cameras: A Comparative Study. *Sensors and Actuators*, **B177**, 500-506. http://dx.doi.org/10.1016/j.snb.2012.11.041

12

Effect of Seed Size, Suspension Recycling and Substrate Pre-Treatment on the CVD Growth of Diamond Coatings

Awadesh Kumar Mallik[1*], Sandip Bysakh[1], Radhaballabh Bhar[2], Shlomo Z. Rotter[3], Joana Catarina Mendes[3]

[1]CSIR-Central Glass & Ceramic Research Institute, Kolkata, India
[2]Department of Instrumentation Science, Jadavpur University, Kolkata, India
[3]Instituto de Telecomunicações, Campus Universitário de Santiago, Aveiro, Portugal
Email: [*]amallik@cgcri.res.in

Abstract

CVD growth of uniform conformal polycrystalline diamond (PCD) coatings over complex three dimensional structures is very important material processing technique. It has been found that the nucleation and subsequent growth period is very critical for successful development of CVD diamond based technologies. There are many methods of enhancing diamond nucleation on foreign substrates-ultrasonic treatment with diamond seed suspension being the best among them. A combination of ultrasonic seeding (US) technique with prior treatment (PT) of the substrate under CVD diamond growth conditions for brief period of time, has found to be very effective in enhancing the diamond nucleation during CVD growth—together they are known as NNP. But successive usage of the same seeding suspension up to ten cycles deteriorates the seeding efficiency. 6th seeding cycle onwards the silicon substrates are barely get covered by diamond crystallites. Five different diamond micron grits were used for seeding the silicon substrates and it is observed that US with the sub-micron particles (0.25 µm) is very effective in efficient nucleation of PCD on Si substrates. PT of the substrate somewhat negates the effect of successive use of the same seeding slurry but it is best to avoid recycling of the same seeding suspension using micron size diamond grits.

Keywords

Microwave Plasma CVD, Seeding, Diamond Micron Grits, Novel Nucleation Process (NNP), Polycrystalline Diamond

---

[*]Corresponding author.

1. Introduction

Diamond is one of the most astonishing materials on earth. Not only in terms of aesthetic significance, which has wondered human beings for ages, but also for its technological superiority over others. But it was very recently in the 1950s that man could synthesise diamond inside laboratories. To achieve this, scientists tried to emulate natural environments under which diamond is formed beneath the earth over millions of years in high pressure and high temperature (HPHT) conditions [1]. In 1955, [2] the first artificial diamond crystals were successfully synthesized in laboratory conditions by HPHT method. However, this is very expensive technology in terms of experimental set ups. One has to pour graphite into high pressure (7 - 10 GPa) anvils and set high enough temperature (1700 - 2000 K) to transform it into diamond phase. Making diamond pieces larger than a few square millimetres is still not possible by this route. In 1949, American researcher William Eversole [3] from Union Carbide Corporation first ever documented low pressure growth of diamond from carbon monoxide gas onto diamond seeds, even before General Electric's HPHT discovery in 1954. Russian researchers independently confirmed results of Eversole in 1969. They used atomic hydrogen which enhanced the growth rate. Spitsyn *et al.* at the Physical Chemistry Laboratory, RAS, Moscow started growing diamond at low pressure of 10^{-6} mm of Hg from CBr_4 and CI_4 vapors in between 800°C - 1000°C [4]-[6], since as early as 1956. J. C. Angus of Case Western University pioneered the role of atomic hydrogen in facilitating growth of diamond and removing graphite under chemical vapour deposition (CVD) conditions [7]. But these experiments proceeded very slowly and were looked at like a suspicious process among scientific community [8]. But it is until recently that in the year 1982 a Japanese group at NIRIM successfully developed a very innovative process of synthesising metastable diamond by CVD technique [9] [10]. This so called metastable CVD growth [11] is actually a stable growth of diamond phase with capillary rise in pressure up to 4 GPa under CVD conditions [12] [13]. The challenges of synthesising artificial diamond inside laboratories are still many, like, faster growth rates to grow bigger crystals, purity of the diamond, n-type doping, large area uniformity, process reproducibility etc. The science and technology of CVD diamond [14] [15] is fast developing but is not yet fully commercialised for serving engineering needs, due to technological obstacles in commissioning capable machines/reactors or sustaining economically viable processes. There is further need of thorough fundamental research in understanding the basic science in solving present technological problems. Nucleation and growth have been the most important phenomenon in CVD growth of diamond and they are still debatable processes among scientific community, which motivates authors to carry out the present work. For example it was the general belief among scientists that atomic hydrogen is the most important factor in depositing diamond phase during CVD of diamond but the discovery at Argonne National Laboratory (ANL), using the same reactor which has been utilised in the present work, has shown that diamond can be formed from Ar gas without the presence of atomic hydrogen. Moreover, it was understood that the CVD diamond growth is a metastable process but recently it has been proved that it is formed under stable region of carbon phase diagram. So it is essential to further understand the nucleation and growth mechanisms of CVD diamond [16] for advancement of knowledge in this field, which can add to the existing database.

The general approach of CVD growth of diamond is to seed the foreign substrate with diamond or diamond like particles to initiate nucleation process [17]-[29]. Otherwise diamond phase is very difficult to appear with very low 10^2 - 10^3 nuclei/cm^2 nucleation densities [30]. It is well understood that to synthesise diamond it is essential to start with diamond seeds [31]. The substrate over which diamond growth can happen is also limited to very few materials, which has to meet the stringent conditions of Hume Rothery rules. Silicon is the most widely used substrate material, since it has the similar diamond cubic structure. The lattice parameters are also not very different which allow diamond to grow over silicon. But closed diamond films cannot be deposited on bare silicon substrate and some surface pre-treatment [32] [33] of the silicon substrate prior to CVD growth is required. A custom procedure is the manual scratching of the smooth silicon surface with some abrasive grits, in order to introduce scratches and grooves on the polished surface. These abrasive powders after scratching are also left over inside the microscopic grooves and pits on the substrate surface. If the abrasive powers are composed of diamond grit, the encrusted particles help in nucleation of the sp3 diamond phase under CVD growth environments. But this technique has some limitations: manual scratching does not produce completely random scratches; it rather creates a systematic pattern on the bare silicon substrate depending on the individual who does the scrubbing. Moreover this technique is only applicable for flat surfaces and, as such, complex 3-d structures cannot be abraded manually. These limitations can be overcome if the scratching of the silicon substrate is performed with the help of ultra-sonication (US) [34]. By immersing the substrate in an abrasive suspension containing diamond grit and using ultrasounds to agitate it, a random and uniform 3-d substrate seeding, *i.e.* a surface pattern (US does not produce scratches; if it is observed, then these are due to the mechanical polishing of the Si wafers, and not due to US seeding) with particles embedded inside can be easily obtained on complex 3-d structures. The concentration of the abrasive suspension is supposed to be an important parameter in ultra-sonication method. Since the abrasive particles leave the diamond particles on the polished surface, it is obvious that the concentration of diamond suspensions would determine the density of diamond seeds embedded on the silicon substrate. This density may depend also on the size of the abrasive particle, which can vary from few hundred microns to hundreds of nanometers. Now, it is interesting to note that once an abrasive suspension is prepared, it is not that the suspension is discarded after single use. The same suspension can be reused several times. After each usage, it is expected that the concentration of the suspension is reduced, since some of the abrasive particles get embedded on the substrate surface and also some of them get entrapped inside the scratches and grooves produced on the polished silicon surface. So the concentration and successive usage of slurries can also be important factors in increasing the density of diamond seeds on the substrate. Earlier, researchers used to use diamond grits of few microns sizes but after the advent of detonation nanodiamond (DND) slurries, it became common to seed the silicon substrate [35]-[37] with these slurries. Such techniques of abrasive scratching or ultrasonic seeding could produce nucleation densities from 10^9 to 10^{11} per cm^2.

Another method of successful nucleation enhancement (10^{11} nuclei/cm^2) is to bombard the negatively biased substrate surface [38] [39] with ionised carbon species, which thereby modify the surface structure favourable for diamond nucleation. However, this process is detrimental for device application, since the energetic plasma species can cause deep and wide wounds on substrate surface [40].

Other technique that has been used to enhance the nucleation density is the deposition of a buffer interlayer on the substrate, prior to the US seeding [41]. This practice has reportedly led to nucleation enhancement of NCD and UNCD films [42]-[45] for coating three-dimensional complex structures [46] [47], which has immense technological significance. Buijnsters *et al.* suggested that metallic nucleation layer might be a solution for developing conformal very thin coatings [48].

Shlomo *et al.* [49]-[52] have shown in 1990s, that a brief period of pre-treatment (PT) [53] of silicon substrate under CVD diamond growth conditions (GC), before any seeding procedure, would help significantly in nucleation and growth of diamond. The steps [54] [55] that were followed are the followings, 1) to keep the bare substrate inside CVD reactor with conditions similar to diamond growth for 30 minutes (PT), and thereafter 2) to ultrasonically seed the pre-treated substrate (USP) with a diamond abrasive suspension for embedding diamond seeds on the substrate surface. This pre-treatment enhances nucleation density by two orders of magnitude compare to standard US seeding only process. The principle behind nucleation enhancement is such that the 10 - 15 nm thin sp3 enriched film [49]) formed during PT supplies an extra amount of carbon during incubation period of diamond nucleation, in addition to the carbon that is provided from the plasma environment. The result is that lateral growth of diamond is promoted immediately after the 3) substrate is exposed to diamond GC. This method also prevents the appearance of the backside crevices which are formed when using typical diamond nucleation processes. The combination of PT, USP and GC steps, are called together novel nucleation process (NNP) [56]-[58], and it can produce 10^{12} nuclei/cm^2 nucleation density, resulting uniform and conformal nanocrystalline diamond coatings, as thin as 60 nm.

Rotter *et al.* [49] reported performing NNP under several sets of experimental conditions. PT was conducted on metal coated and uncoated Si substrates using HFCVD, which was followed by identical USP for all the samples. Then again all the samples were exposed to GC except for one sample with microwave plasma CVD (MPCVD) treatment. Rotter used identical CVD parameters for both the PT and GC steps. Sumant *et al.* [55] reported modified NNP for growing NCD or UNCD. They used MPCVD for both the pre-treatment and growth steps, with an intermediate step of seeding with DND suspension. The only difference between Rotter's earlier work and the work from ANL group was that ANL used higher methane concentration for the pre-treatment step. Lee *et al.* [53] also reported modified (different parameters for PT and GC steps) NNP with DC arc CVD and HFCVD, and compared the results with "seed" only process. They observed that the pre-treatment deteriorated the nucleation density, contrary to general belief. Whereas, Fahner *et al.* [59] noticed improvement in nucleation density when they tried to grow boron doped diamond on pre-treated platinum wires.

In all the earlier work, researchers following NNP, have used either HFCVD or low power 2.45 GHz frequency microwave plasma or direct current arc for CVD growth (GC) experiments. Researchers at ANL might had used this same DT1800 reactor which is being used in the present work, but there, very thin film UNCD was developed using Ar rich plasma and the same MPCVD process was used for PT as well. Here we report the influence of GC on NNP when 915 MHz MWPCVD reactor was used to grow microcrystalline diamonds over large area in hydrogen rich plasma and when, HFCVD was used for PT. For the work by Sumant *et al.*, the grain size of the diamond crystals was considerably smaller than with hydrogen-rich chemistries and the deposited films are usually referred to as being nanocrystaline diamond (NCD) or ultrananocrystalline diamond (UNCD) films.

2. Materials & Methods

A modified NNP method was followed to prepare the polycrystalline diamond samples.

One silicon wafer was cut into smaller (p-type, 0.5 mm thick, one side polished and about 25 mm^2) pieces. Half of the pieces were given PT for 30 minutes in a HFCVD system (1% CH_4/H_2 precursor gas flow, 30 Torr chamber pressure, 600°C substrate temperature).

After PT, the Si samples were paired (one with PT, the other without PT) and were successively seeded with diamond grit suspension inside ultrasonic bath (USP). They were prepared to study the effect of suspension saturation. Five different seeding suspensions were prepared (**Table 1**). Each suspension was used ten times, each time a different pair of silicon samples was seeded, one with PT and the other one without PT. A total number of $2 \times 10 \times 5$ samples were prepared (**Figure 1**).

Table 1. Seeding suspensions.

Diamond Grit Size (μm)	Suspension identification number
0.25	1
6 - 12	2
10 - 20	3
30 - 40	4
40 - 60	5

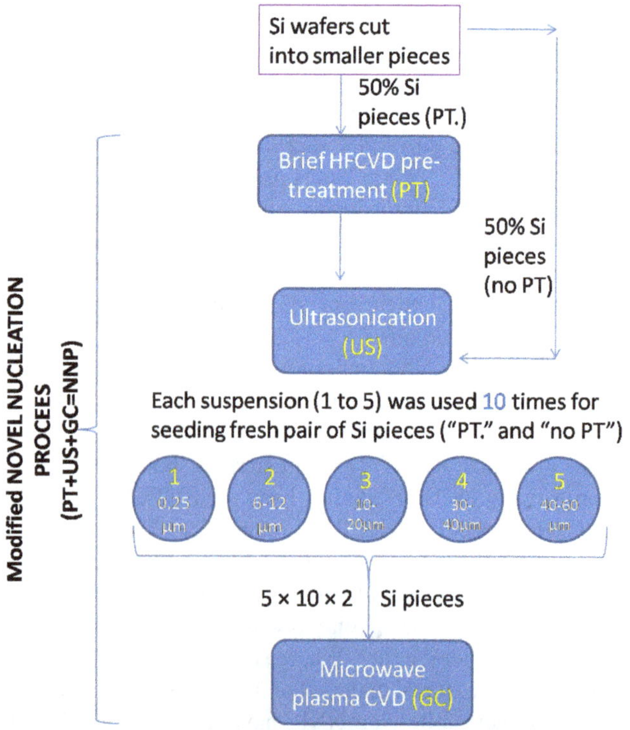

Figure 1. Schematic of the modified novel nucleation process (NNP).

Each sample is identified by 2 numbers; the first number refers to the particular suspension being used (starting from 1 to 5, according to **Table 1**) and the second number refers to the seeding cycle, *i.e.* the number of times each suspension is being used for seeding (from 1 to 10). Moreover, the samples are marked with a dot (.) which underwent PT. For example a sample with identification number 5-5., has to be considered as a NNP grown polycrystalline diamond (PCD) coated silicon sample, which has been seeded with solution No. 5, the solution was reused for 5th successive time and also the sample was pre-treated under HFCVD growth environment before ultrasonic seeding.

After the first two steps of NNP, (PT or without PT, + USP), the samples were kept inside plastic boxes with proper labelling. Once they were shipped from University of Aveiro, Portugal to CSIR-Central Glass and Ceramic Research Institute, India, the final step of the modified NNP was carried out, *i.e.*, MPCVD diamond growth (GC), using a different reactor, 915 MHz DT1800 (Lambda Technologies Inc. USA).

The smaller pieces of samples from Aveiro were kept over one large 100 mm diameter 6 mm thick silicon wafer. Thermal management [60] [61] of plasma heating necessitates usage of such a thick wafer for placing over the thinner silicon sample pieces. Samples were grown using 10 sccm methane gas mixed with 500 sccm of hydrogen and 8 kW microwave input power with 110 Torr pressure. The substrate temperature was maintained at 1050°C.

The PCD coated experimental samples were then seen under optical microscope (Olympus BX 51, country), electron microscope (Phenom Pro X at 5 - 15 kV beam energy) for surface structural evaluation. EDAX data were also taken from them to identify the elements present on the PCD growth surface. Laser Raman spectroscopy (STR500, Technos Instruments, originally from Seki Technotron, Japan) was used to identify quality and stress of the as-grown PCD coatings.

3. Results & Discussion

The PCD coated silicon samples were seen under optical microscope, which showed clear crystalline deposition of coatings with the presence of discontinuity among many of them (**Figure 2**). **Figure 2(a)** and **Figure 2(b)** show the different grain sizes observed when the substrate is pre-treated in NNP. 1-3 is the sample without pre-treatment and the sample 1-3. is the one which underwent HFCVD pre-treatment before USP and GC. **Figure 2(b)** has finer grains compare to **Figure 2(a)**, although both the samples were seeded with 0.25 μm diamond grit suspension. The effect of PT on grain size becomes evident when without PT sample 2-2 shows similar (to **Figure 2(a)**) bigger grains in **Figure 2(c)**. Now, 2-2 was seeded with 6 - 12 μm grit size, whereas sample 3-1 was seeded with bigger 10 - 20 μm grits, but it appears that since none of them were pre-treated, they do not have different grain sizes, but their polycrystals are appearing to be comparable in size. 1-3, 2-2 and 3-1 samples were seeded with gradual increase in grit size (**Table 1**), which are well reflected in the optical micrographs in **Figure 2(a)-(d)** with apparent successive bigger PCD crystal sizes. All the samples discussed so far, are seeded with fresh stock (first 5 USP cycles) of diamond suspension, but when the samples like 4-7. and 5-7. (7th seeding USP cycle of the same suspension) are observed under optical microscope, it was noticed that there was discontinuous PCD growth over silicon substrates. Discontinuity is more prominent for the sample 5-7. than the sample 4-7. It may be because of their grit size difference. Sizes of the PCD grains are apparently bigger in 5-7. than 4-7., which is again may be because of the bigger grit size used in preparing the substrate. It is to be pointed out that both the samples in **Figure 2(e)** and **Figure 2(f)** were pre-treated. It appears that when the old stocks (last 5 USP cycles) of seeding suspension are used even PT cannot become effective to grow fully covered PCD. Moreover, surface of the sample 5-7., although seeded with biggest grit size, appears to be covered with much smaller sized seed particles in **Figure 2(f)**. It can be assumed that the successive use of the ultra-sonication, up to seven times, broke down the macrometer size diamond grit into smaller particle seeds.

So three parameters appear to influence the NNP nucleation and growth characteristics of PCD, and they are abrasive grit size, successive use of same seeding suspension (USP cycles) and substrate pre-treatment (PT). The influences of these three parameters are presented in the following separate sub-sections.

Raman signals (**Figure 3**) were also taken at 50× magnification with 5 micron laser spot size (inset figures). 514.5 nm Ar+ ion laser was used for 10 sec of exposure time, successively for 5 times of signal acquisition. The laser spot was focussed on grain (**Figure 3(a)**), grain boundary (**Figure 3(b)**), scratch marks (**Figure 3(c)**) and blackish surface spots (**Figure 3(d)**) of PCD sample 4-4 (as an example of all the PCD grown), prepared with 30 - 40 μm grit size suspension used for successive 4th time without PT. It is found that the PCD quality marginally

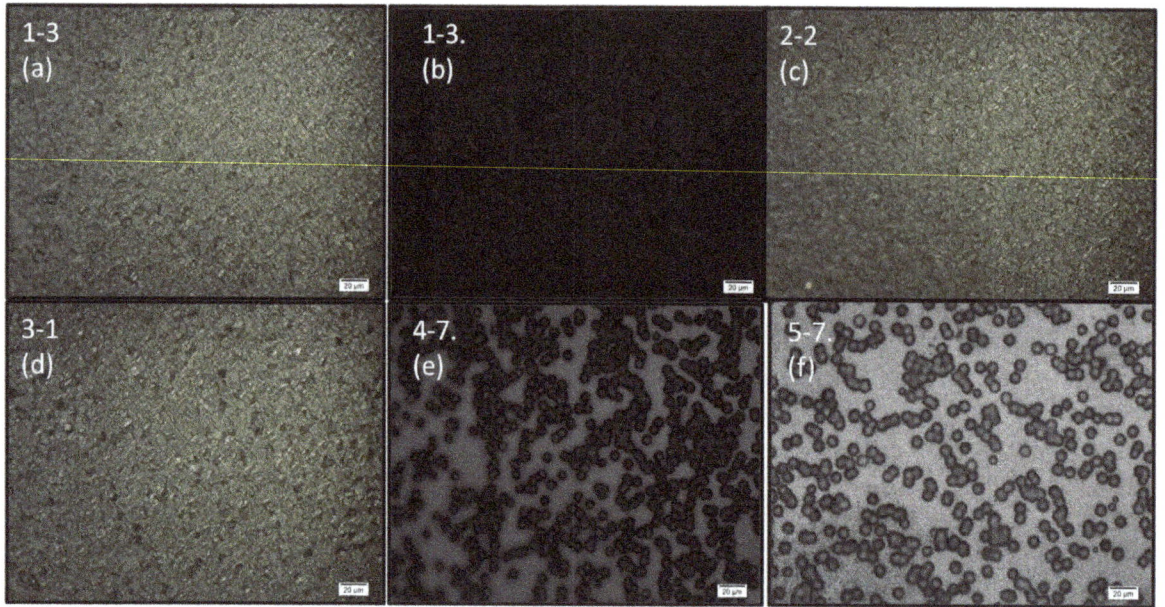

Figure 2. 500× optical microscope images of polycrystalline diamond coatings showing, (a)-(d) full coverage and (e) & (f) discontinuous growth.

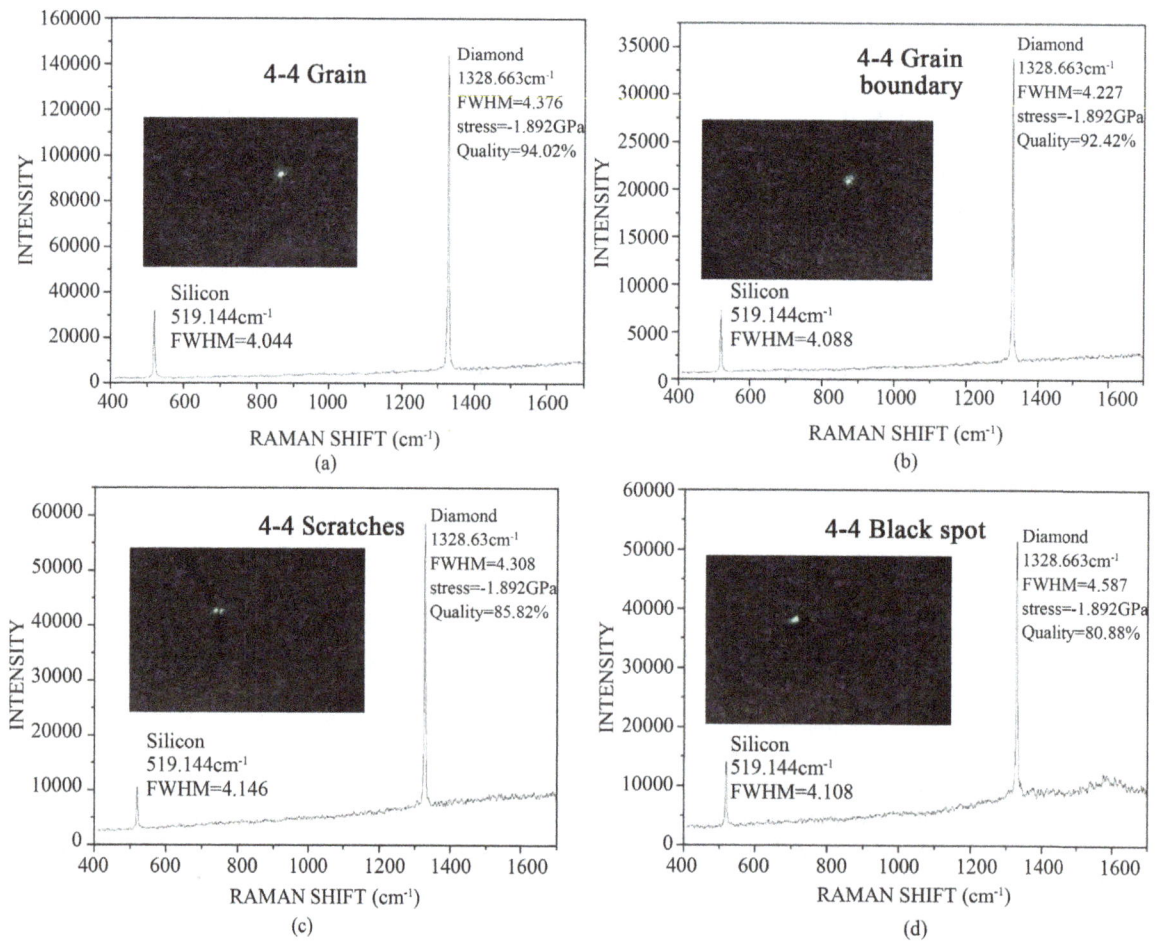

Figure 3. Raman signals of PCD coatings—from different spots on same sample.

deteriorates from 94% to 92%, as the laser point focus changes from grain to grain boundary regions, as expected. Grain size calculation by line intercept method revealed that the average grain size of each crystal for sample 4-4 is 1.21 μm. The size of the laser spot is 5 μm, which implies that at least four or five grains will fall within the laser spot. So it is very difficult to focus the laser spot inside the grain or on its grain boundary. While focussing the Raman laser, care was taken to concentrate the laser point on bigger size grain for generating reliable result on diamond quality. Moreover when the signals are taken from scratch mark (these marks are due to the mechanical polishing of the silicon wafer) or black spot on the PCD surface, the quality further deteriorates to 85% and 80% respectively. But surprisingly FWHM and internal stress were found to be more or less equal for all the regions in **Figure 3**. FWHM values of about 4.5 cm^{-1} indicate good quality PCD crystal. The stress was tensile in nature with values about 1.9 GPa. Signals from base silicon substrate were always present, indicating thin nature of the PCD coating.

3.1. Effect of Suspension Grit Size

Five different diamond abrasive grits were used, starting from sub-micron 0.25 micron grits through some intermediate sizes of 6 - 12, 10 - 20 and 30 - 40 μm, and to very large 40 - 60 μm diamond particles of as described in **Table 1**. Effect of grit size on NNP grown PCD grain size, its quality and stress values are discussed below.

3.1.1. PCD Grain Size

Average PCD grains are found to be more than 2 μm size, irrespective of grit sizes used (**Table 2**). It is observed that as we start decreasing the suspension grit size, the average PCD grain sizes diminishes but does not alter very much (2.26 - 2.84 μm). Sub-micron grit size even after 3rd cycle of seeding (1-3), produces big enough diamond grains. For the sake of comparison between different grit sizes only, samples which were seeded at the beginning of the seeding cycles are shown in **Table 2**. For grit sizes of 5, 4 and 3 - samples prepared with 1st seeding cycle were chosen and found to show definitive trend, whereas, for grit sizes, 2 and 1 - 2nd, 3rd and 5th seeding cycles were chosen to show that lower grit sizes, even after successive uses, show better or similar results, than found for bigger grit size suspensions. It is to be noted that when the sub-micron grit size (1) is reused for the 5th consecutive time, it gives finer diamond grains of 1.21 μm. The electron microscopy images of the **Table 2** samples are shown in **Figure 4**. They show complete coverage of the silicon substrate, as all of them were seeded with relatively fresher stock of suspensions. It is to be noted that none of the **Figure 4** represents sample with pre-treatment (to avoid its influence). **Figure 4** compares the CVD grown PCD microstructures (10 K or 20 K magnifications) based solely on the use of different grit sizes. Even the effect of successive seeding cycle is not taken into account by comparing samples seeded within first few USP cycles. So, the lower value 1.21 μm grain size of 1-5 sample compared to 2.26 μm of 1-3, even after using the same grit size, may be because of successive usage of the same diamond suspension. **Figure 4(c)** (sample 3-1) and **Figure 4(b)** (sample 4-1) show diamond octahedrals with growth steps, whereas, the other micrographs (**Figure 4(e)** for sample 1-3, **Figure 4(d)** for sample 2-2 and **Figure 4(a)** for sample 5-1) show crystals with signs of secondary nucleation on their surfaces. **Figure 2(a)**, **Figure 2(c)** and **Figure 2(d)** are the optical micrographs of the electron microscopy images **Figure 4(c)**, **Figure 4(d)** and **Figure 4(e)**. It was discussed earlier that as we increase the grit size from 0.25 μm (1) to 6 - 12 μm (2) and then to 10 - 20 μm (3) for the USP, the resulting PCD grain size appear to

Table 2. Effect of suspension grit size on Raman peak FWHM, PCD quality, internal stress, grain size.

Laser spot position	Sample name	Diamond peak position (cm^{-1})	Silicon peak position (cm^{-1})	FWHM of diamond peak	Stress (GPa)	Quality (%)	Grain size (μm)
Grain	5-1	1327.70	521.28	12.50	2.44	76.98	2.84
Grain	4-1	1330.58	520.21	3.40	0.80	94.93	2.64
Grain	3-1	1329.62	NOT FOUND	3.52	1.35	94.15	2.60
Grain	2-2	1329.62	521.28	5.90	1.35	94.73	2.40
Grain	1-3	1331.54	520.21	3.88	0.26	93.96	2.26
Grain	1-5	1328.66	519.14	7.40	1.90	89.65	1.21

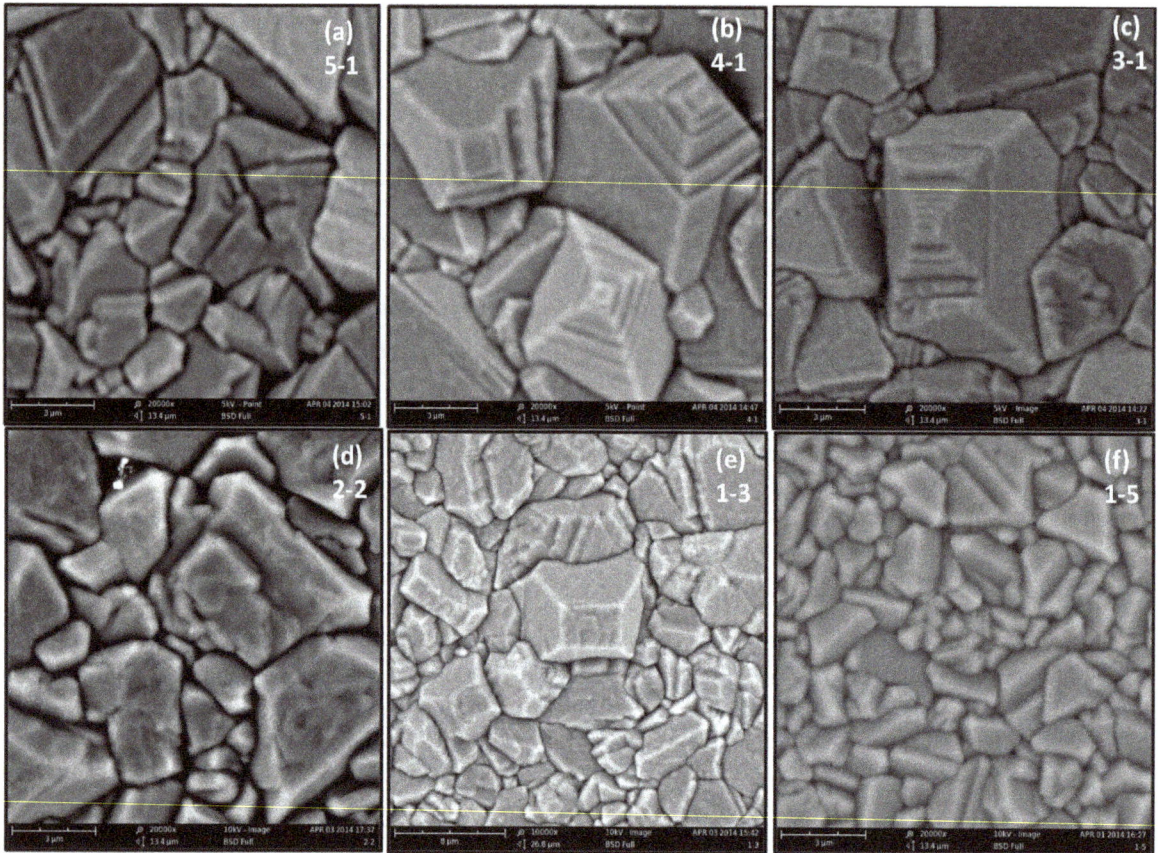

Figure 4. SEM images of fully covered PCD coatings without substrate pre-treatment.

increase gradually. Grain size calculations from SEM pictures by line intercept method show that, indeed the PCD crystal size increase with increase in grit size, although marginally (**Table 2**).

3.1.2. Quality & Stress

Detailed analysis of the Raman signals from inside of the PCD grains (for the samples shown in **Figure 2** and **Figure 4**) are shown in **Table 2**. It is found that use of very large size diamond grit may degrade (~76%) the deposited PCD quality with very high amount of tensile stress (sample 5-1). It will also decrease the crystallinity of the PCD grain with very high FWHM (12.5 cm^{-1}). As we reduce the grit size successively from 30-40 μm down to 0.25 μm submicron particle suspension, the CVD grown PCD quality does not change very much (around 94%). The diamond peak positions are also found to be in and around 1330 cm^{-1}, *i.e.* keeping the internal stress within 0 - 2 GPa. The least amount of stress is found in sample 1-3, whereas, the highest stress is found in sample 5-1. Although there is no definitive correlation between the coating material internal stress and the seed particle size, it appears that using very large diamond grit of 40 - 60 μm deteriorates the properties of CVD grown PCD in terms of internal stress, quality and crystallinity. Sample 1-5 was prepared by seeding the silicon substrate with 5 time's reused sub-micron diamond suspension 1, which produces very fine grain (1.21 μm) microstructure (**Figure 4(f)**). Fine grains are responsible for the loss of crystallinity as well, as **Table 2** shows quite high 7.40 cm^{-1} FWHM for 1-5 sample.

3.2. Effect of Successive Seeding

Now after seeing the effect of suspension grit size on CVD growth of PCD, it is necessary to investigate the effect of successive seeding of the same grit size suspension on the features of PCD coatings. Earlier, samples with initial seeding cycles were only discussed. Now samples which were seeded later on with the reuse of the same diamond grit suspension will be discussed.

3.2.1. Grain Size

10 - 20 μm grit size seeding suspension (**Table 1**, suspension 3) was reused ten times, just like for other seeding suspensions. When the grain sizes of the resultant PCD material are compared, it is observed that there is a gradual decrease of average PCD grain size with successive seeding with the same solution. At the 1st cycle, the sample 3-1 was producing considerable big grains of 2.4, but with the 3rd, 4th and 6th successive cycles, grain size gradually reduces to 2.12 μm (3-3), 1.54 μm (3-4) and 1.18 μm (3-6) respectively (**Table 3**). It may be thought that with repetitive ultra-sonication of the same suspension, the size of the diamond particles decreases with disintegration of their agglomeration etc. So as the number of seeding cycles is increased the particle size of individual seeds decreases, giving rise to finer PCD microstructures (as discussed in sub-section 3.1.1). But with an exception for 3-9, as the 10 - 20 μm grit size particle suspension on use for the 9th successive time, produces 1.5 μm size PCD grains, somewhat similar to sample 3-4. This may be because as we successively use the same suspension the seeding efficiency decreases, (with depletion of seed particles present in suspension) with discontinuous PCD coverage on the silicon substrates. So the individual crystal size may become bigger under CVD growth condition with improper coalescence, which is further discussed in the next sub-section 3.2.2.

3.2.2. Microstructure: Full or Discontinuous Substrate Coverage

EDAX elemental analysis was done on the NNP grown PCD coatings in order to know the chemical nature of the films. It is seen that 3-5 sample (5th seeding cycle) shows almost full coverage (pockets of the base substrate are also visible, as circled in the **Figure 5(a)**) of the silicon substrate, whereas at the 9th seeding cycle, even for

Table 3. Effect of successive use of diamond seeding suspension on CVD grown PCD grain sizes.

Sample	Grain Size
3-1	2.40
3-3	2.12
3-4	1.54
3-6	1.18
3-9	1.50

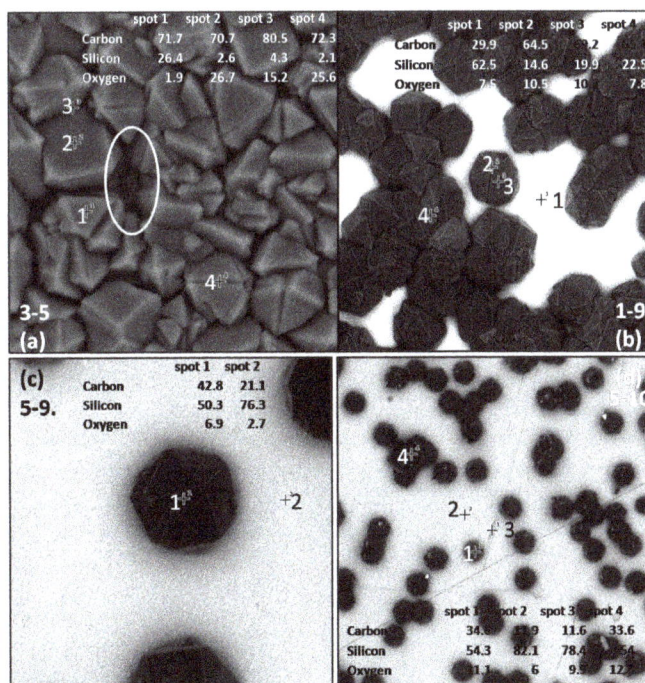

Figure 5. EDAX spot analysis images of PCD coatings, showing discontinuous growth 5th seeding cycle onwards.

the sample seeded with submicron grit size (1 is known for high seeding efficiency), there is discontinuous growth of PCD (**Figure 5(b)**). When the biggest diamond grit size (not known for efficient seeding) of 40 - 60 µm was used in successive seeding for the 9th and 10th consecutive times, even substrate pre-treatment (known for enhancing nucleation density) could not induce full substrate coverage on NNP growth of PCD coatings (**Figure 5(c)** and **Figure 5(d)**). EDAX spot analysis was done for grain, grain boundary, isolated grains, bare silicon substrate regions. EDAX signals from grain and grain boundary regions (1, 2, 3 & 4 spots in **Figure 5(a)**, 2, 3, 4 spots in **Figure 5(b)**) give predominant presence of carbon, whereas, the EDAX peaks from whitish un-covered base region (spot 1 in **Figure 5(b)**; spot 2 in **Figure 5(c)**; spots 2 & 3 in **Figure 5(d)**) show more percentages of silicon. The areas where no diamond grains are visible may be covered with amorphous carbon. Moreover, since the percentage concentration of silicon is higher, it may be inferred that there are phases like silicon carbide. It is interesting to note that some of the isolated PCD grains are producing almost 50% concentration silicon peaks (spot 1 in **Figure 5(c)** and spots 1 & 4 in **Figure 5(d)**). They may reveal the presence of silicon carbide phases.

3.2.3. Quality & Stress
Sub-micron grit suspensions are known as efficient seeding material. Earlier result indicates that sample seeded with 0.25 µm diamond grit produces good quality PCD (**Table 2**, last two rows describing samples 1-3 and 1-5). In order to show the effect of seeding cycle efficiency on PCD quality, internal stress and the Raman peak width at half maxima using suspension 1, few more examples in this set of samples (1-4, 1-8, 1-9 & 1-10) are described in **Table 4**. It is found that with successive reuse of the same slurry 1, after 3rd, 4th, 5th, 6th, 9th and 10th times, the PCD quality gradually deteriorates from 93% to 91%, 89%, 87%, 69% and 60% respectively. Internal stress also successively increases from 0.26 GPa (sample 1-3) to 2.44 GPa (sample 1-9), within the resultant PCD films, exception being the 1-10 sample which has 0.8 GPa stress similar to sample 1-4. Later in the sub-section 3.3.1, it will be further discussed that 1-10 also has deviated grain size value, mainly because of successive seeding.

The Raman signals were taken from within the grain and in case of discontinuous growth (incomplete substrate coverage), grains lying inside the covered region were studied. The crystallinity (FWHM) of the PCD grains also fall with the use of successive seeding, from 3.88 cm^{-1} (sample 1-3) down to 11.36 cm^{-1} (sample 1-9), again with a slight trend reversal for sample 1-10. Silicon peak at 520.5 cm^{-1} is noticed always, as the PCD coatings were thin enough to get Raman signal from the base Si substrate. The general trend is that with successive use of same seeding slurry, resulting PCD becomes poorer.

3.3. Effect of PT
So far the effect of abrasive grit size and the use of successive seeding cycles on the CVD growth of polycrystalline diamond films have been discussed. But the most significant parameter is the substrate surface pre-treatment (PT) on the nucleation enhancement, which makes the process termed as "NNP". Samples marked with (.) were pre-treated inside hot filament CVD chamber for a brief period of time. PT facilitates diamond nucleation during CVD growth (GC). Pre-treatment of bare silicon substrate under diamond growth conditions leaves a carbon film on the surface which supplies carbon material during initial nucleation and growth process of the diamond CVD (GC). PT was done under GC which means that the carbon film that formed under conditions didn't leave any sp2 bonded carbon on the growing surface, as the hydrogen would have etched them away. Initial 10 - 15 nm of carbon film formed during PT was almost 100% sp3 bonded carbon in an amorphous form

Table 4. Effect of successive use of diamond seeding suspension on CVD grown PCD peak width, stress and quality.

Laser spot position	Sample name	Diamond peak position (cm^{-1})	Silicon peak position (cm^{-1})	FWHM of diamond peak	Stress (GPa)	Quality (%)	Grain size (µm)
Grain	1-4	1330.58	520.21	4.72	0.80	91.56	1.28
Grain	1-8	1328.66	520.21	8.03	1.90	87.00	Discontinuous
Grain	1-9	1327.70	519.14	11.36	2.44	69.94	Discontinuous **Figure 6(k)**
Grain	1-10	1330.58	519.14	10.84	0.80	60.97	1.43

[49]. Afterwards, seeding with pure diamond grit was done, avoiding any use of the detonation nanodiamond and by the ultrasonic action the grit left behind small diamond particles, much smaller than the grit size, that were partly embedded inside the sea of the sp3 carbon. The diamond particles under ultrasonic agitation, would interact with the carbon film on silicon substrate, creating activated surface sites (pits, channels, edges, corners etc.) for nucleation. It may be termed as the nucleation film. Then when the GC started, these seeds commenced growing using the PT carbon film in a way transforming it into the crystalline form, in what can be like a re-crystallization action. This is the reason that the incubation time is very short and the growth is done laterally as the seeds grow sidewise until they touch each other. This is also the reason that the interface is very smooth with very little crevices for NNP grown films [55]. Following sub-sections discuss the effect of PT on NNP grown PCD grain size, quality, crystallinity and internal stress.

3.3.1. Grain Size

Submicron seed particles are known for enhancing the nucleation and growth of PCD better than micron size grits. So from the seeding efficiency point of view, the discussion is limited to suspension 1 only. Other particle sizes (2, 3, 4 & 5) are not included in **Table 4** & **Table 5**, in order to remove the influence of the "grit size" parameter on the result. When the 4th, 5th and 10th seeding cycle samples, *i.e.* those prepared with older stock suspension of sub-micron grits, are compared between their pairs (1-4 & 1-4. −1.28 μm & 1.39 μm; 1-5 & 1-5. −1.21 μm & 1.60 μm; 1-10 & 1-10. −1.43 μm & 1.78 μm respectively) as described in **Table 2**, **Table 4** and **Table 5**), it is found that the samples without substrate pre-treatment produce smaller size PCD grains. Although, this trend is not shown for the sample pairs prepared with fresh stock of seeding suspension (*i.e.* between samples 1-3 & 1-3. −2.26 μm & 1.60 μm). It is observed that there is difference in grain sizes between pair of samples prepared with or without PT. Sample pairs when seeded with fresh stock of suspension (3rd cycle) they show decrease in grain size (from 2.26 to 1.60 μm), when the sample is pre-treated (**Figure 2(a)** and **Figure 2(b)** with **Table 2** & **Table 5**). But the reverse is true for NNP sample seeded with older stock suspensions. So it is seen that successive use of seeding suspension influence significantly the grain sizes of the PCD samples, may be it has stronger influence than PT effect.

In the earlier sub-section 3.2.1, it was noticed that as the cycle number is increased for suspension 3, application of more ultrasonic force disintegrates the diamond seed particle in the suspension, which produces finer PCD microstructure (**Table 3**). Similar trend is also observed for suspension 1 (**Table 2** & **Table 4**). The grain sizes for sample 1-3 (2.26 μm), 1-4 (1.28 μm) and 1-5 (1.21 μm) gradually diminish with successive use of ultrasonic force. On the other hand just like suspension 3, suspension 1 also has reverse trend of increased grain size (3-9 shows 1.5 μm and 1-10 shows 1.43 μm) for higher seeding cycles. It seems that the recycling of the same suspension has very strong effect on the final PCD characteristics. Although some kind of correlation between successive seeding cycles and final grain size is noticed for without PT samples, but for PT samples no noticeable trend is observed with more or less similar grain sizes (1.39 μm—1-4., 1.41 μm—1-5., 1.60 μm—1-3., 1.78 μm—1-10.). In other words, pre-treatment of substrate surface may overshadow the trends found in successive seeding.

3.3.2. Quality & Stress

Table 2, **Table 4** and **Table 5** also show the Raman results of the PCD coating prepared with suspension 1, both with PT and without PT. It is observed that pre-treatment enhances (comparing each pair, with PT or without PT from **Table 2**, **Table 4** and **Table 5**) the diamond coating quality, significant one being the samples prepared

Table 5. Raman and grain size data for pretreated (PT) PCD samples seeded.

Laser spot position	Sample name	Diamond peak position (cm^{-1})	Silicon peak position (cm^{-1})	FWHM of diamond peak	Stress (GPa)	Quality (%)	Grain size (μm)
Grain	1-3.	1328.66	520.21	5.46	1.89	94.52	1.60
Grain	1-4.	1329.62	520.21	7.16	1.35	93.22	1.39
Grain	1-5.	1329.66	520.21	4.45	1.35	90.38	1.41
Grain	1-8.	1327.70	519.14	4.75	2.44	94.91	1.40
Grain	1-10.	1326.74	519.14	6.19	2.98	90.31	1.77

using 10 successive seeding where PT enhances the quality from 60% (1-10) to 90% (1-10.). More than 90% is the average quality of the PCD coating of the PT samples. But one interesting trend is that the surface pre-treatment enhances the internal stress to a significant extent (exception 1-5 pair). The explanation may be that the nature of stress present is found to be tensile. Tensile stress arises due to generation of defects during growth. The defects may be points, vacancy, dislocation, grain boundary etc. It is to be understood that the pre-treated surface produces diamond films with more growth defects. PT produces soft carbon film before diamond CVD growth. This carbon film supplies carbon atoms during diamond crystal formation in the early stages of nuclea-tion and growth. Now when a crystal is forming from an amorphous source of carbon atoms, it is already creat-ing vacancy in the underlying carbon film, which also influences Raman signal to significant level. If we con-sider the effect of successive seeding on PT results, it is found that fresh stock of suspensions (1-3 & 1-3. -3.88 cm^{-1} & 5.46 cm^{-1}; 1-4 & 1-4. -4.72 cm^{-1} & 7.16 cm^{-1}) increase the FWHM values (decrease crystallinity) for samples with PT, whereas, the older stocks of seeding suspension (1-5 & 1-5. -7.4 cm^{-1} & 4.45 cm^{-1}; 1-8 & 1-8. -8.03 cm^{-1} & 4.75 cm^{-1}; 1-10 & 1-10. -10.84 cm^{-1} & 6.19 cm^{-1}) decrease the FWHM values (better crystals) between pair of samples with PT. General understanding is that PT enhances nucleation or helps in faster initial growth or coverage of the silicon substrate by creating a soft carbon film which supplies carbon atoms during nucleation and initial CVD growth of diamond. Sample pairs 1-5, 1-8, 1-10 seem to show expected result of im-proving quality and crystallinity with PT, whereas the sample pairs 1-3 and 1-4 showed the reverse trend. Here, the role of successive seeding is supplementing the influence of PT on diamond nucleation and growth.

3.4. NNP Grown PCD Microstructure and Quality-Coalescence

NNP grown PCD samples were seen under microscope as depicted in **Figure 6**. Samples are shown at the lowest possible magnifications (**Figure 6(a)**, **Figure 6(e)** and **Figure 6(g)**) in order to understand the shape and size of the individual silicon samples. Red square on each sample denotes the area under next higher magnifications. **Figure 6(b)** shows magnified view of the smaller highlighted red square in **Figure 6(a)**. Similarly **Figure 6(c)** and **Figure 6(d)** show successive higher magnifications of the sample 4-9., which was prepared following all the three NNP steps, namely PT, USP and GC respectively. The grit size used was 30 - 40 µm and the seeding sus-pension was used successively for 9 times. Such exhausted big particle slurry is found to be ineffective in pro-ducing continuous growth over susbstrate but PT seems to have helped them in more or less complete coverage, since the sample 4-10 (for the sake of argument it is mentioned that authors found 4-9 and 4-10 samples were almost identical from coalescence perspective, but 4-10 was chosen for SEM sample preparation) prepared with same suspension but without susbstrate pretreatment is found to be completly uncovered as shown in **Figure 6(e)** and **Figure 6(f)**. Comparing **Figure 6(a)** and **Figure 6(e)**, it can be conclusively stated that PT helps in effective film coverage on siliocn substrates. Similar to 4-9., sample 5-9., which was prepepared with larger grit size of 40 - 60 µm in 9th successive seeding cycle and with PT, is found to be discontinuous. The degree of discontinuity (compare **Figure 6(a)** and **Figure 6(g)**) is more for 5-9. than 4-9., may be because of the inefficiency of seeding by bigger grit sizes. Some circular patch marks are seen for the samples 4-10 and 5-9. in **Figure 6(e)** and **Figure 6(g)**. The sizes of the patch marks are bigger and more covered, in **Figure 6(g)** and **Figure 6(h)** than **Figure 6(e)**, may be because of the combination of both the effects, bigger grit size and PT. So, PT seems to facilitate substrate coverage and helps to overcome the negative effect of using bigger seed particles.

If the samples prepared with comparatively less exhausted seeding suspensions are compared, it is seen that sample seeded with smaller grit size (sample 3-6 in **Figure 6(i)**), even without preptreatment, gives more susbtrate covereage than **Figure 6(a)-(h)**. **Figure 6(i)** shows distinct linear seeding patterns which had been im-printed on the substrate during the mechanical polishing of the silicon substrate and was followed on during CVD growth (GC) of diamond. The size of the isolated PCD crystal is found to be around 7 µm (**Figure 6(j)**). It may be somewhat larger considering the fact that the individual nucleation site grows on its own without com-petition from the neighbouring crystallites. Use of smaller seed particles help in getting faster substrate coverage. Successive seeding may be the most detrimental factor in effective diamond seeding for NNP nucleation and growth. From 5th seeding cycle (**Figure 5(a)**) onwards the seeding solution appears to become ineffective (may be termed as old solution) in prodcuing PCD with complete substrate coverage. But usage of sub-micron grit size may significantly overcome the drawback of successive seeding as shown in **Figure 6(k)** and **Figure 6(l)**, where even after 9 times use of the same slurry, NNP without PT produces almost complete coverage. It was just a matter of time had the susbstrate been given sufficient deposition period, it would have been completely covered.

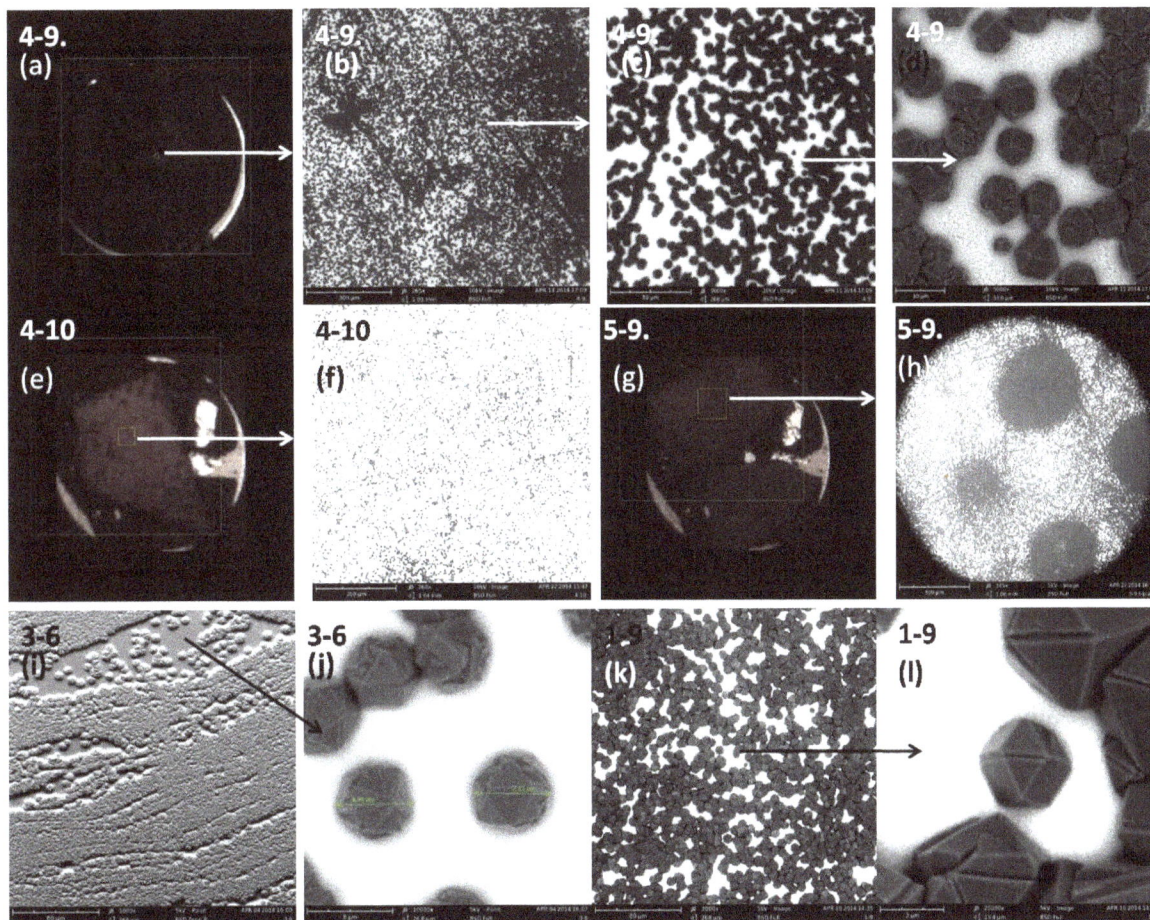

Figure 6. SEM images of PCD coatings showing discontinuous growth with older (>5th cycles) suspension stocks.

4. Conclusions

A large batch of samples were systematically prepared to study the effect of substrate surface pretreatment (PT), reuse of seeding suspension (USP cycles) and suspension grit size on the nucleation and growth behavior of polycrystalline diamond coatings on siliocn substrate. Novel nucleation process (NNP) was compared with normal ultrasonication seeding procedure by diamond grit suspensions and followings are the experimental outcomes:

1) Decreasing the seeding suspension grit size (from 40 - 60 μm to 0.25 μm) diminishes the average PCD grain sizes (2.84 μm - 2.26 μm) marginally.

2) Quality and crystallinity of PCD coatings are satisfactorily high (~94%) with usage of different grit sizes, but usage of the biggest seed particle size considerably degrades the quality (76%) and its crystallinity (FWHM 12.5 cm^{-1}) with introduction of large internal stress (2.44 GPa).

3) Successive reuse of the same diamond suspension (1st, 2nd, 3rd and so on) for seeding silicon substrates, progressively decreases the resulting PCD grain sizes, may be because, successive use of the ultrasonic force disintegrates the large diameter grits into smaller ones.

4) Quality of the resulting PCD coatings suffers very much (1-10 shows 60%) with the reuse of same diamond suspensions for seeding.

5) 6th seeding cycle onwards, discontinuous growth (1-8 & 1-9) of PCD occurs, which can be improved with PT of substrate (1-8. & 1-9.).

6) PT of the silicon substrate before USP, appears to help in improving diamond film quality from 60% to 90% (1-10 & 1-10.).

7) Successive use of the same diamond suspension is found to be very much detrimental in terms of continuous growth of PCD over the silicon substrates. PT can somewhat overshadow that effect but not to a great extent.

8) Since the diamond seeding suspensions are not in nanometer dimension (no detonation nano diamond slurry was used), NNP is observed to be very sensitive to recycling of the same suspension for successive seeding.

9) However, reducing seed particle size helps in grain coalescence wherein sub-micron grit size is found to be the most efficient seeding material.

This work is different from earlier findings [55] which were for growing UNCD film with argon rich plasma chemistry. Here we report the effectiveness of HFCVD pre-treatment [49] with diamond abrasive seeding procedure for growing polycrystalline diamond coatings using 915 MHz CVD reactor. It is found that the NNP process of combining PT with sub-micron diamond particle seeding by ultrasonication is very effective in enhancing CVD nucleation and growth of PCD coatings.

Acknowledgements

The financial support was from the Council of Scientific and Industrial Research (CSIR) network project MTDDC, PSC0101 and Department of Science and Technology (DST) sponsored international Indo-Russian S&T program GAP0246. This work was also funded by FCT/MEC through Portuguese funds and when applicable co-funded by FEDER-PT2020 partnership agreement under the project UID/EEA/50008/2013. J. C. Mendes also acknowledges the FCT for the grant BPD/90306/2012. Authors acknowledge the technical assistance of Mr. Nandadulal Dandapat and Ms Anuradha Jana. Prof. Arun Kumar Pal guided us in understanding many research concepts. Dean, Faculty of Science, Jadavpur University and Director, CSIR-CGCRI are also acknowledged for their permissions.

References

[1]   Angus, J.C. (2014) Diamond Synthesis by Chemical Vapor Deposition: The Early Years. *Diamond & Related Materials*, **49**, 77-86. http://dx.doi.org/10.1016/j.diamond.2014.08.004

[2]   Bundy, F.P., Hall, H.T., Strong, H.M. and Wentorf Jr., R.J. (1955) Manmade Diamond. *Nature*, **176**, 51-54. http://dx.doi.org/10.1038/176051a0

[3]   Eversole, W. (1962) Synthesis of Diamond. US Patent 3,030,188.

[4]   Derjaguin, B.V., Fedoseev, D.V., Lukyanovich, V.M., Spitzin, B.V. and Lavrentyev, A.V. (1968) Filamentary Diamond Crystals. *Journal of Crystal Growth*, **2**, 380-384. http://dx.doi.org/10.1016/0022-0248(68)90033-X

[5]   Spitsyn, B.V., Bouilov, L.L. and Derjaguin, B.V. (1981) Vapor Growth of Diamond on Diamond and Other Surfaces. *Journal of Crystal Growth*, **52**, 219-226. http://dx.doi.org/10.1016/0022-0248(81)90197-4

[6]   Deryagin, B.V., Dzevitsky, B.E., Kochkin, D.A. and Spitsyn, B.V. (1972) Producing Diamonds Synthetically. US Patent 3705937 A.

[7]   Angus, J.C., Will, H.A. and Stanko, W.S. (1968) Growth of Diamond Seed Crystals by Vapor Deposition. *Journal of Applied Physics*, **39**, 2915. http://dx.doi.org/10.1063/1.1656693

[8]   Angus, J.C. (2002) Diamond Films Handbook. Meckel Dekker Inc., New York, 17-26.

[9]   Matsumoto, S., Sato, Y., Tsutsumi, M. and Setaka, N. (1982) Growth of Diamond Particles from Methane-Hydrogen Gas. *Journal of Materials Science*, **17**, 3106-3112. http://dx.doi.org/10.1007/BF01203472

[10]   Kamo, M., Sato, Y., Matsumoto, S. and Setaka, N. (1983) Diamond Synthesis from Gas Phase in Microwave Plasma. *Journal of Crystal Growth*, **62**, 642-644. http://dx.doi.org/10.1016/0022-0248(83)90411-6

[11]   De Vries, R.C. (1987) Synthesis of Diamond under Metastable Conditions. *Annual Review of Materials Science*, **17**, 161-187. http://dx.doi.org/10.1146/annurev.ms.17.080187.001113

[12]   Wang, C.X. and Yang, G.W. (2005) Thermodynamics of Metastable Phase Nucleation at the Nanoscale. *Materials Science and Engineering R*, **49**, 157-202. http://dx.doi.org/10.1016/j.mser.2005.06.002

[13]   Zhang, C.Y., Wang, C.X., Yang, Y.H. and Yang, G.W. (2004) A Nanoscaled Thermodynamic Approach in Nucleation of CVD Diamond on Nondiamond Surfaces. *Journal of Physical Chemistry B*, **108**, 2589-2593. http://dx.doi.org/10.1021/jp036887d

[14]   Nemanich, R.J., Carlisle, J.A., Hirata, A. and Haenen, K. (2014) CVD Diamond—Research, Applications, and Challenges. *MRS Bulletin*, **39**, 490-494. http://dx.doi.org/10.1557/mrs.2014.97

[15]   Schwander, M. and Partes, K. (2011) A Review of Diamond Synthesis by CVD Processes. *Diamond & Related Materials*, **20**, 1287-1301. http://dx.doi.org/10.1016/j.diamond.2011.08.005

[16]   Mallik, A.K., Mendes, J.C., Rotter, S.Z. and Bysakh, S. (2014) Detonation Nanodiamond Seeding Technique for Nucleation Enhancement of CVD Diamond—Some Experimental Insights. *Advances in Ceramic Science and Engineer-

ing, **3**, 36-45. http://dx.doi.org/10.14355/acse.2014.03.005

[17] Lee, S.T., Lin, Z. and Jiang, X. (1999) CVD Diamond Films: Nucleation and Growth. *Materials Science and Engineering R*, **25**, 123-154. http://dx.doi.org/10.1016/S0927-796X(99)00003-0

[18] Smolin, A.A., Ralchenko, V.G., Pimenov, S.M., Kononenko, T.V. and Loubnin, E.N. (1993) Optical Monitoring of Nucleation and Growth of Diamond Films. *Applied Physics Letters*, **62**, 3449-3451. http://dx.doi.org/10.1063/1.109045

[19] Liu, H. and Dandy, D.S. (1995) Studies on Nucleation Process in Diamond CVD: An Overview of Recent Development. *Diamond and Related Materials*, **4**, 1173-1188. http://dx.doi.org/10.1016/0925-9635(96)00297-2

[20] Popovici, G. and Prelas, M.A. (1992) Nucleation and Selective Deposition of Diamond Thin Films. *Physica Status Solidi*, **132**, 233-252. http://dx.doi.org/10.1002/pssa.2211320202

[21] Kromka, A., Potocký, S., Čermák, J., Rezek, B., Potměšil, J., Zemek, J. and Vaněček, M. (2008) Early Stage of Diamond Growth at Low Temperature. *Diamond & Related Materials*, **17**, 1252-1255. http://dx.doi.org/10.1016/j.diamond.2008.03.035

[22] Chavanne, A., Barjon, J., Vilquin, B., Arabski, J. and Arnault, J.C. (2012) Surface Investigations on Different Nucleation Pathways for Diamond Heteroepitaxial Growth on Iridium. *Diamond & Related Materials*, **22**, 52-58. http://dx.doi.org/10.1016/j.diamond.2011.12.005

[23] Naguib, N.N., Elam, J.W., Birrell, J., Wang, J., Grierson, D.S., Kabius, B., Hiller, J.M., Sumant, A.V., Carpick, R.W., Auciello, O. and Carlisle, J.A. (2006) Enhanced Nucleation, Smoothness and Conformality of Ultrananocrystalline Diamond (UNCD) Ultrathin Films via Tungsten Interlayers. *Chemical Physics Letters*, **430**, 345-350. http://dx.doi.org/10.1016/j.cplett.2006.08.137

[24] Mallik, A.K., Binu, S.R., Satapathy, L.N., Narayana, C., Seikh, M.M., Shivashankar, S.A. and Biswas, S.K. (2010) Effect of Substrate Roughness on Growth of Diamond by Hot Filament CVD. *Bulletin of Materials Science*, **33**, 251-255. http://dx.doi.org/10.1007/s12034-010-0039-3

[25] Lin, S.J., Lee, S.L., Hwang, J., Chang, C.S. and Wen, H.Y. (1992) Effects of Local Facet and Lattice Damage on Nucleation of Diamond Grown by Microwave Plasma Chemical Vapor Deposition. *Applied Physics Letters*, **60**, 1559-1561. http://dx.doi.org/10.1063/1.107250

[26] Das, D. and Singh, R.N. (2007) A Review of Nucleation, Growth and Low Temperature Synthesis of Diamond Thin Film. *International Materials Reviews*, **52**, 29-64. http://dx.doi.org/10.1179/174328007X160245

[27] Bogdanowicz, R., Śmietana, M., Gnyba, M., Gołunski, L., Ryl, J. and Gardas, M. (2014) Optical and Structural Properties of Polycrystalline CVD Diamond Films Grown on Fused Silica Optical Fibres Pre-Treated by High-Power Sonication Seeding. *Applied Physics A*, **116**, 1927-1937. http://dx.doi.org/10.1007/s00339-014-8355-x

[28] Pobedinskas, P., Janssens, S.D., Hernando, J., Wagnera, P., Nesládek, M. and Haenen, K. (2011) Selective Seeding and Growth of Nanocrystalline CVD Diamond on Non-Diamond Substrates. *MRS Proceedings*, **1339**, 04-02. http://dx.doi.org/10.1557/opl.2011.992

[29] Fox, O.J.L., Holloway, J.O.P., Fuge, G.M., May, P.W. and Ashfold, M.N.R. (2009) Electrospray Deposition of Diamond Nanoparticle Nucleation Layers for Subsequent CVD Diamond Growth. *MRS Proceedings*, **1203**, J17-J27.

[30] Iijima, S., Aikawa, Y. and Baba, K. (1990) Early Formation of Chemical Vapor Deposition Diamond Films. *Applied Physics Letters*, **57**, 2646-2648. http://dx.doi.org/10.1063/1.103812

[31] Liu, X., Yu, T., Wei, Q., Yu, Z. and Xu, X. (2012) Enhanced Diamond Nucleation on Copper Substrates by Employing an Electrostatic Self-Assembly Seeding Process with Modified Nanodiamond Particles. *Colloids and Surfaces A: Physicochemical and Engineering Aspects*, **412**, 82-89. http://dx.doi.org/10.1016/j.colsurfa.2012.07.020

[32] Arnault, J.C., Demuynck, L., Speisser, C. and Normand, F.L. (1999) Mechanisms of CVD Diamond Nucleation and Growth on Mechanically Scratched Si(100) Surfaces. *The European Physical Journal B—Condensed Matter and Complex Systems*, **11**, 327-343. http://dx.doi.org/10.1007/s100510050943

[33] Podesta, A., Salerno, M., Ralchenko, V., Bruzzi, M., Sciortino, S., Khmelnitskii, R. and Milani, P. (2006) An Atomic Force Microscopy Study of the Effects of Surface Treatments of Diamond Films Produced by Chemical Vapor Deposition. *Diamond & Related Materials*, **15**, 1292-1299. http://dx.doi.org/10.1016/j.diamond.2005.10.005

[34] Anger, E., Gicquel, A., Wang, Z.Z. and Ravet, M.F. (1995) Chemical and Morphological Modifications of Silicon Wafers Treated by Ultrasonic Impacts of Powders: Consequences on Diamond Nucleation. *Diamond & Related Materials*, **4**, 759-764. http://dx.doi.org/10.1016/0925-9635(94)05301-4

[35] Shenderova, O., Hens, S. and McGuire, G. (2010) Seeding Slurries Based on Detonation Nanodiamond in DMSO. *Diamond & Related Materials*, **19**, 260-267. http://dx.doi.org/10.1016/j.diamond.2009.10.008

[36] Williams, O.A., Douheret, O., Daenen, M., Haenen, K., Osawa, E. and Takahashi, M. (2007) Enhanced Diamond Nucleation on Monodispersed Nanocrystalline Diamond. *Chemical Physics Letters*, **445**, 255-258. http://dx.doi.org/10.1016/j.cplett.2007.07.091

[37] Ralchenko, V., Saveliev, A., Voronina, S., Dementjev, A., Maslakov, K., Salerno, M., Podesta, A. and Milani, P. (2005) Nanodiamond Seeding for Nucleation and Growth of CVD Diamond Films; Synthesis, Properties and Applications of Ultrananocrystalline Diamond. *NATO Science Series*, **192**, 109-124. http://dx.doi.org/10.1007/1-4020-3322-2_9

[38] Chen, Y.C., Zhong, X.Y., Konicek, A.R. and Grierson, D.S. (2008) Synthesis and Characterization of Smooth Ultrananocrystalline Diamond Films via Low Pressure Bias-Enhanced Nucleation and Growth. *Applied Physics Letters*, **92**, Article ID: 133113. http://dx.doi.org/10.1063/1.2838303

[39] Stoner, B.R., Ma, G.H.M., Wolter, S.D. and Glass, J.T. (1992) Characterization of Bias-Enhanced Nucleation of Diamond on Silicon by Invacuo Surface Analysis and Transmission Electron Microscopy. *Physical Review B*, **45**, 11067-11084. http://dx.doi.org/10.1103/PhysRevB.45.11067

[40] Maillard-Schaller, E., Kuttel, O.M., Groning, O., Agostino, R.G., Aebi, P., Schlapbach, L., Wurzinger, P. and Pongratz, P. (1997) Local Heteroepitaxy of Diamond on Silicon (100): A Study of the Interface Structure. *Physical Review B*, **55**, 15895. http://dx.doi.org/10.1103/PhysRevB.55.15895

[41] Lu, P., Gomez, H., Xiao, X., Lukitsch, M., Durham, D., Sachdeve, A., Kumar, A. and Chou, K. (2013) Coating Thickness and Interlayer Effects on CVD-Diamond Film Adhesion to Cobalt-Cemented Tungsten Carbides. *Surface & Coatings Technology*, **215**, 272-279. http://dx.doi.org/10.1016/j.surfcoat.2012.08.093

[42] Chen, H.C., Liu, K.F., Tai, N.H., Pong, W.F. and Lin, I.N. (2010) On the Mechanism of Enhancing the Nucleation Behaviour of UNCD Films by Mo-Coating. *Diamond & Related Materials*, **19**, 134-137. http://dx.doi.org/10.1016/j.diamond.2009.11.020

[43] Buijnsters, J.G., Vázquez, L., Galindo, R.E. and Meulen, J.J. (2010) Molybdenum Interlayers for Nucleation Enhancement in Diamond CVD Growth. *Journal of Nanoscience and Nanotechnology*, **10**, 2885-2891. http://dx.doi.org/10.1166/jnn.2010.1392

[44] Li, Y.S., Tang, Y., Yang, Q., Maley, J., Sammynaiken, R., Regier, T., Xiao, C. and Hirose, A. (2010) Ultrathin W-Al Dual Interlayer Approach to Depositing Smooth and Adherent Nanocrystalline Diamond Films on Stainless Steel. *ACS Applied Materials & Interfaces*, **2**, 335-338. http://dx.doi.org/10.1021/am9007159

[45] Chen, L.J., Tai, N.H., Lee, C.Y. and Lin, I.N. (2007) Effects of Pretreatment Processes on Improving the Formation of Nanocrystalline Diamond. *Journal of Applied Physics*, **101**, Article ID: 064308.

[46] Sumant, A.V., Auciello, O., Carpick, R.W., Srinivasan, S. and Butler, J.E. (2010) Ultrananocrystalline and Nanocrystalline Diamond Thin Films for MEMS/NEMS Applications. *MRS Bulletin*, **35**, 281-288. http://dx.doi.org/10.1557/mrs2010.550

[47] Remes, Z., Kozak, H., Rezek, B., Ukraintsev, E., Babchenko, O., Kromka, A., Girard, H.A., Arnault, J.C. and Bergonzo, P. (2013) Diamond-Coated ATR Prism for Infrared Absorption Spectroscopy of Surface-Modified Diamond Nanoparticles. *Applied Surface Science*, **270**, 411-417. http://dx.doi.org/10.1016/j.apsusc.2013.01.039

[48] Buijnsters, J.G., Vázquez, L., Dreumel, G.W.J., Meulen, J.J., Enckevort, W.J.P. and Celis, J.P. (2010) Enhancement of the Nucleation of Smooth and Dense Nanocrystalline Diamond Films by Using Molybdenum Seed Layers. *Journal of Applied Physics*, **108**, Article ID: 103514. http://dx.doi.org/10.1063/1.3506525

[49] Rotter, S.Z. and Madaleno, J.C. (2009) Diamond CVD by a Combined Plasma Pretreatment and Seeding Procedure. *Chemical Vapor Deposition*, **15**, 209-216. http://dx.doi.org/10.1002/cvde.200806745

[50] Edelstein, R.S., Gouzman, I., Folman, M., Rotter, S. and Hoffman, A. (1999) Surface Carbon Saturation as a Means of CVD Diamond Nucleation Enhancement. *Diamond & Related Materials*, **8**, 139-145. http://dx.doi.org/10.1016/S0925-9635(98)00261-1

[51] Gouzman, I., Richter, V., Rotter, S. and Hoffman, A.J. (2000) Study of Chemical Vapor Deposition Diamond Film Evolution from a Nanodiamond Precursor by C13 Isotopic Labeling and Ion Implantation. *Journal of Vacuum Science & Technology A*, **18**, 2997. http://dx.doi.org/10.1116/1.1319677

[52] Feng, Z., Komvipoulos, K., Brown, I.G. and Bogy, D.B. (1993) Effect of Graphitic Carbon Films on Diamond Nucleation by Microwave-Plasma-Enhanced Chemical-Vapor Deposition. *Journal of Applied Physics*, **74**, 2841. http://dx.doi.org/10.1063/1.354636

[53] Lee, H.J., Jeon, H. and Lee, W.S. (2011) Ultrathin Ultrananocrystalline Diamond Film Synthesis by Direct Current Plasma-Assisted Chemical Vapor Deposition. *Journal of Applied Physics*, **110**, Article ID: 084305. http://dx.doi.org/10.1063/1.3652752

[54] Butler, J.E. and Sumant, A.V. (2008) The CVD of Nanodiamond Materials. *Chemical Vapor Deposition*, **14**, 145-160. http://dx.doi.org/10.1002/cvde.200700037

[55] Sumant, A.V., Gilbert, P.U.P.A., Grierson, D.S., Konicek, A.R., Abrecht, M., Butler, J.E., Feygelson, T., Rotter, S.S. and Carpick, R.W. (2007) Surface Composition, Bonding, and Morphology in the Nucleation and Growth of Ultra-Thin, High Quality Nanocrystalline Diamond Films. *Diamond & Related Materials*, **16**, 718-724. http://dx.doi.org/10.1016/j.diamond.2006.12.011

[56] Philip, J., Hess, P., Feygelson, T., Butler, J.E., Chattopadhyay, S., Chen, K.H. and Chen, L.C. (2003) Elastic, Mechanical, and Thermal Properties of Nanocrystalline Diamond Films. *Journal of Applied Physics*, **93**, 2164-2171. http://dx.doi.org/10.1063/1.1537465

[57] Metcalf, T.H., Liu, X., Houston, B.H., Baldwin, J.W., Butler, J.E. and Feygelson, T. (2005) Low Temperature Internal Friction in Nanocrystalline Diamond Films. *Applied Physics Letters*, **86**, 81910. http://dx.doi.org/10.1063/1.1868065

[58] Sekaric, L., Parpia, J.M., Craighead, H.G., Feygelson, T., Houston, B.H. and Butler, J.E. (2002) Nanomechanical Resonant Structures in Nanocrystalline Diamond. *Applied Physics Letters*, **81**, 4455-4457. http://dx.doi.org/10.1063/1.1526941

[59] Fhaner, M., Zhao, H., Bian, X., Galligan, J.J. and Swain, G.M. (2011) Improvements in the Formation of Boron-Doped Diamond Coatings on Platinum Wires Using the Novel Nucleation Process (NNP). *Diamond & Related Materials*, **20**, 75-83. http://dx.doi.org/10.1016/j.diamond.2010.11.003

[60] Mallik, A.K., Pal, K.S., Dandapat, N., Guha, B.K., Datta, S. and Basu, D. (2012) Influence of the Microwave Plasma CVD Reactor Parameters on Substrate Thermal Management for Growing Large Area Diamond Coatings inside a 915 MHz and Moderately Low Power Unit. *Diamond & Related Materials*, **30**, 53-61. http://dx.doi.org/10.1016/j.diamond.2012.10.001

[61] Mallik, A.K., Bysakh, S., Pal, K.S., Dandapat, N., Guha, B.K., Datta, S. and Basu, D. (2013) Large Area Deposition of Polycrystalline Diamond Coatings by Microwave Plasma CVD. *Transactions of the Indian Ceramic Society*, **72**, 225-232. http://dx.doi.org/10.1080/0371750X.2013.870768

13

Cell-Phone Recycling by Solvolysis for Recovery of Metals

Lorena Eugenia Sánchez Cadena[1], Zeferino Gamiño Arroyo[2],
Mario Alberto González Lara[2], Q. Demetrio Quiroz[1]

[1]Departamento de Ingeniería Civil, Universidad de Guanajuato, DI, Av. Juárez 77, CP 36000, Guanajuato, Gto., México
[2]Departamento de Ingeniería Química, Universidad de Guanajuato, DCNyE, Noria Alta s/n, CP 36050, Guanajuato, Gto., México
Email: hau10@hotmail.com, hau10@hotmail.com

Abstract

Mobile phones represent a significant and growing problem with respect to electrical waste and electronic equipment (WEEE). Nevertheless, they are perhaps one of the most valuable electronic products, since they are an important resource for the recovery of metals in terms of mass and volume. In this research a chemical recycling of mobile phones by solvolysis was investigated. The processing was performed by comminution in a hammer mill followed by screening to obtain mesh-4 sized flakes. Flakes were subjected to solvolysis. Different reaction conditions were tested. A reaction time between 2 - 7 hours and a temperature between 150°C - 300°C were the optimum conditions to dissolve the polymer contained in mobile phones. Metals were separated by filtration. Chemical analyses (ATR FT-IR, UV) were carried out on the solvent and the mobile phone flakes before and after solvolysis. A SEM study was carried out, before and after solvolysis, but only to the mobile phone flakes. Thermal transitions of mobile phone flakes were determined by DSC. Chemical results showed that some aromatic species migrate from mobile phones flakes to the solvent, due to the solvolysis reaction. Thermal analysis showed that the Tg, (glass transition temperature) of mobile phone flakes after solvolysis was different to Tg of the polymer before solvolysis, this is due to chemical changes in the molecule. A comparative SEM study revealed that, after solvolysis, the polymer contained in mobile phone flakes is more elastomeric. After solvolysis, solvent was recovered by means of a rotatory evaporator, so that it can be used again. The results obtained in this research showed that solvolysis is an alternative for metal recovery from mobile phones.

Keywords

Cell-Phones, Solvolysis, Recycling

1. Introduction

Over the last few decades, the polymer industry has been very important, specifically in the field of electronics and communication, such as in cellular phones. At the beginning of the 2000's, there were 500 million cell phone users, in 2011 this number had increased to 5750 million world-wide. Generally, cell phones are not recycled in Mexico [1]. This provokes the generation of waste that should not be thrown away with domestic trash, as they are composite materials that contain conducting liquids and metals in the battery that cause soil and water pollution by lixiviation and air pollution when combusted. The above in addition to the fact that its degradation is very slow.

Over time, physical and structural changes in cell phones have been notable. The preference for the latter in this technology is part of modern youth psychology. This contributes to the considerable and growing number of cell phones in disuse. The presence of metals in cell phones creates the expectation of being able to obtain them in solid form, in order to do so, the polymer section should be removed [2]. Different kinds of recycling may be applied (physical, chemical, and incineration) [3]. However, the physical method cannot separate the two components (change in crystalline structure; metallic solutions). We therefore favor chemical recycling, a catalytic solvolysis reaction. Glycols are good solvents due to its hydrogen bridges that can establish, between one molecule and another.

Method Different reactions were carried out where variables were tested such as; temperature (150°C - 220°C), solvent quantity (glycol), reaction time, catalyst quantity, and agitation.

2. Method

1) Collected cell phones were opened to remove the batteries and metal plates.
2) The cell phones were then pulverized and sieved.
3) The system was set-up, at a small or large scale, correctly sealing to prevent leaks.
4) The initial quantity of pulverized cell phone was measured.
5) Glicol was added until the level achieved was slightly higher than the cell phone powder.
6) A nitrogen atmosphere was applied.
7) The solvent is separated from the metals.
8) The solvolysis liquid, which is the solvent, is recovered in a rotary evaporator.
9) Different instrumental analyses are applied; FT-IR ATR, UV-Vis, DSC and SEM, with the sample liquids and solids, before and after the solvolysis.

3. Results and Discussions

3.1. FT-IR ATR Analysis

A comparative study was carried out between the FT-IR ATR spectra of the polymer before and after treatment with the solvent.

This study shows how aromatic species with a C-H bond that are found outside the plane on (751 cm^{-1}) present on the black cell phone before the solvolysis, migrate towards the solvent (758 cm^{-1}) (**Figure 1** and **Figure 2**).

Additionally, the signal of 1586 cm^{-1} related to aromatics di and tri substitutes appear in the solvent (polymer + glycol), it is clear that this signal comes from the polymer as the solvent is not aromatic.

The peak at 1066 cm^{-1} present on the column of the black cell phone without treatment indicates the presence of phenols, said signal appears again in the solvent, further, between 1241 - 1267 cm^{-1}, the presence of aromatic ethers are noted, in the untreated black cell phone and in the solvent we can see similar signs.

The observed peak at 1720 cm^{-1} associated to a carboxyl group, present in the untreated black cell phone disappeared in the solvent spectra [4].

According to the above, we can assert that the black cell phone polymer has been chemically attacked, as certain chemical species pass to the solution.

3.2. UV Analysis

A study of the UV spectrum was carried out using ethanol to dilute the previously filtered solvent, results are shown in **Figure 3**. As may be observed, a signal in the 200 to 300 nm range is detected, indicating the presence of phenol aromatic compounds [5]. These are the products of secondary reactions from the dissolution process.

3.3. SEM Study

A study in a scanning electron microscope was carried out, coupled with X-ray dispersion equipment for element

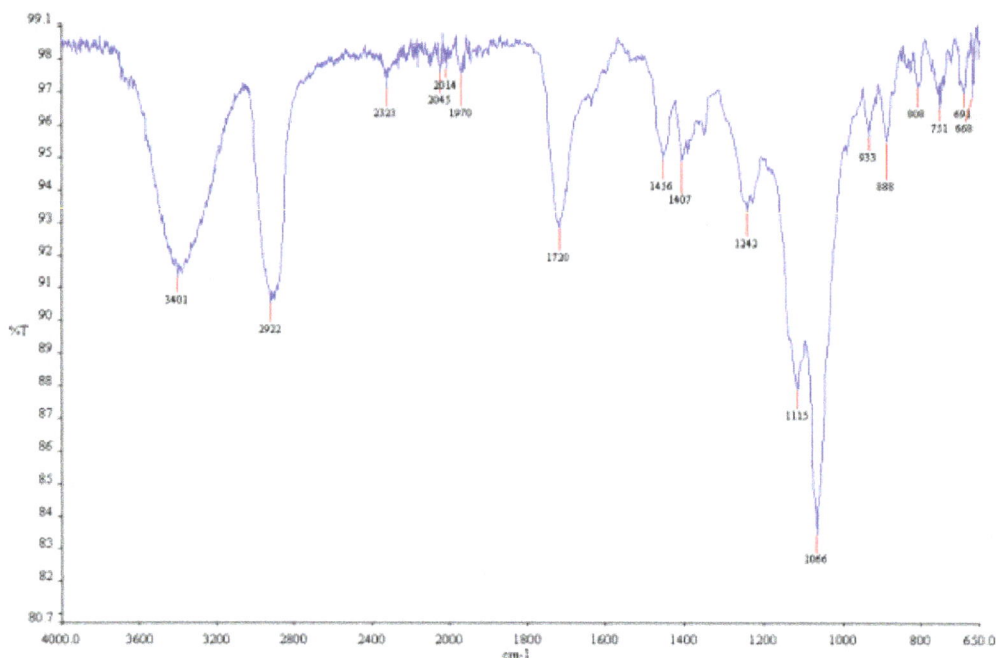

Figure 1. ATR spectra of the cell phone before treatment.

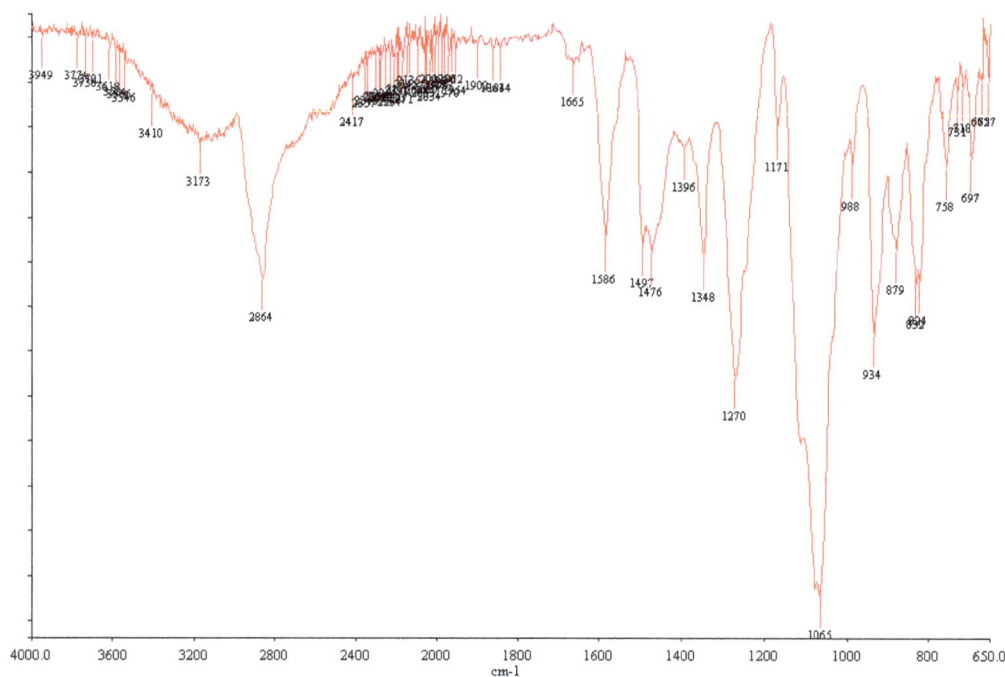

Figure 2. FT-IR ATR spectrum of the solvent after treatment.

analysis. The results are shown in **Figure 4**. As can be seen, in the photo on the left, the polymer before the treatment is more rigid and after the treatment becomes more elastic, such as the photo on the right. It is normal to break macromolecules, some small molecular species have greater movement.

The increased pore size provokes the detachment of micro-particles, favoring migration towards the glycol. The SEM analysis with the coupled X-ray dispersion equipment, allowed us to know the elements present in the sample which are: Si, Pb, C, and Al. The presence of metals is associated with small metal filaments that are left behind.

Espectro UV usando Perkin Elmer Lambda 35; blanco TEG-Etanol

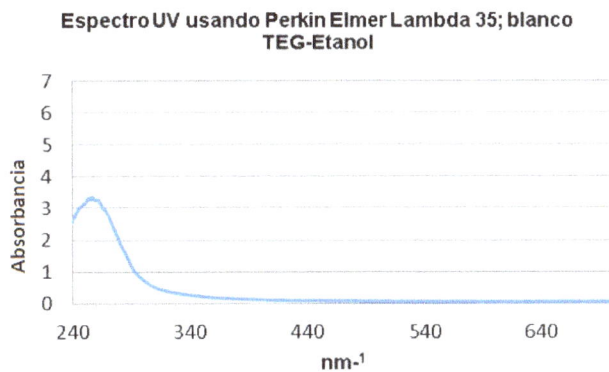

Figure 3. UV spectrum of the solvent that remains after treatment.

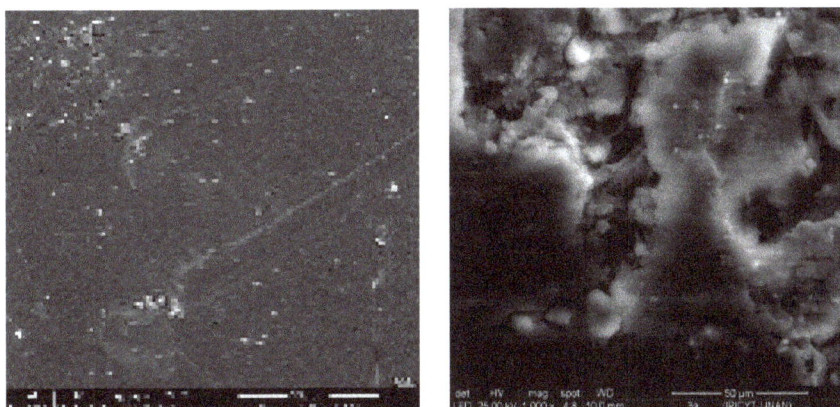

Figure 4. Cell photo before and after treatment.

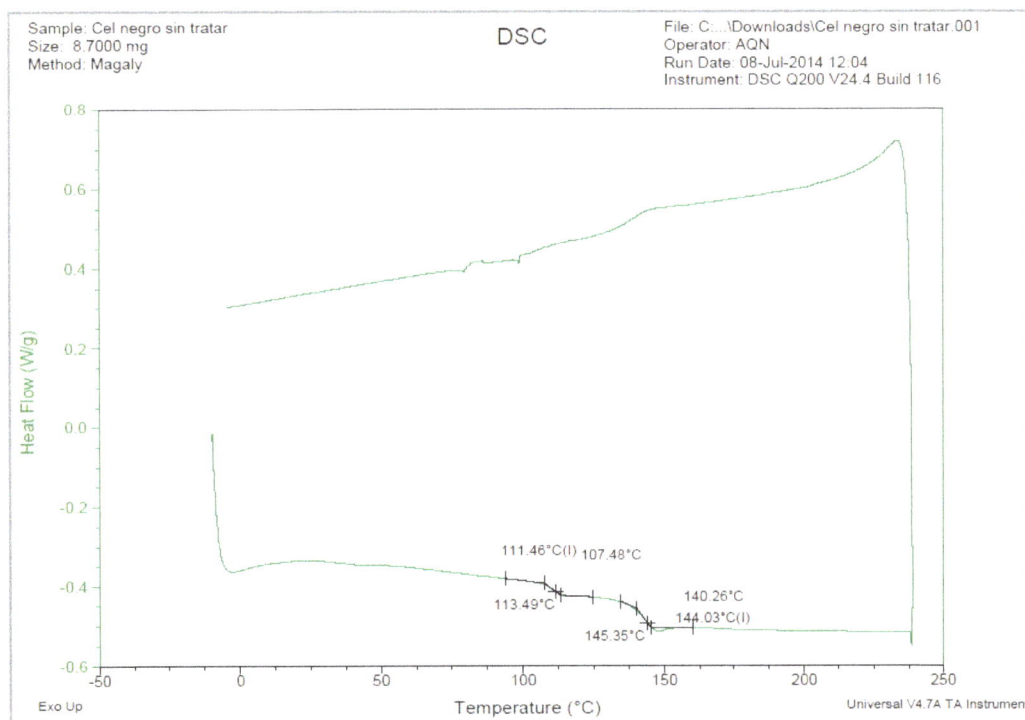

Figure 5. Untreated DSC sample.

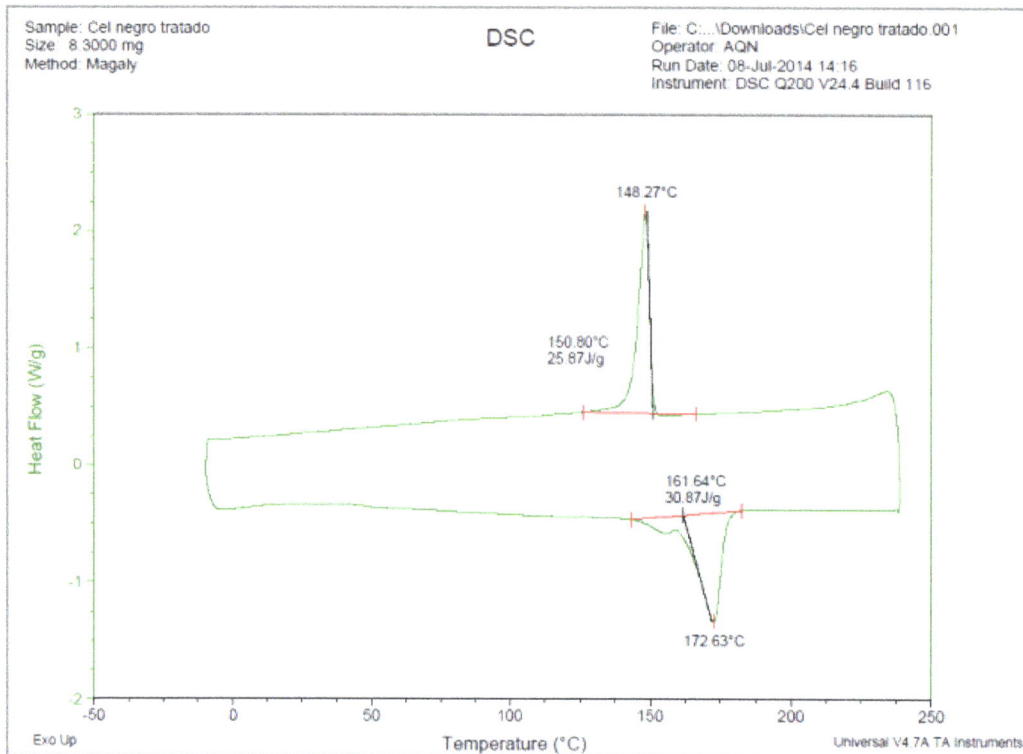

Figure 6. Treated DSC sample.

3.4. DSC Analysis

The differential sweeping calorimetry analysis on polymer samples of black cell phones, one before treatment, and one after treatment. The thermograms obtained are shown in **Figure 5** and **Figure 6**.

The comparative study shows that the glass transition temperature in each sample is different. During the warming of the sample without treatment two temperatures were observed, 17.48°C and 144.03°C, however, in the treated polymer only one temperature is observed: 161.64°C. In the case of cooling in the treated sample, no change was detected from 148.27°C.

The aforementioned results proves that the chemical organization of the material has been modified, changing the structure of the mixed solid (amorphous and crystalline, completely modifying the mechanical properties of the material [6].

4. Conclusions

This study allows us to conclude that cell phone recycling by solvolysis of the polymer part to recuperate metals is possible, showing how well glycol works as a solvent, and a reaction of 150°C to 220°C, with agitation and with a basic catalyst, and reaction times varying from 1 to 7 hours.

Through spectrophotometric evidence of the presence of multiple compounds present in the solvent from the polymer, it has been verified that the macromolecule has been chemically altered. More generally, the instrumental analysis (FT-IR ATR, UV, DSC and SEM) show that de-polymerization has taken place.

The majority of solids left after the reaction are metals, but there are still some polymer particles, and we therefore suggest prolonging the reaction time. These solids are more resistant to solvolysis, due to the fact that they are linear saturated polymers or contain tertiary carbons. The presence of a basic atmosphere increases the de-polymerization. The absence of oxygen prevents hydroperoxides from forming (these compounds are easily decomposed to give macro-radicals).

Acknowledgements

DAIP, University of Guanajuato, and IPICYP.

References

[1]   www.milenio.com/negocios/mexico-reciclan-mil-celulares-mes

[2]   Vinh, H.H. (2010) Thiosulfate Leaching of Gold from Waste Mobile Phones. *Journal of Hazardous Materials*, **178**, 1115-1119.

[3]   Sheirs, J. (1998) Polymer Recycling Science Technology and Applications. John Wiley and Sons, 381-406.

[4]   Koening, L. (1992) Spectroscopy of Polymers. ACS Professional Reference Book, USA.

[5]   Silverstein, Basler, Morrill (1998) Identification Spectroscopique de composes organiques; de boeck.

[6]   Sanchez Cadena, L.E., Alvarado Tenorio, B., Romo Uribe, A., Campillo, B., Flores, O. and Huifen, Y. (2013) Hot-Pressed Boards Based on Recycled High-Density Polyethylene Tetrapack: Mechanical Properties and Fracture Behavior. Reinforced Plastics and Composites.

14

Synthesis of CaO-SiO$_2$ Compounds Using Materials Extracted from Industrial Wastes

Nobuaki Yamaguchi[1], Yoshiko Masuda[2], Yoshishige Yamada[2], Hideaki Narusawa[1], Cho Han-Cheol[1], Yukimichi Tamaki[3*], Takashi Miyazaki[1]

[1]Department of Conservative Dentistry, Division of Biomaterials & Engineering, Showa University School of Dentistry, Tokyo, Japan
[2]Department of Conservative Dentistry, Division of Endodontology, Showa University School of Dentistry, Tokyo, Japan
[3]Department of Dental Materials Science, Asahi University School of Dentistry, Gifu, Japan
Email: *tamaki@dent.asahi-u.ac.jp

Abstract

Mineral trioxide aggregate (MTA) cement is an attractive material in endodontic dentistry. The purpose of this study was to produce calcium silicate, which is a major component of MTA, from waste materials. A dental alginate impression gel and used chalks were selected and mixed in a suitable ratio (Code: EXP). As a control, CaCO$_3$ and a commercial diatomite were used (Code: CON). Each powder was heated to 850°C and 1000°C, and then kneaded with water. TG-DTA, compressive tests, SEM observations, elemental mapping analyses, and XRD analyses were performed. TG-DTA indicated that weight reduction of CaCO$_3$ started at 600°C, and it completely decomposed on heating at 850°C. The strength was affected by the temperature. After heating, CaCO$_3$ was transformed into CaO and/or Ca$_2$SiO$_4$, and Ca(OH)$_2$ was formed by mixing with water. There were no differences between EXP and CON. These data suggested that recycled wastes might be promising MTA sources.

Keywords

Recycling, Calcium Silicate, Diatomite, Alginate Impression Material, Calcium Hydroxide

1. Introduction

Most used materials generally become wastes, and dental materials such as waxes, impression materials, gyp-

---
*Corresponding author.

sum, investment materials, and acrylic resins are also handled as industrial wastes; however, they may also contain some reusable materials and/or recyclable waste [1]-[3]. In particular, inorganic substances such as oxides, nitrides, hydroxides, and sulfides are valuable resources. Recently, some oxide materials have been shown to be promising for metal-free restoration because they have sufficient strength, no flammability, high stiffness, a natural tooth-color, and excellent bio-inertness [4]-[6]. Such materials are used in clinical restoration as composites with monomers. Because of the aesthetic outcome and their safety for use in humans, these ceramics are valuable in dentistry.

Mineral trioxide aggregate (MTA) cement is an attractive material in endodontic dentistry, especially for teeth with severe and/or long-term root canal treatment [7]-[10]. Root canal treatment usually involves delicate work because of the limited view and area. As a result, the possibility of accidents increases, and, unfortunately, this may lead to tooth extraction. It has been reported that MTA consists mainly of various calcium compounds such as calcium silicates (Ca_3SiO_5, Ca_2SiO_4, $CaSiO_3$), calcium aluminate ($CaAl_2O_4$), and calcium sulfate ($CaSO_4 \cdot 2H_2O$) [9] [11]. Calcium silicates in particular have been reported to have excellent antibacterial, hard tissue regeneration, and bone conductivity properties [12] [13]. Specifically, dicalcium silicate [11] or tricalcium silicate [13]-[15] produces calcium hydroxide in chemical reactions with water, and hardens MTA [9]. Excellent properties are obtained as a result of the strong alkalinity of calcium hydroxide [11] [16] [17]. Calcium silicate plays a pivotal role in these characteristics and predominates in MTA [9]-[11]. However, as is well known, MTA is more expensive than other dental products for root canal treatment.

The industrial syntheses of calcium silicates are represented as follows [9]:

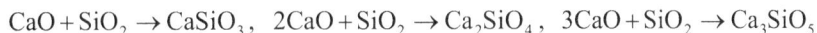

$$CaO + SiO_2 \rightarrow CaSiO_3, \quad 2CaO + SiO_2 \rightarrow Ca_2SiO_4, \quad 3CaO + SiO_2 \rightarrow Ca_3SiO_5$$

Calcium silicate is produced by the reaction between SiO_2 and CaO. In dentistry, SiO_2 is a very popular compound and is frequently used not only as a refractory in dental investment materials (quartz or cristobalite), but also as a component of dental materials such as porcelain powder, model resins for veneering crowns, filling resin composite materials, and glass ionomer cements. CaO is not used as a raw material in dentistry; however, several dental products contain CaO. Gypsum is a model CaO-containing compound. CaO can be derived from such compounds using various procedures. For instance, when a molten metal at above 1000°C comes into contact with a gypsum-bonded material, CaO is produced by decomposition of the gypsum.

The purpose of this study was to investigate the possibility of obtaining MTA-like cement products from waste materials. To do this, it is necessary to produce calcium silicate from these wastes. First, we investigated the synthesis of calcium silicate from recycled waste materials, and then generated calcium hydroxide by kneading with water. Generally, commercial alginate impression products contain more than 60 wt% diatomite [18]. Diatomite consists of diatomaceous fossils, originally algae, and its chemical formula is SiO_2. It is known that SiO_2 is allotropic and is divided into quartz, tridymite, cristobalite, and amorphous silica glass, with different crystal structures. However, quartz and cristobalite have poor reactivities with other compounds because of their stabilities at high temperature. Diatomite has a porous structure and is relatively reactive at low temperature.

2. Materials and Methods

2.1. Powder Preparation

To extract silicon dioxide, a commercial alginate impression product (Algiace Z, DENTSPLY International Inc., Tokyo, Japan), which includes 70 wt% diatomite (SiO_2), was selected. On the assumption that the taking of impressions from patients had been completed, the powder was mixed with water and then gelled. The obtained gel was heated in a ceramic dish in an electric furnace at 800°C. The residual pieces obtained by firing were milled using a pestle and a mortar, and then retrieved. This procedure is shown in **Figure 1**.

Although it is not impossible to extract CaO from gypsum products, as described above, an electric furnace with a high-power source is needed, for firing at 1200°C [19]. We therefore investigated another compound, which easily generates CaO. Chalk is a calcium-based compound that is used routinely in elementary schools. Chalk consists of calcium carbonate, and used chalk usually becomes waste. Also, the decomposition temperature of $CaCO_3$ is lower than that of gypsum [20]. To synthesize calcium silicate, both an alginate impression gel and chalk powder were prepared.

Used chalk (Dustless Chalk, Nihon Rikagaku Industry Co., Ltd., Kawasaki, Japan) was similarly pulverized

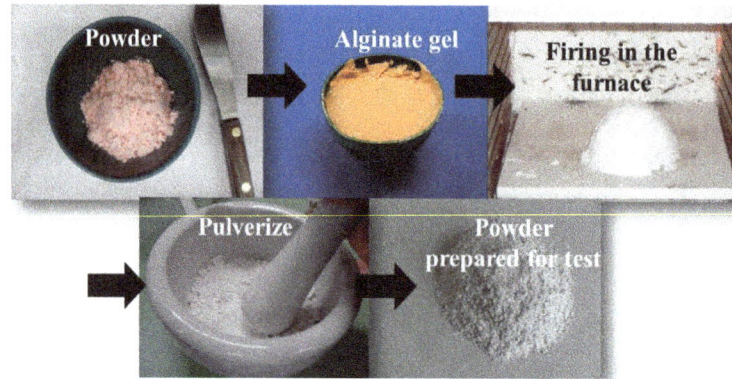

Figure 1. Preparation of diatomite extracted from alginate impression gel.

and retrieved.

Reagent-grade calcium carbonate (Wako Pure Chemical Industries, Ltd., Osaka, Japan) and a commercial diatomite (Wakkanai Green Factory Co., Ltd., Wakkanai, Japan) were used as references to check the validity of the experimental results.

2.2. Synthesis of Calcium Silicate

The diatomite/$CaCO_3$ ratio was calculated based on molecular weights, using the following chemical formula:

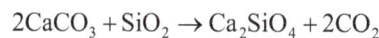

$$2CaCO_3 + SiO_2 \rightarrow Ca_2SiO_4 + 2CO_2$$

Calcium carbonate (3 g) and diatomite (10 g) were used to produce Ca_2SiO_4.

$CaCO_3$ was blended with diatomite in the above quantities, and the mixture was placed in a heat-resistant dish and uniformly kneaded with distilled water at a water (mL)/powder (g) ratio (W/P) of 0.4 for 30 s. The mixture consisting of both alginate impression material and dustless chalk was the experimental sample (Code: EXP). After sufficient drying, the mixture was heated at 10°C/min in an automatic electric furnace (KDF-009, Yoshida Co., Ltd., Tokyo, Japan). The limit of the heating temperature was 850°C or 1000°C. As a control (Code: CON), a mixture consisting of commercial ingredients (diatomite and $CaCO_3$) was prepared for comparison with EXP. All the materials prepared for the experiments are listed in **Table 1**.

2.3. Thermal Analysis

In this study, thermo-decomposition of calcium carbonate was required for the synthesis of Ca_2SiO_4. Moreover, the calcium oxide formed during heating needed to react with silicon dioxide. The thermal behaviors of two $CaCO_3$ powders and two mixtures (EXP and CON) were therefore measured using thermo-mechanical analysis equipment (Thermo Plus TMA 8310, Rigaku Co., Ltd., Tokyo, Japan). Thermogravimetric/differential thermal analysis (TG-DTA) was performed on each of the prepared powders. Around 30 mg of sample were placed in a platinum pan, heated up to 1500°C at 10°C/min, and then heat-soaked for 10 min. Changes in the weights of the samples were monitored along with the exothermic or endothermic nature of the reaction.

2.4. X-Ray Diffraction (XRD) Analysis

The compositions of the powders after heating were evaluated using XRD (XRD 6100, Shimadzu Corp., Kyoto, Japan). Sample powders were placed in the glass holder and analyzed using Cu-Kα radiation, at 40 kV and 30 mA, with a scanning speed of 2°/min and a scanning range of 10° - 70°. The XRD analyses were performed on the following samples:

1) four kinds of powders prior to mixing (SiO_2: fired alginate gel and commercial diatomite; $CaCO_3$: chalk and the reagent);
2) EXP and CON blended with water;
3) EXP and CON fired at 850°C and 1000°C;
4) EXP and CON blended with water after firing at 850°C and 1000°C.

Table 1. Materials tested in this study.

EXP: As used materials,
- Dental alginate impression product (SiO$_2$: Diatomite)
  (Algiace Z, DENTSPLY International Inc., Tokyo, Japan)
- Chalk (CaCO$_3$)
  (Dustless chalk, Nihon Rikagaku Industry Co., Ltd., Kawasaki, Japan)

CON: As a control,
- Diatomite (SiO$_2$)
  (Wakkanai green factory Co., Ltd., Wakkanai, Japan)
- CaCO$_3$reagent
  (Calcium carbonate, Wako Pure Chemical Industries, Ltd., Osaka, Japan)

2.5. Scanning Electron Microscopy (SEM) Observations

The fired alginate gel and commercial diatomite were observed using SEM (SEM 8000, Hitachi Co., Ltd., Tokyo, Japan) and their morphologies were compared. In addition, SEM observations were performed for EXP after firing at 850°C and 1000°C. Prior to placing in the sample holder, specimens were coated with gold.

2.6. Elemental Mapping Analyses

Using the same equipment as for the SEM observations, elemental mapping analysis was carried out for the fired alginate impression material to identify included SiO$_2$. The elements mapped were silicon and calcium. To estimate the elemental distribution and/or morphological changes after firing at high temperature, the fired EXP was also examined.

2.7. Compressive Strength

The green compressive strengths of EXP and CON (ø 6 mm, height 12 mm) after mixing for 24 h were measured using a universal testing machine (Instron MD-1125, Instron Japan Co., Ltd., Kawasaki, Japan), with a cross-head speed of 1.0 mm/min.

The lumps obtained after firing at 850°C and 1000°C were milled. The obtained powders were kneaded with distilled water at a water/powder ratio of 0.5 on a glass plate. Five specimens were prepared for each experiment and the average value was calculated. The obtained data were statistically evaluated by two-way analysis of variance at a significance level of $\alpha = 0.05$.

3. Results

The thermal analysis results are shown in **Figure 2** and **Figure 3**. There were no clear differences between reagent CaCO$_3$ in **Figure 2** and EXP, shown in **Figure 3**. In both TG-DTA curves, a large endothermic peak was observed at around 600°C, and then the sample mass started to decrease abruptly. This decrease continued until nearly 800°C.

In contrast, there were obvious differences between EXP and CON. Although the endothermic reaction with the decomposition of CaCO$_3$ was almost the same, a sharp peak at around 100°C was clearly observed in the CON curve.

Figures 4-7 show the results of the XRD analysis. XRD patterns of the four powders prepared for the experiments are shown in **Figure 4**. The peaks of chalk and calcium carbonate were completely in agreement. Diatomite was the dominant peak in both cases; however, the dental waste had a calcium sulfate peak. A weak quartz peak was detected from the commercial diatomite. The peaks of EXP and CON kneaded with distilled water are shown in **Figure 5**. These consisted of diatomite and CaCO$_3$, and we confirmed that new compounds formed by reaction with water appeared. After firing at 850°C, EXP and CON had strong CaO peaks originating from CaCO$_3$, residual CaCO$_3$, and a weak diatomite peak, as shown in **Figure 6**. However, Ca$_2$SiO$_4$ was not formed. In contrast, peaks from Ca$_2$SiO$_4$ were detected at 31° - 32° in EXP and CON after firing at 1000°C. The peaks of the samples fired at 1000°C were slightly broader than those in the patterns of the samples fired at 850°C, When powders fired at 850°C were kneaded with water, strong peaks for calcium hydroxide [Ca(OH)$_2$] were identified at 18° and 34°. In contrast, the powders fired at 1000°C showed weaker peaks from Ca(OH)$_2$ (**Figure 7**).

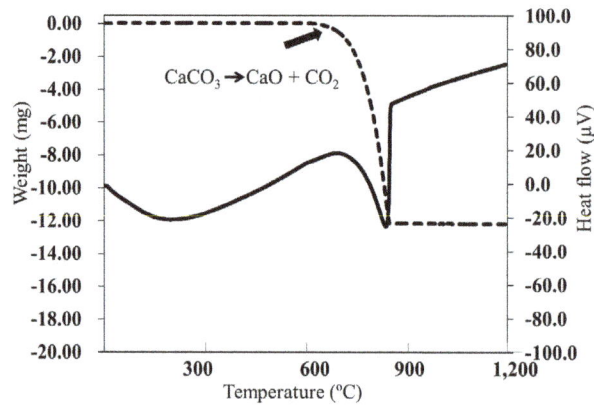

Figure 2. TG-DTA curve for calcium carbonate.

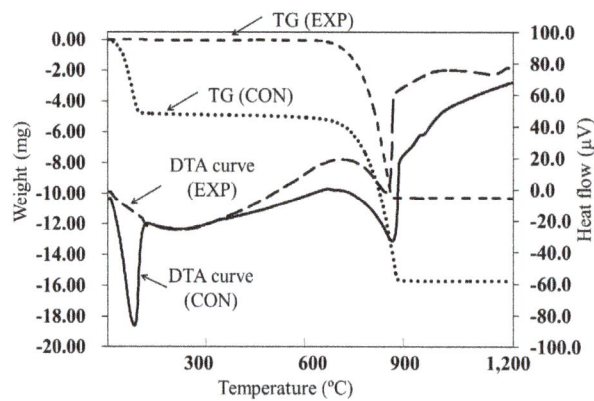

Figure 3. TG-DTA curves for EXP and CON.

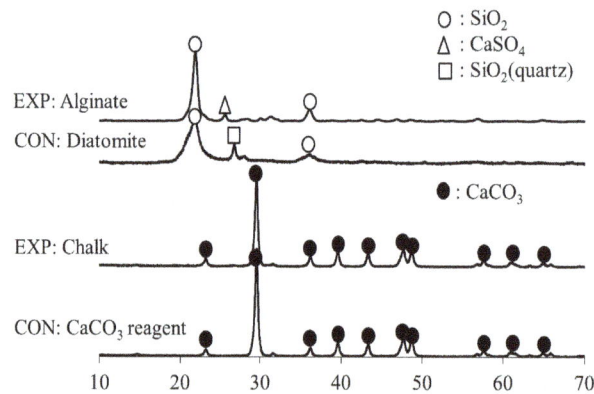

Figure 4. XRD analyses of four materials tested in this study.

SEM images of the diatomites and fired EXP samples are shown in **Figure 8**. A peculiar diatomite crystal structure was distinctly observed for the fired alginate gel at 850°C. There were numerous ordered spaces. In contrast, the structure of the commercial diatomite presented a common grain shape, and there were no characteristic patterns like that from the alginate gel. The crystal shape of the diatomite in EXP fired at 850°C was retained. Partially fused diatomite and $CaCO_3$ was found for the EXP fired at 1000°C, indicated by the black arrow in the SEM image in **Figure 8**.

An elemental map of Si and Ca is shown in **Figure 9**. Silicon existed densely on the diatomite, with a unique structure, and was separate from Ca. In the case of the sample fired at 1000°C, the diatomite was fused with CaO and uniform.

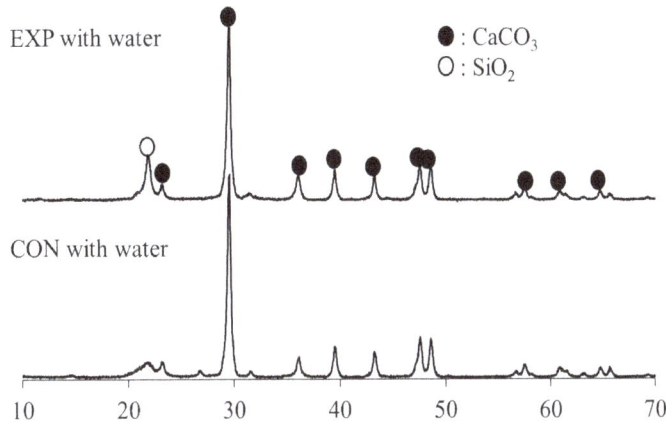

Figure 5. XRD analyses of EXP and CON after kneading with water prior to firing.

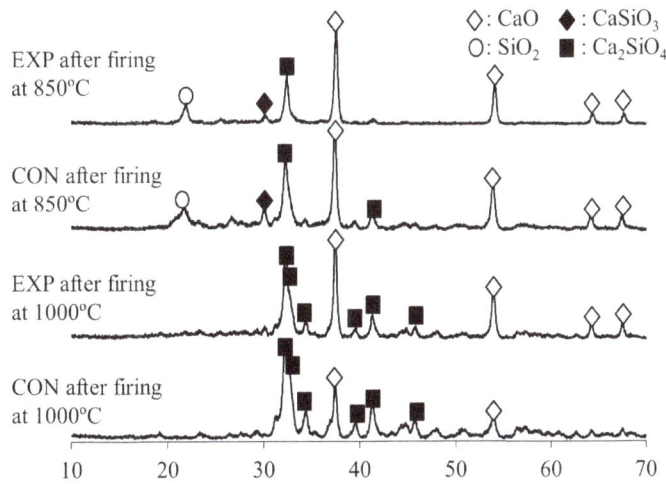

Figure 6. XRD analyses of EXP and CON after firing at 850°C and 1000°C.

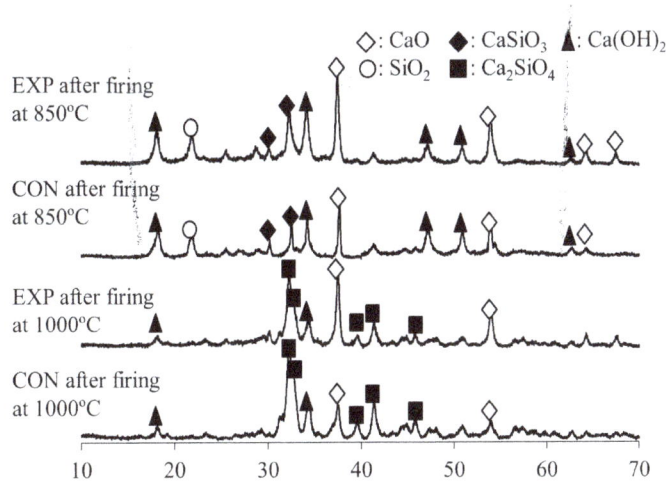

Figure 7. XRD analyses of EXP and CON kneaded with water after firing at 850°C and 1000°C.

Fired alginate impression gel   **Commercial diatomite**

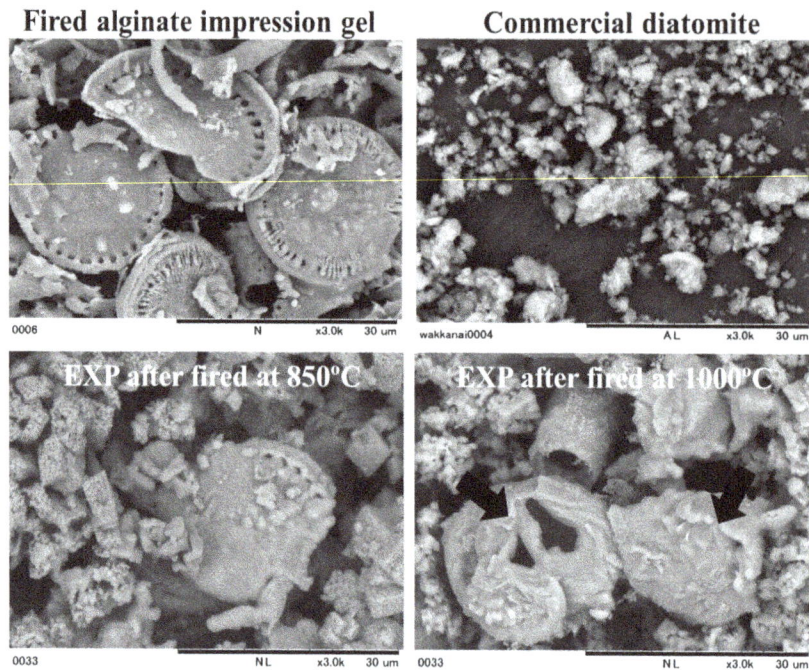

Figure 8. SEM images: upper right: commercial diatomite, upper left: diatomite from alginate gel; lower right: EXP fired at 1000°C, lower left: EXP fired at 850°C.

Figure 9. Elemental mapping of EXP fired at 1000°C. upper: SEM image, lower left: Si distribution, lower right: Ca distribution.

Compressive strength data are shown in **Figure 10**. The values for EXP and CON were 0.30 ± 0.05 and 0.40 ± 0.09 (MPa), respectively, at 850°C, and 0.62 ± 0.23 and 0.50 ± 0.08 (MPa), respectively, at 1000°C. There was no relationship between strength and the type of powder; however, it was found that the maximum firing temperature significantly affected the strength.

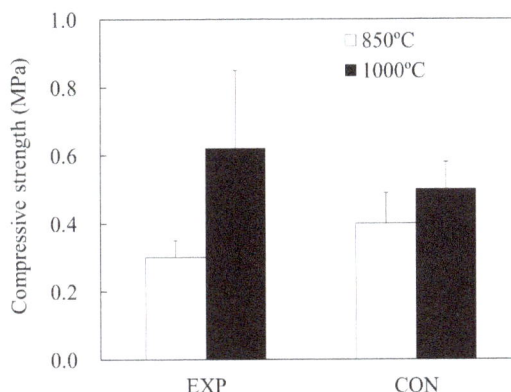

Figure 10. Green compressive strengths of EXP and
CON after firingat 850°C and 1000°C.

4. Discussion

It is important to take recycling into consideration because conservation of finite resources has been practiced worldwide since the end of the 20th century. There have been several studies of recycling of dental alloys [21]-[23], investment materials [1]-[3], and gypsum [24] in dentistry. However, these were concerned with reuse, and recycling in other ways was not considered to any great extent. Our aim was to manufacture new products by extracting ingredients from used products.

To synthesize Ca_2SiO_4, which is one of the major components of MTA [7]-[9], we initially explored possible ways of supplying both SiO_2 and CaO. Two types of waste, namely used chalk for $CaCO_3$ and alginate impression gel for SiO_2 were used.

From the results of thermal analysis, the firing temperature was set at 850°C because weight loss from decomposition of $CaCO_3$ was complete below 850°C. The decomposition of $CaCO_3$ as a result of heat-stress was prominent in the TG-DTA curves, shown in **Figure 2** and **Figure 3**. The rapid decrease at 100°C in the TG-DTA graph of CON was thought to be the result of evaporation of water in the powder. As a result, the total decreases for EXP and CON were different.

Figure 4 showed that the waste materials used in this study had almost the same compositions as reagents and commercial product. However, the shape of the diatomite in the alginate impression material was different from that in the commercial product, as shown in **Figure 8**. To homogenize the powder prior to firing, each of the mixtures (EXP and CON) was kneaded with water. Since the two different peaks in **Figure 4** are both present in **Figure 5**, it was found that water did not influence the $CaCO_3$ content of the mixture.

Based on TG-DTA results, the following two-step chemical reaction was predicted on firing at 850°C:

$$CaCO_3 \rightarrow CaO + CO_2, \quad 2CaO + SiO_2 \rightarrow Ca_2SiO_4$$

The results after firing at 850°C showed that $CaCO_3$ had decomposed. However, Ca_2SiO_4 peaks were not detected in the XRD pattern. This meant that the firing temperature was not high enough for the second of the above reactions, based on the finding that the SiO_2 peak remained after firing. It is reasonable to suppose that a higher temperature than 850°C is needed. Under the experimental conditions for the TG-DTA analysis, only a small quantity of EXP or CON was used for the measurements. In order to activate the decomposition of $CaCO_3$, the samples needed to be fired at a temperature above 850°C.

The notable features of the XRD patterns after firing at 1000°C were the disappearance of the SiO_2 peak and broadening of the peak at 33°. It is possible that several peaks overlapped. In fact, Ca_2SiO_4 has many different crystal types, and these peaks are concentrated at around 32° - 33°. It was therefore considered that firing at 1000°C helps to form Ca_2SiO_4. The formation of Ca_2SiO_4 can be explained based on the SEM images in **Figure 8** and the mapping analysis results.

As a final step, EXP and CON were milled after firing and kneaded with water. The experimental results are shown in **Figure 7**. The following findings were obtained: $Ca(OH)_2$ was mainly generated from CaO, and this phenomenon was marked at 850°C, but not at 1000°C. It is likely that the samples after firing at 850°C had insufficient CaO to form Ca_2SiO_4. Initially, we assumed that $Ca(OH)_2$ would also be produced by reaction of water with Ca_2SiO_4, based on the following equation:

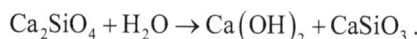

$$Ca_2SiO_4 + H_2O \rightarrow Ca(OH)_2 + CaSiO_3 .$$

Unexpectedly, it seems that the above reaction does not occur easily. It is therefore reasonable to suppose that the $Ca(OH)_2$ in the sample fired at 1000°C, shown in **Figure 7**, was also derived from CaO. It should also be added that the generation of Ca_2SiO_4 was time-dependent since we detected strong $Ca(OH)_2$ peaks in the XRD pattern after 24 h.

Another explanation for the formation of Ca_2SiO_4 with firing at 1000°C is provided by the compressive strength results. As shown in **Figure 7**, CaO and/or Ca_2SiO_4 were converted to $Ca(OH)_2$ by kneading with water, regardless of the firing temperature. Nevertheless, the difference between the firing temperatures was significant, as shown in **Figure 10**. It could be inferred from the higher compressive strength obtained from samples fired at 1000°C that the formed Ca_2SiO_4 plays an important role as a filler.

On the other hand, there are some problems with this study. For instance, the compressive strength is still controversial. Although Portland cement is involved in the hardening of MTA, it was not added to EXP. The green strength value of EXP was therefore inferior to that of MTA previously reported. The hardening mechanism also needs to be improved. In order to improve the mechanical properties and manipulation of EXP, it will be necessary to examine additives and/or reconsider the powder preparation. Additionally, detailed evaluations in cells or organisms have not yet been performed. We will carry out further investigations in the future.

5. Conclusion

In summary, these experimental results show that the synthesis of Ca_2SiO_4 was possible by the firing at 1000°C of a mixture containing both $CaCO_3$ and diatomite. Moreover, the fired mixtures produced $Ca(OH)_2$ on mixing with water. This result was in agreement with a previous report [9]. The EXP sample prepared in this study almost corresponded to the CON sample, which consisted of a reagent and a commercial product, except in the case of the thermal analysis results. In conclusion, it was expected that it would be possible to successfully produce $CaSiO_3$-based compounds with properties like those of MTA by recycling used products.

References

[1] Aida, Y., Zhang, Z., Yagi, S. and Tamaki, Y. (2009) Experimental Critobalite Investments with Reused Glass Powder as Binder Materials Are Available for Reuse of Casting. *Dental Medicine Research*, **29**, 139-147. http://dx.doi.org/10.7881/dentalmedres.29.139

[2] Zhang, Z., Tamaki, Y., Hotta, Y., Miyazaki, T. (2005) Recycling of Used Commercial Phosphate-Bonded Investments with Additional Mono-Ammonium Phosphate. *Dental Materials Journal*, **24**, 14-18. http://dx.doi.org/10.4012/dmj.24.14

[3] Yagi, S., Zhang, Z., Aida, Y., Hotta, Y., Tamaki, Y. and Miyazaki, T. (2011) Soda-Lime Glass as a Binder in Reusable Experimental Investment for Dental Castings. *Dental Materials* Journal, **30**, 611-615. http://dx.doi.org/10.4012/dmj.2011-040

[4] Vichi, A., Louca, C., Corciolani, G. and Ferrari, M. (2011) Color Related to Ceramic and Zirconia Restorations: A Review. *Dental Materials*, **27**, 97-108. http://dx.doi.org/10.1016/j.dental.2010.10.018

[5] Bottino, M.A., Salazar-Marocho, S.M., Leite, F.P., Vásquez, V.C. and Valandro, L.F. (2009) Flexural Strength of Glass-Infiltrated Zirconia/Alumina-Based Ceramics and Feldspathic Veneering Porcelains. *Journal of Prosthodontics*, 18, 417-420. http://dx.doi.org/10.1111/j.1532-849X.2009.00462.x

[6] Miyazaki, T. and Hotta, Y. (2011) CAD/CAM Systems Available for the Fabrication of Crown and Bridge Restorations. *Australian Dental Journal*, **56**, 97-106. http://dx.doi.org/10.1111/j.1834-7819.2010.01300.x

[7] Aeinehchi, M., Eslami, B., Ghanbariha, M. and Saffar, A.S. (2003) Mineral Trioxide Aggregate (MTA) and Calcium Hydroxide as Pulp-Capping Agents in Human Teeth: A Preliminary Report. *International Endodontic Journal*, **36**, 225-231. http://dx.doi.org/10.1046/j.1365-2591.2003.00652.x

[8] Dammaschke, T., Gerth, H.U.V., Zuchner, H. and Schafer, E. (2005) Chemical and Physical Surface and Bulk Material Characterization of White ProRoot MTA and Two Portland Cements. *Dental Materials*, **21**, 731-738. http://dx.doi.org/10.1016/j.dental.2005.01.019

[9] Darvell, B.W. and Wu, R.C.T. (2011) "MTA"—An Hydraulic Silicate Cement: Review Update and Setting Reaction. *Dental Materials*, **27**, 407-422. http://dx.doi.org/10.1016/j.dental.2011.02.001

[10] Han, L. and Okiji, T. (2011) Uptake of Calcium and Silicon Released from Calcium Silicate-Based Endodontic Materials into Root Canal Dentine. *International Endodontic Journal*, **44**, 1081-1087. http://dx.doi.org/10.1111/j.1365-2591.2011.01924.x

[11] Chiang, T.Y. and Ding, S.J. (2010) Comparative Physicochemical and Biocompatible Properties of Radiopaque Dicalcium Silicate Cement and Mineral Trioxide Aggregate. *Journal of Endodontics*, **36**, 1683-1687. http://dx.doi.org/10.1016/j.joen.2010.07.003

[12] Liu, X., Morra, M., Carpi, A. and Li, B. (2008) Bioactive Calcium Silicate Ceramics and Coatings. *Biomedicine & Pharmacotherapy*, **62**, 526-529. http://dx.doi.org/10.1016/j.biopha.2008.07.051

[13] Gandolfia, M.G., Ciapettib, G., Taddeic, P., Perutb, F., Tinti, A., Cardosod, M.V., Meerbeek, B.V. and Prati, C. (2010) Apatite Formation on Bioactive Calcium-Silicate Cements for Dentistry Affects Surface Topography and Human Marrow Stromal Cells Proliferation. *Dental Materials*, **26**, 974-992. http://dx.doi.org/10.1016/j.dental.2010.06.002

[14] Liu, W.-N., Chang, J., Zhu, Y.-Q. and Zhang M. (2011) Effect of Tricalcium Aluminate on the Properties of Tricalcium Silicate-Tricalcium Aluminate Mixtures: Setting Time, Mechanical Strength and Biocompatibility. *International Endodontic Journal*, **44**, 41-50. http://dx.doi.org/10.1111/j.1365-2591.2010.01793.x

[15] Camilleri, J. (2011) Characterization and Hydration Kinetics of Tricalcium Silicate Cement for Use a Dental Biomaterial. *Dental Materials*, **27**, 836-844. http://dx.doi.org/10.1016/j.dental.2011.04.010

[16] Marao, H.F., Panzarini, S.R., Aranega, A.M., Sonoda, C.K., Poi, W.R., Esteves, J.C. and Silva, P.I.S. (2012) Periapical Tissue Reactions to Calcium Hydroxide and MTA after External Root Resorption as a Sequera of Delayed Tooth Replantation. *Dental Traumatology*, **28**, 306-313. http://dx.doi.org/10.1111/j.1600-9657.2011.01090.x

[17] Pelisser, F., Steiner, L.R. and Bernardin, A.M. (2012) Recycling of Porcelain Tile Polishing Residue in Portland Cement. Environmental Science & Technology, **21**, 2368-2374. http://dx.doi.org/10.1021/es203118w

[18] Phillips, R.W. (1991) Skinner's Science of Dental Materials, Chapter 8 Elastic Impression Materials: Alginate. 9th Edition, W.B. Saunders, Philadelphia, 123-133.

[19] Matsuya, S. and Yamane, M. (1981) Decomposition of Gypsum Bonded Investments. *Journal of Dental Research*, **60**, 1418-1423. http://dx.doi.org/10.1177/00220345810600080501

[20] Rodríguez, N., Alonso, M., Grasa, G. and Abanades, J.C. (2008) Process for Capturing CO_2 Arising from the Calcination of the $CaCO_3$ Used in Cement Manufacture. *Environmental Science & Technology*, **42**, 6980-6984. http://dx.doi.org/10.1021/es800507c

[21] Reisbick, M.H. and Brantley, W.A. (1995) Mechanical Property and Microstructural Variations for Recast Low-Gold Alloy. *Journal of Prosthodontics*, **8**, 346-350.

[22] Isaac, L., Joseph, M., Bhat, S. and Shetty, P. (2000) Stress Variations in Recast Ni-Cr alloy—A Finite Element Analysis. *Indian Journal of Dental Research*, **11**, 27-32.

[23] Yilmaz, B., Ozcelik, T.B., Johnston, W.M., Kurtulmus-Yilmaz, S. and Company, A.M. (2012) Effect of Alloy Recasting on the Color of Opaque Porcelain Applied on Different Dental Alloy Systems. *The Journal of Prosthetic Dentistry*, **108**, 362-369. http://dx.doi.org/10.1016/S0022-3913(12)60193-0

[24] Ibrahim, R.M., Seniour, S.H. and Sheehab, G.I. (1995) Recycling of Calcium Sulphatedihydrate. *Egyptian Dental Journal*, **41**, 1253-1256.

15

Progress in Recycling of Composites with Polycyanurate Matrix

Christian Dreyer, Dominik Söthje, Monika Bauer

Fraunhofer Research Institution for Polymeric Materials and Composites PYCO, Teltow, Germany
Email: christian.dreyer@pyco.fraunhofer.de

Abstract

Thermoset based composites are used increasingly in industry for light weight applications, mainly for aircraft, windmills and for automobiles. Fiber reinforced thermoset polymers show a number of advantages over *conventional* materials, like metals, especially their better performance regarding their strength-to-weight ratio. However, composite recycling is a big issue, as there are almost no established recycling methods. The authors investigate the recyclability of polycyanurate homo- and copolymers with different recycling agents under different conditions. Also the influence of the recycling process on the most important reinforcement fibers, *i.e.* carbon-, glass-, aramid-, and natural-fiber is investigated. The authors find that: the recycling speed is not only dependent on the temperature, but also is significantly influenced by the particular recycling agents and the polycyanurate formulation. Hence, the stability against the recycling media can be adjusted over a broad range by *adjusting* the polymer composition. Furthermore, the authors find that the inorganic reinforcement fibers (carbon and glass) are almost unaffected by neither recycling agent at either temperature. Aramid-fibers degrade, depending on the particular recycling agent, from slightly up to extremely strong. This leaves one with the possibility to find a combination of matrix resin and recycling agent, which does not affect the aramid-fiber significantly. In the case of natural fibers, the dependence on the particular recycling media is very strong: some media do not affect the fiber significantly; others reduce the mechanical properties (tensile strength and elongation at break) significantly, and still others even improve both mechanical properties strongly. From the Recyclate, the authors synthesize and subsequently characterize a number of new polyurethane thermosets (foamed and solid samples) with different contents of recyclate, exhibiting T_g in the range of 60°C to 128°C.

Keywords

Polycyanurate, Composite Recycling, Thermoset, Cyanate Ester Resins, High Performance Polymers, Reinforcement Fiber, Carbon Fiber, Natural Fiber

1. Introduction

Composites are combinations of two (or more) materials, where all individual components retain their separate identities. They are neither dissolved into each other, nor merged together. They also do not react chemically with each other (reactions of functional groups at the fiber-surface not taken into regard). The materials are *mixed* homogeneously, but they are still physically discrete and mechanically separable. Typically, the mechanical properties of the composites are superior to the properties of the individual components [1]. When mentioning composites in this paper, the authors refer to fiber reinforced polymers, in particular with a thermoset matrix.

Composites based on fiber reinforced polymers are used in a wide range of applications these days, in particular for light-weight structures. They are predominantly found in aviation (e.g. for the A380) but also in structures for wind turbines in the building sector and increasingly in the automotive field [2].

Light-weight construction with fiber-reinforced polymers has been known in the production of automobiles already since the 1950s [3]. Since the late 1980s, the average weight of a car has grown steadily, mainly caused by an increasing amount of active and passive safety systems and due to a rising demand for comfort [4]. Especially for electrically driven vehicles, the need to reduce weight is high, as a reduction of the weight of a middle-class E-car by 200 kg results in a range increase of 20 - 30 km, which is quite significant when taking a typical range of around 150 km into regard [5]. In aircraft, the percentage of composites is increasing strongly, especially since the introduction of the A380 with more than 20 mass-% of composites in the structure [6].

Composites show a number of advantages over *conventional* materials like metals: They have a higher performance per weight. Excellent strength- and stiffness-to weight-ratios are achievable. Using unidirectional laminates, direction-dependent strength and stiffness are possible and composites with polymer matrix typically show excellent resistance to corrosion and chemical attack. Of course, composites also show some disadvantages over the *classical* materials. Composites are usually more brittle than metals. They exhibit lower thermostability and lower flame resistance compared to most metals. Furthermore, they show a different impact behavior than metals, which can be deformed better plastically. The repair of the composites is also more complex than the repair of metals. For most metals, efficient recycling processes are established, which is not the case for composites (see Section 1.2).

1.1. Polycyanurates

From the second half of the 19th century, unsuccessful attempts to synthesize cyanates, the monomeric units of polycyanurates took place [7]. More than a hundred years later, in 1964, almost simultaneously in three workgroups, the main synthetic routes for the synthesis of cyanates were developed [8]-[10].

The polycyanurate ester resins obtained from (at least) difunctional Cyanate monomers were originally developed as matrix resins for laminated printed circuit boards, due to their excellent performance, especially regarding their high thermostability and low dielectric constant, their use extended to composites for avionics (air ducts, plenums etc.) rad domes and also for space applications [11].

In contrast to phenolic resins, which form water during their polycondensation reaction, cyanates react with themselves via cyclotrimerization (an additional hardener like in the case of epoxides is not required). The reaction scheme for the formation of a dense crosslinked polycyanurate network is shown in **Figure 1**.

Due to the high network density and the high content of aromatic rings, glass transition temperatures of up to 400°C are possible for polycyanurates [12]. In addition to a high thermal stability they also show a relatively high intrinsic flame retardancy [13] and good chemical stability [11]. Their good transparency, especially in the infrared wavelength region, makes them suitable for use in integrated optical devices [14]-[16].

Cyanates not only react with themselves, they react with various compounds, which gives one the possibility to develop polymers with adjusted property profiles for different applications. Furthermore, the various reaction routes give one the possibility to accelerate the reaction speed, making these polymers interesting for industries where fast curing is needed, like in the automotive industry [17]. **Figure 2** gives an overview about the manifold reactions of cyanates.

Last but not least, the different resulting structural components, including the cyanurate-structure, which is responsible for the recyclabilities are described quantitatively for coreaction with phenolics and epoxides in [11], pp. 58-86.

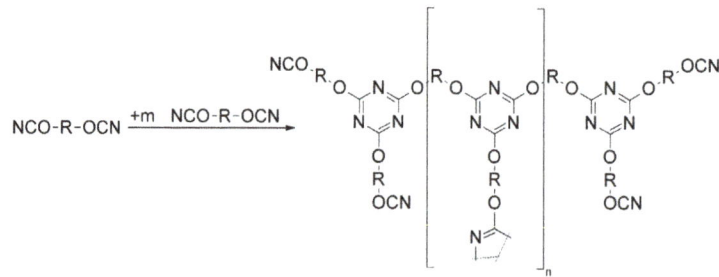

Figure 1. Polycyclotrimerization reaction of cyanates.

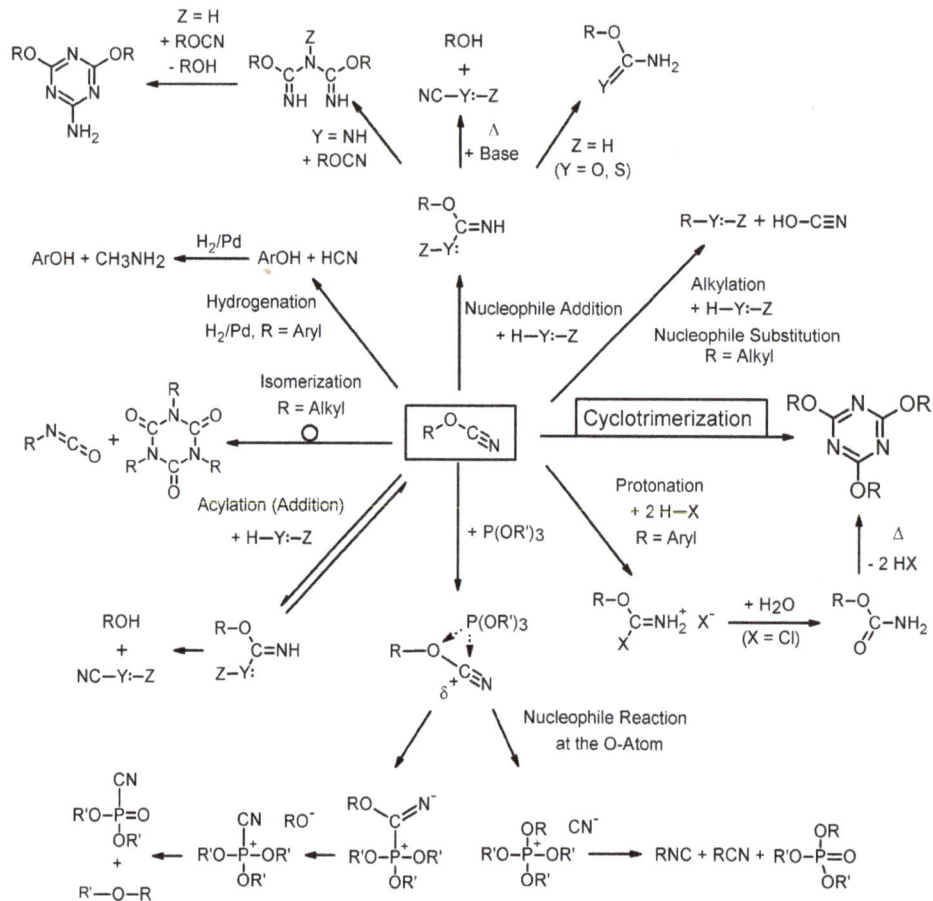

Figure 2. Reactions of cyanates with various compounds.

1.2. State-of-the-Art Recycling of Composites

Due to (expected) future governmental regulations, not only a high percentage of a car needs to be recycled at its end-of-life, also this will be demanded for aircraft in the future [18] [19]. With increasing use of composites the need for efficient recycling technologies will grow.

The major amount of composites has thermoset polymers as a matrix. As mentioned above, these polymers are in contrast to thermoplastic polymers [20]-[22] neither soluble nor meltable and therefore the main disposal route is landfilling, but also burning as well as milling and subsequent recovery as fillers (in small amounts). Not only the matrix polymer, but also the reinforcement fibers, which can be very price-intensive (e.g. in the case of carbon fibers), get lost by this procedure. Only a few publications are dealing with recycling and repair of epoxide resins, e.g. [23]-[25]. Furthermore pyrolysis, gasification, hydrolysis via subcritical resp. hypercritical water, and hydrogenation are under development for the recycling of composites [25]-[31].

Most of these common recycling methods do not reuse the matrix resin. A chemical recycling of epoxides is described in [24] [32]-[34]. But in this case, the recycling can only done for epoxides with a specific hardener, resulting in cured epoxides with a maximum T_g of around 100°C. The recycling is carried out under extremely mild conditions, having the big disadvantage of low media stability resulting therefrom.

Recycling of amine-cured epoxide thermosets by nitric acid leads to reusable glass fibers as well as reusable matrix material, nevertheless the recycling time is very long with approx. 400 hours [23].

2. Recyclabilityof Polycyanurates

Initial research on the recycling of polycyanurate thermosets was carried out by Bauer and Bauer already in the 1990s [35]. At that time, intense research was not done, because the need to recycle thermosets was relatively low. Lately, the interest in composite recycling is increasing, correlated to the extensive use in aircraft, windmills, trains and latterly in the automotive industry.

The recycling, *i.e.* chemical decomposition, of polycyanurates can be done under relatively mild conditions and within relatively short times.

The products of the cyanurate recycling are mainly triazine-containing structures and phenols. The particular recycling products depend of course on the original composition and the used recycling agent(s). **Figure 3** depicts the principle reaction scheme of the chemical decomposition of a polycyanurate network. The recycling process can be done with a low amount of recycling agent and with low energy consumption, especially compared to pyrolysis [36].

The chemical nature of cyanates, to react with a variety of compounds, not only lets one adjust the property-profile of the resulting polymer, but also permits to change the recycling behavior. The incorporation of mono-functional cyanates and/or phenols results in an acceleration of the recycling process (under same conditions). Of course, the structure of the recycling agent as well as the temperature influences the recycling speed significantly. Detailed research on the recycling of cyanates is reported in section 3.2.1 of this paper.

3. Experimental

In this section, recycling experiments of polycyanurate homo- and copolymers, the investigations of the recycling processes' influence on different reinforcement fibers, and the synthesis of new PUR-thermosets are described.

3.1. Materials

3.1.1. Recycling of Polycyanurates and Polycyanurate Copolymers

As a *model*-thermoset for this research the commercially available Dicyanatobisphenol A (DCBA, B10), LONZA AG, Switzerland was selected. Sample plates with a thickness of approx. 6 mm were manufactured by curing the cyanate monomer in vertical aluminum molds under a defined temperature regime up to a final curing

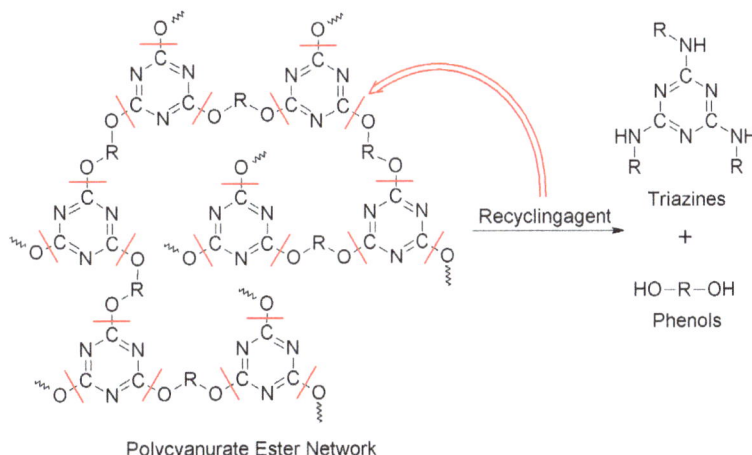

Figure 3. Chemical decomposition of polycyanurates.

temperature of 250°C, resulting in transparent brownish plates, showing a glass transition temperature (T_g) of about 287°C (DMA, tanδ).

The synthesis of L10- and PT15-homopolymers and the different polycyanurate-copolymers were carried out analogously to the B10-synthesis. The curing regimes were modified for the particular formulations.

Twelve different recycling agents were used for decomposing the resin sample plates. In the following chapter they are continuously named from Agent A to Agent M.

3.1.2. Effect of the Recycling Process on the Reinforcement Fibers

For the research regarding the influence of different recycling agents on the reinforcement fibers four types of different fibers—for composite materials well known—were selected. As carbonfiber a 3 k carbonfiber roving from SGL was chosen. Furthermore glass fiber rovings with approx. 680 dtex, Aramid fibers with approx. 1730 dtexand natural fiber yarns (flax), with approx. 530 dtex were selected.

3.1.3. New Polyurethane Thermosets from Recycled Polycyanurates

For the polyurethane-synthesis Glycerine (hydroxyl value: 1828 g·kg^{-1} KOH) and 1,4-Butanediol (hydroxyl value: 1240 g·kg^{-1} KOH) were selected as hydroxyl-compounds in addition to the OH-group containing recyclate.

As bio-based polyol castor oil (hydroxyl value: 160 g·kg^{-1} KOH) was chosen. As isocyanate components Desmodur N100 (1,6-hexamethylene-1,6-diisocyanate based aliphatic polyisocyanate, HDI-Biuret, NCO-group content of 22.0% ± 0.3%) and Desmodur N3300 (hexamethylene-1,6-diisocyanate based trimer, HDI-Trimer, NCO-group content of 21.8% ± 0.3%) both obtained from Bayer MaterialScience and ISO 134/3 (polymeric, aromatic diphenylmethanediisocyanate, P-MDI, NCO-group content of 24.6%) obtained from BASF, were selected. As a blowing agent water was added to the corresponding formulations. The above mentioned chemicals for the synthesis of the new polyurethane thermosets are depicted in **Table 1**.

3.2. Synthesis and Recycling Procedure

3.2.1. Decomposition of Polycyanurates and Polycyanurate Copolymers

The recycling process was carried out in a 50 ml flask equipped with a condenser for temperatures >60°C. For experiments at lower temperatures, closed Duran® bottles were used. Details regarding recycling and work up can be found in reference [37].

3.2.2. Effect of the Decomposition Process on the Reinforcement Fibers

The media stability experiments on the reinforcement fibers were carried out in 50 ml flasks equipped with a condenser.

3.2.3. Synthesis of Polyurethanes from Recyclate

The recycling process was carried out in a 500 ml flask equipped with magnetic stirrer and condenser. Agent Hwas used in excess as recycling agent at the boiling point (172°C). The recycling time was 16 h and 2 h respectively. The excess of Agent H was removed via distillation. The dark brown highly viscous residue was used without further purification. Details regarding recycling and work up can be found in reference [37].

The different formulations were all targeted to obtain a 90% conversion of the OH-groups. Formulations P1 - P3 were synthesized using recyclate 1, P4 - P6 were synthesized using recyclate 2. 1,4-Butanediol, respectively castor oil acted as a solvent and a chain extender. Glycerin was used as a crosslinker in formulations P1-P3 and P6. **Table 2** gives an overview about the polyurethane-formulations used.

The components were mixed in a Speedmixer™ from the company Hauschild, poured into molds, precured for 24 h at RT, followed by a curing step for 1 h at 100°C. To obtain a complete conversion, a postcure for 1 h at 150°C was done.

For the formulations P1-P3 foams with different densities were obtained. P4-P6 produced massive samples. Some of the samples are shown in **Figure 4** (*left*: massive PUR, *right*: foams from formulations P1-P3).

3.3. Characterization Methods

3.3.1. Determination of Hydroxyl-Values

For the recyclate after 16 h (recyclate 1) a hydroxyl-value of 486 g·kg^{-1} KOH was estimated (ideal network

Table 1. Chemicals for synthesis of new thermosets.

Name	Chemical Structure
Bisphenol A	
Dicyanatobisphenol A (B10, DCBA)	
Dicyanatobisphenol E (L10, DCBE)	
PT15	
1,4-Butandiol	
Glycerin	
Castor oil	
HDI-Biuret	
HDI-Trimer	
P-MDI	

Table 2. Polyurethane formulations.

Formulation	KOH-value [g·kg^{-1} KOH]	Chain extender	Isocyanate	Cross-linking agent	Blowing agent	$\omega_{Recyclate}$ [wt-%]
P1	486*	1,4-Butandiol	HDI-Biuret	Glycerin	Water	13.2
P2	486*	1,4-Butandiol	P-MDI	Glycerin	Water	14.3
P3	486*	Castor oil	HDI-Biuret	Glycerin	Water	22.4
P4	653**	1,4-Butandiol	HDI-Trimer	-	-	10.0
P5	653**	1,4-Butandiol	HDI-Trimer	-	-	20.0
P6	653**	1,4-Butandiol	HDI-Trimer	Glycerin	-	20.0

*Estimated hydroxyl-value; **Hydroxyl values determined in accordance with DIN 53240.

Figure 4. Massive (left) and foam (right) PUR-samples from polyurethane recyclate on a carbon-fiber reinforced polycyanurate-laminate.

formation and ideal decomposition assumed). The hydroxyl-value of 653 g·kg^{-1} KOH for the recyclate after 2 h (recyclate 2) was determined in accordance with DIN 53240 [38].

3.3.2. SEC-Chromatography
The completeness of the decomposition was determined via size exclusion chromatography (SEC), using Tetrahydrofuran (THF) as an eluent at a flow rate of 1 ml·min^{-1}.

3.3.3. DSC-Analysis
The melting behavior of the crude recyclate was determined via Dynamic Scanning Calorimetry (DSC) (10 K·min^{-1}, 25°C - 200°C), exhibiting a melting range of 80°C - 100°C.

The complete curing of the new polymers was determined by DSC from 25°C to 150°C with a heating rate of 10 K·min^{-1}.

DSC-measurements were performed on a Mettler-Toledo DSC 821.

3.3.4. DMA-Measurements
Dynamic Mechanical Analysis (DMA) was performed in torsion with an ARES Rheometer from TA Instruments. Storage (G') and loss (G") modulus as well as tanδ were determined as a function of temperature. The samples were heated from 30°C to 150°C with a heating rate of 4 K·min^{-1} at a torsion frequency of 1 Hz.

3.3.5. Mechanical Measurements
To determine the mechanical properties of the treated and untreated fibers a tensile strength measurement machine from Instron was used. For each experiment regime 5 to 10 tensile tests were carried out.

4. Results and Discussion

In this section, the results of the experimental work on the recycling of polycyanurate homo- and copolymers, using different recycling agents under different temperatures, will be discussed. The second part covers the investigations concerning the influence of the recycling process on different reinforcement fibers. The third and last part gives an overview about the synthesized new PUR-thermosets.

4.1. Recycling of Polycyanurates and Polycyanurate Copolymers

Multiple factors influence the recyclability of polycyanurate ester resins and their copolymers: The recyclability of different fully cured polycyanuarate homopolymers under same conditions is significantly different, as shown in **Figure 5**.

Here the mass loss of the polycyanurate sample plates (due to the chemical decomposition in the recycling agent) is plotted against the recycling time. It can be seen clearly, that the reaction rate from B10 via L10 to PT

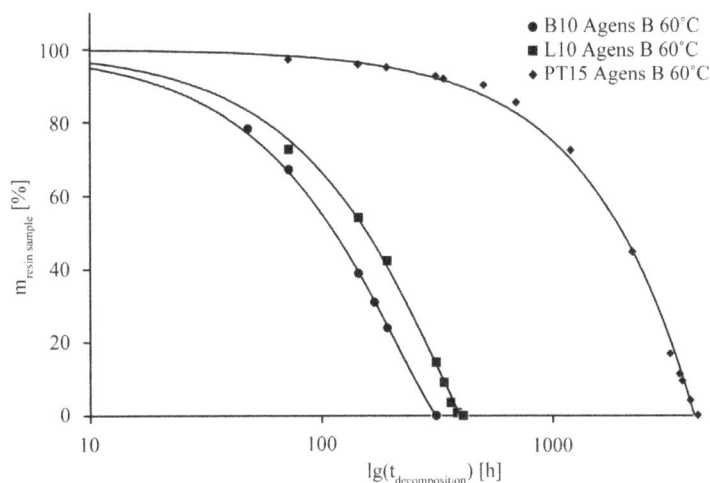

Figure 5. Recycling behavior of cured B10, L10 and PT15 (**Table 1**) in recycling agent B at 60°C (accuracy ± 12 h).

15 is decreasing. A complete decomposition of B10 is reached after about 300 h, compared to 400 h for L10 and more than 4000 h for PT15. The extreme difference in recycling time between the polymers obtained from the two difunctional cyanate-monomers B10 and L10 and the polyfunctional PT15 is not really surprising, because to release the polyphenol-units from the PT15-network a significant higher number of cyanurate bonds need to be decomposed, than to release the bisphenol-units in case of B10 and L10.

The influence of different recycling agents on the recycling behavior is shown in **Figure 6**. It can be seen, that the recycling times for the same Polycyanurate homopolymer vary between 48 h and 816 h (for a complete decomposition, *i.e.* complete dissolution of the polycyanurate in the recycling agent).

Figure 7 shows impressively the big differences in the recycling times beginning from approx. 48 h up to approx. 816 h for cured B10 samples depending on the particular recycling agent.

An increase of the reaction temperature results also for the chemical recycling of the polycyanurates—in an increase of the decomposition rate—as it is usual for a typical organic reaction. **Figure 8** shows the decomposition times of the three cured polycyanurates B10, L10 and PT15 in recycling agent B at 60°C and under refluxing conditions. One can see the dramatic reduction of the recycling times by a factor of at least 10. It is remarkable, that the decomposition of B10 at 60°C with 312 h is faster than that of L10 (408 h). Under refluxing conditions this effect is reversed so that L10 (18 h) is faster decomposed than B10 (36 h).

Under ambient pressure, the highest temperature which can be applied is the boiling temperature of the recycling agent. It can be seen in **Figure 9**, that the recycling times can be reduced down to approx. 1 h when doing the recycling under reflux. The columns are correlated to the recycling times in hours, the solid squares are correlated to the boiling point of the respective agents, which are identical with the recycling temperatures (with agent L and M as exceptions, where a recycling temperature of 195°C was used due to their high boiling points). **Figure 7** and **Figure 9** cannot be compared directly, as the mass-ratio of polycyanurate to recycling agent is 1:10 for **Figure 7** and 1:4 for **Figure 9**, resulting in a higher recycling time for agent C at boiling point (138h) than at 60°C (72 h). A correlation between the boiling point resp. the recycling temperature and the recycling speed cannot be seen directly, as the efficiencies of the particular agents are obviously strongly different as can be seen in **Figure 7**. When comparing similar chemical structures (as for agents A, B, and C, which differ only in their chain length) the recycling efficiency seems to be correlated with the recycling temperature (**Figure 9**).

A further attempt to investigate the recycling processes was done by using 3-Aminophenol dissolved in an inert solvent as recycling agent (mass-ratio 20:80). It can be seen in **Figure 10** that the 3-Aminophenol in n-Butanol, having a boiling-point of 118°C decomposes approx. 6 times faster compared to 3-Aminophenol in THF (boiling point 66°C) or in Ethanol (boiling point 78°C). Between the two lower boiling solvents there is no significant difference observable; the corresponding trendlines are almost identical.

As already mentioned in section 1.1, the property profile of polycyanurate ester resins can be adjusted over a broad range due to the various reactions which cyanates can undergo. The recyclability of the cyanates of course also depends on the particular cyanate, as explained in **Figure 5**. Due to the possibility to copolymerize the cya-

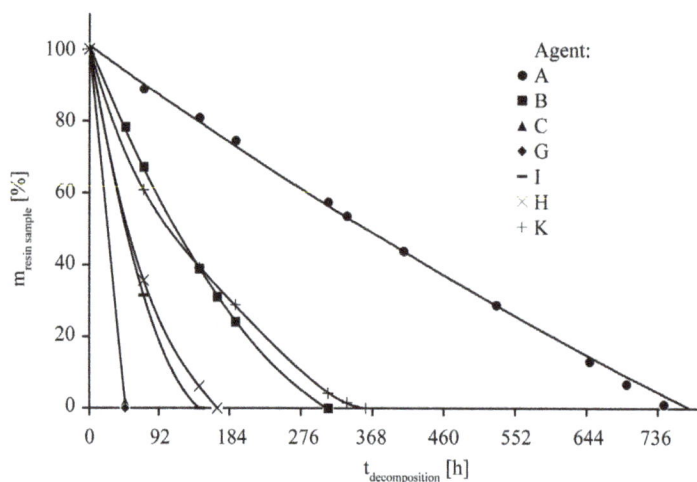

Figure 6. Recycling behavior of cured B10 sample plates in different recycling agents at 60°C (mass ratio B10:agent = 1:10) (accuracy ± 12 h).

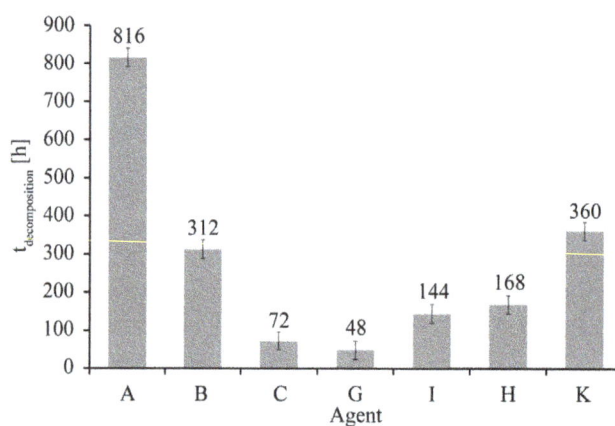

Figure 7. Recycling times of cured B10-Polycyanurate samples in seven different recycling agents at 60°C (mass ratio B10:agent = 1:10).

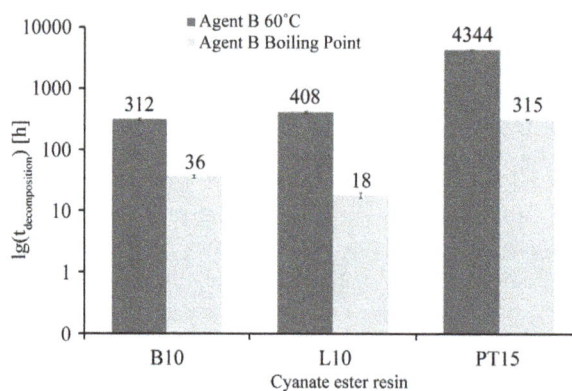

Figure 8. Recycling times of cured Polycyanurate samples at 60°C and under refluxing conditions in recycling agent B.

nates with different coreactants, the recyclability of a cyanate can either be increased or decreased, depending on the nature of the coreactant(s). **Figure 11** shows the decomposition curves of L10 copolymerized with two dif-

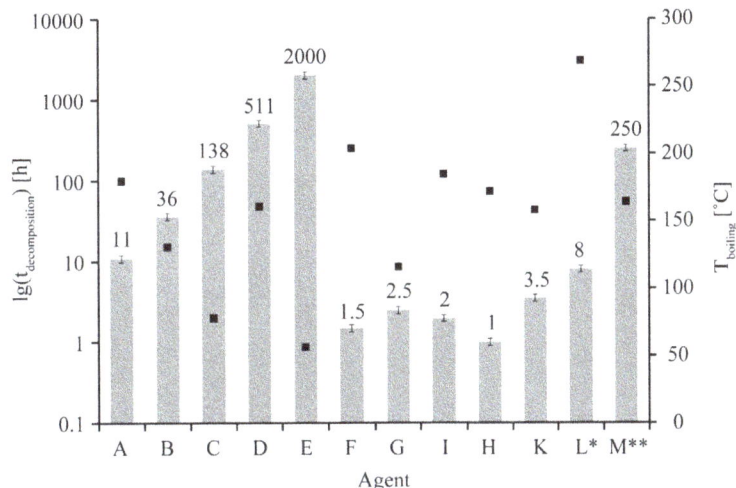

Figure 9. Recycling times of cured B10-Polycyanurate samples in 12 different recycling agents at their respective boiling point (mass ratio B10:agent = 1:4; columns heights are correlated to the recycling time, solid squares to the boiling point; $^*T_{decompositon} = 195°C$; $^{**}T_{decompositon} = 195°C$ and $T_{boiling} = 164°C$ at 15 mbar).

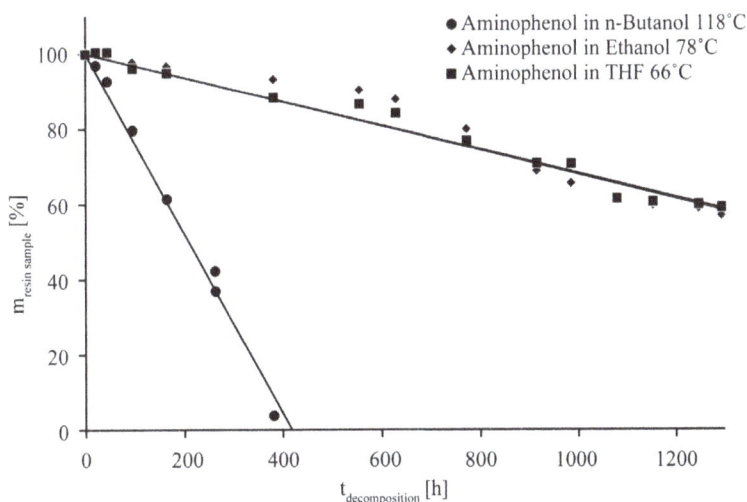

Figure 10. Recycling behavior of cured B10-Polycyanurate samples in solutions of 3-Aminophenol in n-Butanol ●, Ethanol ◆, and THF ■ (mass-ratio 20:80) (accuracy ± 12 h).

ferent comonomers A and B in different mass ratios in agent H at 40°C. The decomposition curves of unmodified (cured) L10 and B10 are also given for comparison. The copolymerization of L10 with Comonomer A results in a reduction of the network density, causing a faster decomposition in the used recycling agent H. The copolymerization of L10 with Comonomer B forms network links, which cannot be decomposed by the recycling agent, which can be clearly seen in **Figure 11**, as the recycling time increases with the mass-content of Comonomer B. The copolymer with a mass ratio of 50:50 is not recyclable within 7000 hours under these conditions.

The influence of the temperature on the reaction rate is also given for the cyanate copolymers. **Figure 12** shows the decomposition curves of L10, L10 + 10 wt-% Comonomer B and L10 + 20 wt-% Comonomer A in recycling agent H, at 20°C and 40°C for each formulation. The decomposition speed is increasing for each polymerformulation, when raising the temperature from 20°C to 40°C. Interestingly the reaction rate is increased by a factor of 6 to 7 irrespective of the particular formulation.

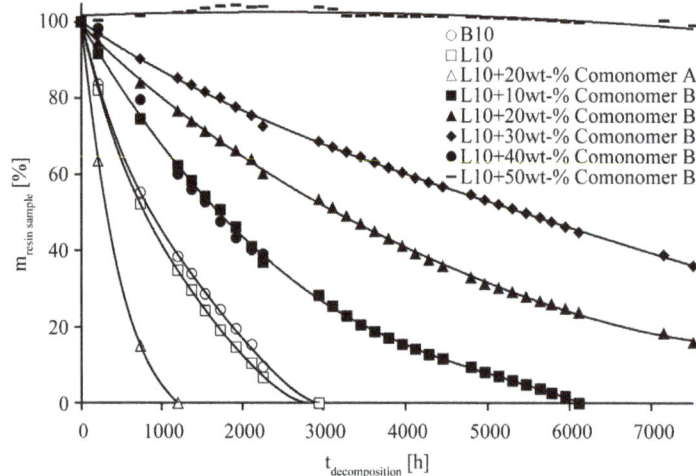

Figure 11. Recycling behavior of cured L10-copolymersamples with different coreactants and in different mass-ratios in recycling agent H at 40°C (neat cured L10 and B10 are depicted for comparison, accuracy ± 12 h).

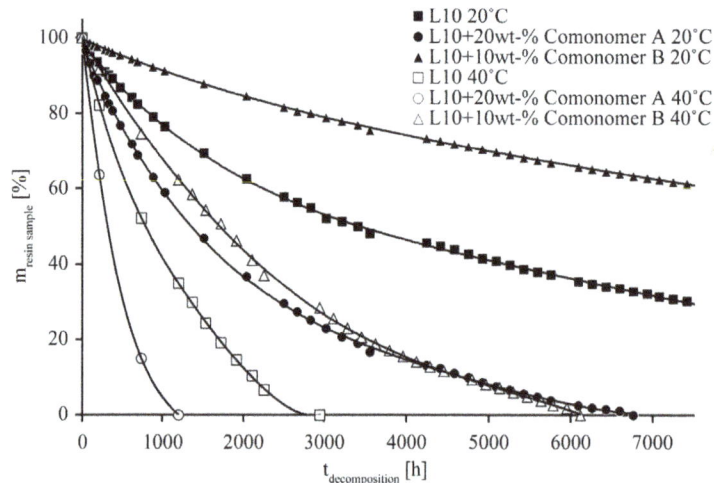

Figure 12. Recycling behavior of cured L10-Homo-and-Copolymer samples in recycling agent H at 20°C and 40°C (accuracy ± 12 h).

4.2. Effect of the Recycling Process on the Reinforcement Fibers

Another major issue for the recycling is the influence of the recycling agents on the reinforcement fibers. The authors investigated the influence of seven different recycling agents on carbon-, glass-, aramid-, and natural fibers at 60°C. The respective recycling times were selected in such a way, that the model-polycyanurate matrix, cured B10, completely decomposes under these conditions. The results of the tensile tests of the treated fibers can be seen in **Figure 13**. The respective recycling time is marked with black squares. Due to the high differences in tensile strengths of the different fibers, the values for tensile strength and elongation at break were normalized to 100% for the untreated fibers. Each column represents the mean value of 5 to 10 single measurements. The normalized tensile strength values are shown in **Figure 13(a)**. One can see, that carbon- and glass-fibers are not affected significantly by the recycling process, independent from the respective recycling agent. More evident is the influence on the aramid-fiber: Depending on the recycling agent the tensile strength is reduced down to approx. 30% (agent H). Using agents B and C, the fiber retains approx. 90% of its initial tensile strength. Interestingly, there is no direct correlation between the recycling time and the reduction of the tensile strength for aramid, e.g. treatment with agent I for 144 h results in a remaining tensile strength of less than 30%, while the aramid-fiber retains approx. 80% of its tensile strength after 816 h in agent A.

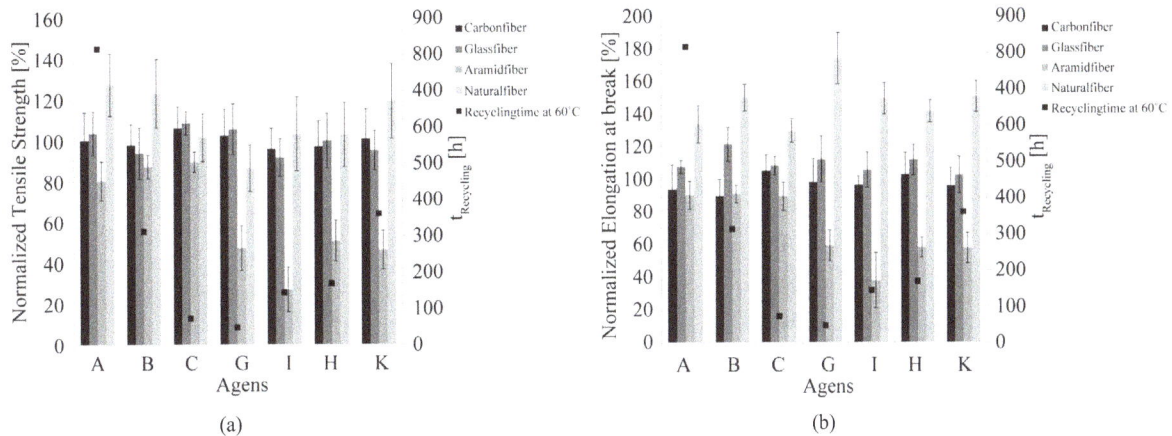

Figure 13. Normalized tensile strengths (a) and elongations at break (b) of reinforcement fibers after treatment in seven different recycling agents at 60°C (Initial values = 100%). The recycling times (represented by ■) are identical with the times needed to decompose the polycyanurate matrix (in this example cured B10).

The natural-fibers are negatively influenced only in case of agent A (approx. 90% remaining tensile strength). It is remarkable, that for agents A and B the tensile strength increases to values > 120%, which indicates a chemical modification by the recycling agent.

The values for the normalized elongation at break are affected relatively little for carbon- and glass-fibers (**Figure 13(b)**). In case of the glass-fibers in principle a slight increase of the elongation at break can be observed; nevertheless the low absolute values of glass must be taken into account (a elongation from 1.5% up to 1.8% is not necessarily significant). Remarkable is the increase of the natural fiber values (between approx. 130% to approx. 170%). An increase of the elongation at break values for all agents with only one exception (agent G) was observed, meaning, that the modification of natural fibers causes (with exception of G) an improvement of the mechanical properties tensile strength and elongation at break. Also remarkable is the increase of the elongation at break for the natural fiber in agent G to approx. 170% by only a slight decrease of the tensile strength down to approx. 80% of the initial value.

The normalized elongations at break of the aramid fibers after the recycling process are directly correlated to the tensile strength values. To be more precise: The agents A, B, and C reduce the mechanical values only marginally, thus affecting the aramid not significantly (reduction of approx. 10%), whereas the other agents yield a reduction of approx. 50%, for agent I of even more than 60%.

The outcome of more detailed examinations regarding the influence of agents A, G and H for 1 h at 80°C, 100°C, and at the chemicals' boiling points on the fibers is displayed in **Figure 14**.

The carbon-fibers are essentially affected by neither agent nor temperature. There is no significant change in the mechanical values, besides a slight increase of the tensile strength under the regime at the boiling point of agent H. The glass-fibers, too, are relatively unaffected; a trend to a slight decrease (less than 10%) of the tensile strength with agent A and a slight increase of the tensile strength with agents G and H is to be visible. The values of elongation at break seem to increase slightly (10% - 20%) compared to the initial point.

The effect on aramid-fibers is strongly depending on the particular agent. Agent A reduces the mechanical values only marginally, whereas G and H reduce both, tensile strength and elongation at break, especially at the boiling point extremely.

Very interesting are the effects on the natural-fibers: Here we get for almost all scenarios an increase of the tensile strength up to approx. 25%; furthermore the elongation at break shows for A and H only a low increase, but for agent G the elongation at break is 150% - 200% *higher* than the initial values.

Because only the aramid-fibers were affected negatively under the above mentioned conditions, further experiments were carried out: A comparative graph of carbon-, glass-, and natural-fibers treated with agents A, G and H for 24 hours at boiling point is given in **Figure 15**.

One can see that the different agents have only a minor effect on the carbon- and glass-fibers (approx. ±10% to the initial values for agent A). The influence on natural-fibers is strong: A tensile strength reduction of approx. 10% (agent H) resp. approx. 30% (agent A) can be found, but most interesting is the increase of +310% for the

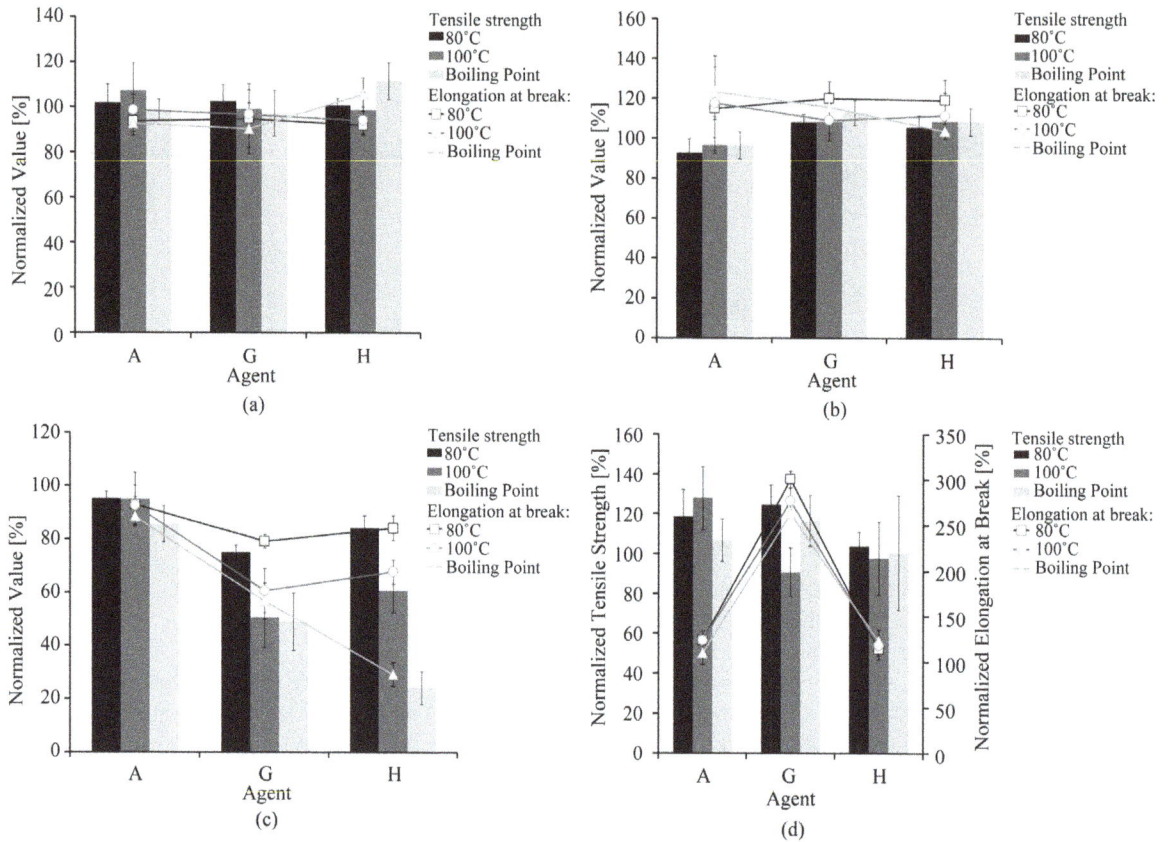

Figure 14. Influence of the agents A, G and H on the four different fiber-species. Treatment for 1 h at 80°C, 100°C, and the particular boiling point. The columns correspond to the tensile strength values and the points to the elongation at break values after treatment; initial values are normalized to 100%.

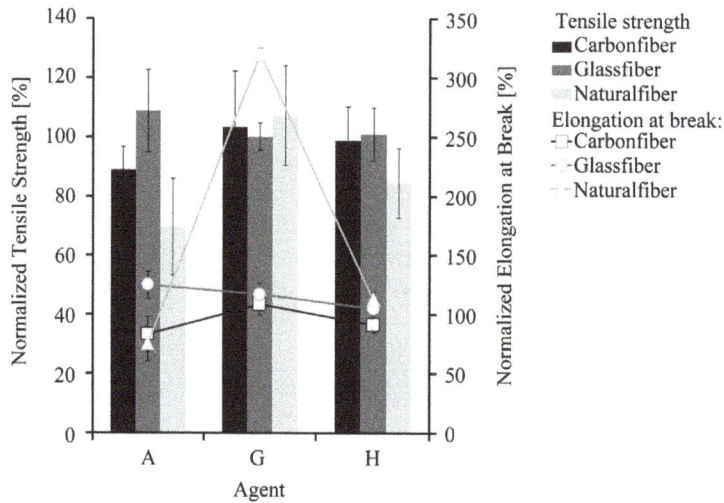

Figure 15. Influence of the agents A, G and H on carbon-, glass-, and natural-fiber. Treatment for 24 h is at boiling point. The columns correspond to the tensile strength values and the points to the elongation at break values after treatment; initial values are normalized to 100%.

elongation at break while maintaining the tensile strength.

Concluding, it can be said about the fiber-treatment with the recycling agents, that the influence on carbon-

and glass-fibers is relatively low. Aramid-fibers are, to some extent very labile against the chemicals, however combinations of recycling agents and decomposition regimes can be found under which the polycyanurate matrix can be dissolved without affecting the aramid-fibers.

4.3. New Polyurethane Thermosets from Recycled Polycyanurates

The new thermosets described in Section 3.4 exhibit glass transition temperatures of up to 128°C, by using 14.3% of Recyclate with P-MDI as Isocyanate component (**Table 3**).

The HDI-containing polymers exhibit lower T_gs. For the HDI-containing foam a T_g of 75°C (13.2 wt-% of recyclate) was found. The DMA-curves for formulations P1-P3 can be found in **Figure 16**.

Massive PUR-materials were obtained with the HDI-Trimer (P4 - P6), having T_gs ranging from 61°C to 80°C. A more *green* version of polyurethane is formulation P3, containing besides 22.4 wt-% of recyclate additional 19.1 wt-% of castor oil as renewable component, making the resulting foam 41.5 wt-% *green*. **Figure 17** shows an exemplary foam structure of formulation P1.

Massive samples were obtained by formulating water-free mixtures. An increase of T_g from 66°C to 80°C was obtained (**Table 3**, **Figure 18**) by rising the amount of recyclate from 10 wt-% to 20 wt-%, probably as a result of the high content of aromatic structures in the recyclate. Adding Glycerine as a crosslinker to the formulation with 20 wt-% of recyclate a decrease of the T_g from 80°C to only 61°C was found.

Epoxide [37] and acrylate based polymers were also synthesized using the recycled polycyanurates, details will be reported in the future.

Actual research regarding additional characterization of the newly developed resins, as well as on the structure elucidation of the polycyanuratere cyclates is in progress.

5. Conclusions and Outlook

The authors investigate the recyclability of polycyanurate homo- and copolymers with different recycling

Table 3. Glass transition temperatures of new polyurethane formulations.

Formulation	KOH-value [g·kg^{-1} KOH]	Chain extender	Isocyanate	Cross-linking agent	Blowing agent	$\omega_{Recyclate}$ [wt-%]	$T_{g, \tan \delta}$ [°C]
P1	486*	1,4-Butandiol	HDI-Biuret	Glycerin	Water	13.2	75
P2	486*	1,4-Butandiol	P-MDI	Glycerin	Water	14.3	128
P3	486*	Castor oil	HDI-Biuret	Glycerin	Water	22.4	94
P4	653**	1,4-Butandiol	HDI-Trimer	-	-	10.0	66
P5	653**	1,4-Butandiol	HDI-Trimer	-	-	20.0	80
P6	653**	1,4-Butandiol	HDI-Trimer	Glycerin	-	20.0	61

*Estimated hydroxyl-value; **Hydroxyl values determined in accordance with DIN 53240.

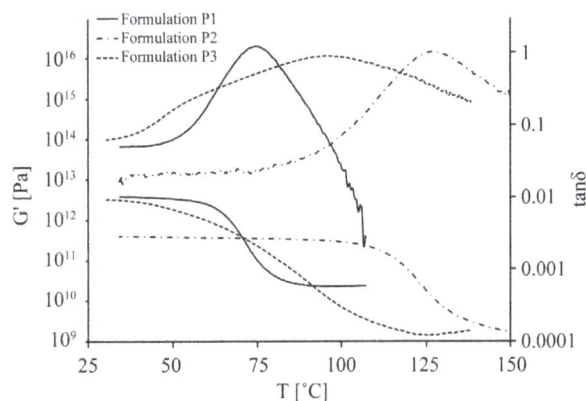

Figure 16. DMA measurements of the new PUR-foam formulations P1-P3.

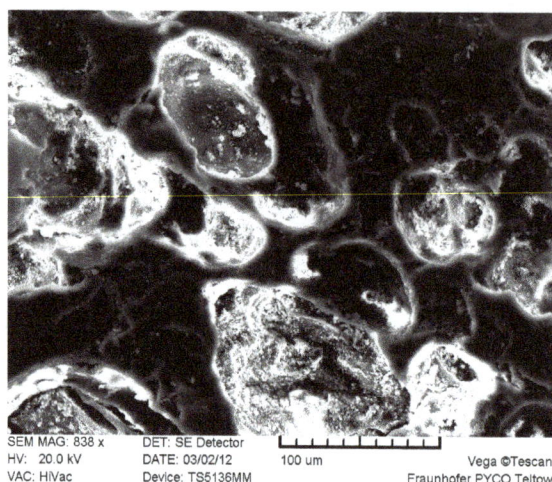

Figure 17. SEM picture of formulation P1, showing the foam structure (magnification 838×).

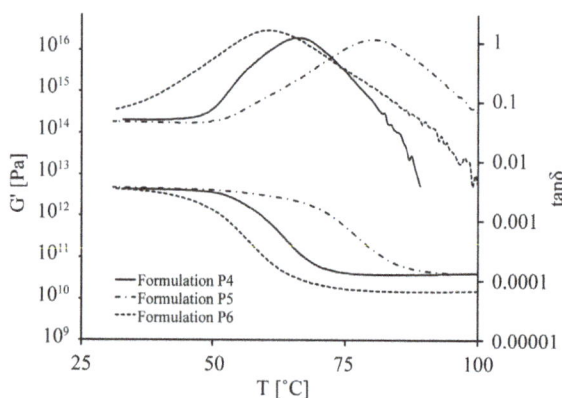

Figure 18. DMA measurements of the new massive PUR formulations P4-P6.

agents under different conditions. Also the influence of the recycling process on the most important reinforcement fibers, *i.e.* carbon-, glass-, aramid-, and natural-fiber is investigated. The authors find that: the recycling speed is not only dependent on the temperature, but is also significantly influenced by the particular recycling agents and the polycyanurate formulation. Hence, the stability against the recycling media can be adjusted over a broad range by *adjusting* the polymer composition. Furthermore, the authors find that the inorganic reinforcement fibers (carbon and glass) are almost unaffected by neither recycling agent at either temperature. Aramid-fibers degrade, depending on the particular recycling agent, from slightly up to extremely strong. This leaves one with the possibility to find a combination of matrix resin and recycling agent, which does not affect the aramid-fiber significantly. In the case of natural fibers, the dependence on the particular recycling media is very strong: some media do not affect the fiber significantly; others reduce the mechanical properties (tensile strength and elongation at break) significantly, and still others even improve both mechanical properties strongly. From the Recyclate, the authors synthesized and characterized a number of new polyurethane thermosets (foamed and solid samples) with different contents of recyclate, exhibiting T_g in the range of 60°C to 128°C.

Experiments on the synthesis of epoxide resins, acrylate resins and other polymer classes from the recycled polycyanuratesand experiments on the repair of polycyanurate laminates are in progress and will be reported elsewhere.

Acknowledgements

We thank Mrs. Katarina Padaszus, Mrs. Katharina Becker, Mr. Felix Herrmann, Mr. Tobias Hoffmann and Mr.

Hadrien Schmitt, all Fraunhofer PYCO, for measurements and assistance in the lab. We also thank Dr. Juergen Schneider, Fraunhofer PYCO, and Dagmar Jones for the active support.

References

[1]   Armstrong, K.B., Bevan, L.G. and Cole, II, W.F. (2005) Care and Repair of Advanced Composites. 2nd Edition, SAE International, Warrendale.

[2]   Witten, E. (2014) Handbuch Faserverbundkunststoffe, Composites: Grundlagen, Verarbeitung, Anwendungen/AVK—Industrievereinigung Verstärkte Kunststoffe e.V. 4th Edition, Springer Vieweg, Wiesbaden. http://dx.doi.org/10.1007/978-3-658-02755-1

[3]   Dreyer, C., Söthje, D., Klauke, K. and Bauer, M. (2013) Chemisches Recycling von Composites. *MaschinenMarkt: MM Compositesworld*, **118**, 26-28.

[4]   Drechsler, K. and Middendorf, P. (2012) New Composite Technologies for Automotive Applications. *3rd International IQPC-Congress*, Wiesbaden, 3-5 December 2012, 45-56.

[5]   Meilhan, N. (2013) Strategies of Global OEMs to Reduce Future Car Weight. *Proceedings of the Transportation Weight Loss Diet Conference*, Stuttgart, 5-6 June 2013, 40.

[6]   Marsh, G. (2013) Composites Poised to Transform Airline Economics. *Reinforced Plastics*, **57**, 18-24. http://dx.doi.org/10.1016/S0034-3617(13)70088-2

[7]   Martin, D. and Bacaloglu, R. (1980) Organische Synthesen Mit Cyansäureestern. 1st Edition, Akademie-Verlag, Berlin.

[8]   Martin, D. (1964) Phenyl Cyanate. *Angewandte Chemie International Edition*, **3**, 311. http://dx.doi.org/10.1002/anie.196403112

[9]   Jensen, K.A. and Holm, A. (1964) Formation of Monomeric Alkyl Cyanates by the Decomposition of 5-Alkoxy-1,2,3,4-Thiatriazoles. *Acta Chemica Scandinavica*, **18**, 826-828. http://dx.doi.org/10.3891/acta.chem.scand.18-0826

[10] Grigat, E. and Pütter, R. (1964) Chemie der Cyansäureester, I. Cyansäureester aus Hydroxylverbindungen und Halogencyan. *Chemische Berichte*, **97**, 3012-3017. http://dx.doi.org/10.1002/cber.19640971107

[11] Hamerton, I. (1994) Chemistry and Technology of Cyanate Ester Resins. 1st Edition, Chapman & Hall, Glasgow. http://dx.doi.org/10.1007/978-94-011-1326-7

[12] Productflyer (2009) LONZA Primaset Cyanate Esters: The Next Generation of High Performance Thermosets. Switzerland. http://www.lonza.com/products-services/materials-science/high-performance-materials/primaset-cyanate-esters.aspx

[13] Bauer, M., Decker, D., Richter, F. and Gwiazda, M. (2010) Hybrid Polymers Made of Cyanates and Silazanes, Method for the Production and Use Thereof. Patent EP Patent 2408846 B1.

[14] Dreyer, C., Schneider, J., Göcks, K., Beuster, B., Bauer, M., Keil, N., Yao, H., Zawadzki, C. and Radmer, O. (2002) Polymere für Anwendungen in der integrierten Optik—Ein Überblick mit ausgewählten Beispielen. *e & i Elektrotechnik und Informationstechnik*, **120**, 178-185.

[15] Dreyer, C.J., Bauer, M., Bauer, J., Keil, N., Yao, H.H. and Zawadzki, C. (2001) Polycyanurate Ester Resins with Low Loss and Low Birefringence for Use in Integrated Optics. *Linear and Nonlinear Optics of Organic Materials*, **4461**, 188-199. http://dx.doi.org/10.1117/12.449847

[16] Dreyer, C., Bauer, M., Bauer, J., Keil, N., Yao, H.H. and Zawadzki, C. (2002) Reduction of the Optical Loss and Optimization of Polycyanurate Thermosets Used in Integrated Optics. *Microsystem Technologies*, **7**, 229-238, http://dx.doi.org/10.1007/s005420100102

[17] Söthje, D., Dreyer, C., Bauer, M., Schulze, C., Riedel, U. and Eppinger, A. (2013) Mikrowellenhärtbareduromere Matrixharze für Automobilanwendungen: Entwicklung Reparatur- und Recyclingfähiger Leichtbaumaterialien. *Konstruktion*, **65**, IW 5-7.

[18] ACARE (2013) Roadmap for Cross-Modal Transport Infrastructure Innovation. http://www.acare4europe.org/sites/acare4europe.org/files/document/ETP%20Roadmap.pdf

[19] Directive 2000/53/EC of the European parliament and of the Council (2000) End-of Life Vehicles.

[20] Xiao, X., Hoa, S.V. and Street, K.N. (1994) Repair of Thermoplastic Resin Composites by Fusion Bonding. In: Damico, D.J., Wilkinson, Jr., T.L. and Niks, L.F.S., Eds., *Composites Bonding*, **1227**, 30-44.

[21] Davies, P., Cantwell, W.J., Jar, P.Y., Bourban, P.E., Zysman, V. and Kausch, H.H. (1991) Joining and Repair of Carbon Fiber-Reinforced Thermoplastics. *Composites*, **22**, 425-431. http://dx.doi.org/10.1016/0010-4361(91)90199-Q

[22] Varadan, V.K. and Varadan, V.V. (1991) Microwave Joining and Repair of Composite Materials. *Polymer Engineer-*

ing & Science, **31**, 470-486. http://dx.doi.org/10.1002/pen.760310703

[23] Dang, W., Kubouchi, M., Sembokuya, H. and Tsuda, K. (2005) Chemical Recycling of Glass Fiber Reinforced Epoxy Resin Cured with Amine Using Nitric Acid. *Polymer*, **46**, 1905-1912. http://dx.doi.org/10.1016/j.polymer.2004.12.035

[24] Liang, B., Pastine, S.J. and Qin, B. (2012) Novel Agents for Reworkable Epoxy Resins. WO Patent 2012/071896 A1.

[25] Williams, P., *et al.* (2003) Recycling of Automotive Composites—The Pyrolysis Process and its Advantages. *Materials World*, **11**, 24-26.

[26] Holl, R. (1982) Process to Increase the Yield of Pyrolysis Oil and to Shorten the Duration of the Pyrolysis. Patent GB 2129009 A.

[27] Giehr, A., Höver, H., Keim, K., Joachim K. and Neuwirth, O. (1984) Verfahren zur Aufarbeitung von Kohlenstoff enthaltenden Abfällen. Patent DE 3442506 C2.

[28] Tippmer, K. (1994) Verfahren zur thermischen Spaltung von Abfall-Kunststoff-Gemischen. Patent DE 4412941 A1.

[29] Bräutigam, K.R., Kupsch, C., Reßler, B. and Sardemann, G. (2003) Analyse der Umweltauswirkungen bei der Herstellung, dem Einsatz und der Entsorgung von CFK-bzw. Aluminiumrumpfkomponenten. Research Report FZKA 6879, Forschungszentrum Karlsruhe, Karlsruhe.

[30] Braun, D., Gentzkow, W. and Rudolf, A. (1998) Recycling von Duroplastwerkstoffen. Patent DE 19839083.

[31] Zia, K.M., Bhatti, N.H. and Bhatti, I.A. (2007) Methods for Polyurethane and Polyurethane Composites, Recycling and Recovery: A Review. *Reactive and Functional Polymers*, **67**, 675-692. http://dx.doi.org/10.1016/j.reactfunctpolym.2007.05.004

[32] Kaeufer, H. and Bollin, H. S. (1996) Verfahren für das Recycling von Epoxidharz enthaltenden Erzeugnissen. Patent WO001996016112A1.

[33] (2013) http://www.connoratech.com/technology/recyclable-epoxy-data/

[34] El Gersifi, K. Durand, G. and Tersac, G. (2006) Solvolysis of Bisphenol A Diglycidyl Ether/Anhydride Model Networks. *Polymer Degradation and Stability*, **91**, 690-702. http://dx.doi.org/10.1016/j.polymdegradstab.2005.05.021

[35] Bauer, J., Bauer, M. and Göcks, K. (1997) Method of Decomposing Polycyanurate-Containing Materials. US Patent No: 000005691388A.

[36] Dreyer, C. Söthje, D. and Bauer, M. (2013) Progress in Recyclable Cyanate Resins. *Thermosets* 2013 *from Monomers to Components Proceedings of the 3rd International Conference on Thermosets*, Berlin, 18-20 September 2013, 159-162.

[37] Söthje, D. Dreyer, C. and Bauer, M. (2013) Advanced Possibilities in Thermoset Recycling. *Thermosets* 2013 *from Monomers to Components Proceedings of the 3rd International Conference on Thermosets*, Berlin, 18-20 September 2013, 219-223.

[38] (2013) Determination of Hydroxyl Value. German Industry Standard, DIN 53240.

16

High Performance Recycling of Polymers by Means of Their Fluorescence Lifetimes*

Heinz Langhals#, Dominik Zgela, Thorben Schlücker

Department of Chemistry, LMU University of Munich, Munich, Germany
Email: #Langhals@lrz.uni-muenchen.de

Abstract

Technical polymers could be identified by means of their remarkably strong auto fluorescence. The time constants of this fluorescence proved to be characteristic for the individual polymers and can be economically determined by integrating procedures. The thus obtained unequivocal identification is presented for their sorting for recycling. Furthermore, polymeric materials were doped with fluorescent dyes allowing a fine-classification of special batches.

Keywords

Recycling, Polymers, Fluorescence Lifetime, Labelling, Fluorescence Spectroscopy

1. Introduction

The recycling of organic polymers obtains an increasing interest both in research and technology. There is a necessity for the development of efficient processes because of increasing environmental pollution by polymers ("plastic planet"). Moreover, their recycling may open an economic source for organic materials. The majority of technical polymers are thermoplasts and melt and moulding again is attractive for their easy re-use. However, the immiscibility and incompatibility of organic polymers are therefore the main obstacles because lacking uniformity as low as 5% lowers the value of polymers appreciably and an even higher uniformity is required for high performance materials. Pure polymers for recycling may be collected in polymer-processing manufactories, however, the majority of collected material forms mixtures where an efficient sorting is required before processing. The machine-based recognition of polymers is a prerequisite for such processes where methods using the density or electrostatic properties were described [1]-[3]. Optical methods are more attractive because of simple, stable and efficient technology where fluorescence is advantageous [4]-[8] because of unproblematic

---

*Dedicated to Prof. Ch. Rüchardt on the Occasion of His 85th Birthday.
#Corresponding author.

light path and detection. The doping of polymers with fluorescent markers [9] and their re-identification by the spectral resolution of their fluorescence in combination with a binary coding was described in preceding papers [10] [11]. This demonstrated the efficiency of the application of fluorescence. However, there are two topics for a fundamental improvement: 1) Only doped material can be recycled where the recycling has to be already targeted in the production of final products; undefined wastes cannot be recycled in this way; 2) the spectral resolution for every flake for recycling costs appreciable efforts for detection and signal processing. Optical processes for the sorting of undoped material would bring about an appreciable progress and would even allow working up deposited material.

2. Experimental

2.1. Spectroscopy

UV/Vis absorption spectra: Varian Cary 5000; fluorescence spectra: Varian Cary Eclipse; fluorescence lifetimes: Edinburgh Analytical Instruments CD900, nF900.

2.2. Materials

The technical polymers Luran® (styrene, polyacrylonitile copolymer from BASF), Delrin® (polyoxymethylene from DuPont) and Ultramid® (polyamide with glass fibre from BASF) were applied without further treatment. The fluorescence labels **1** (PTIE) [15], **2** (S-13) [16], and **3** (S-13TBI) [17] [18] were prepared according to the literature. Spectroscopic grade solvents were applied.

3. Results and Discussion

3.1. Auto Fluorescence of Polymers

The identification of polymers was concentrated to the technical high performance products Luran®, Delrin® and Ultramid®. We found an appreciable strong auto fluorescence of these technical materials with standard optical exciting at 365 nm where mercury lamps may be applied as a light source; see **Figure 1**. Slight variations of the wavelengths of excitation do not alter the spectra.

The investigated polymers exhibit individual shapes of their auto fluorescence spectra; see **Figure 1**. We preferred a fluorescence excitation at 365 nm where intense light sources are available. A slight variation of the wavelength of excitation does not influence the fluorescence. The spectra may be used for the identification and sorting of polymers by means of methods of pattern search. Thus, even undoped material can be sorted, however, this requires still an appreciable effort of calculation.

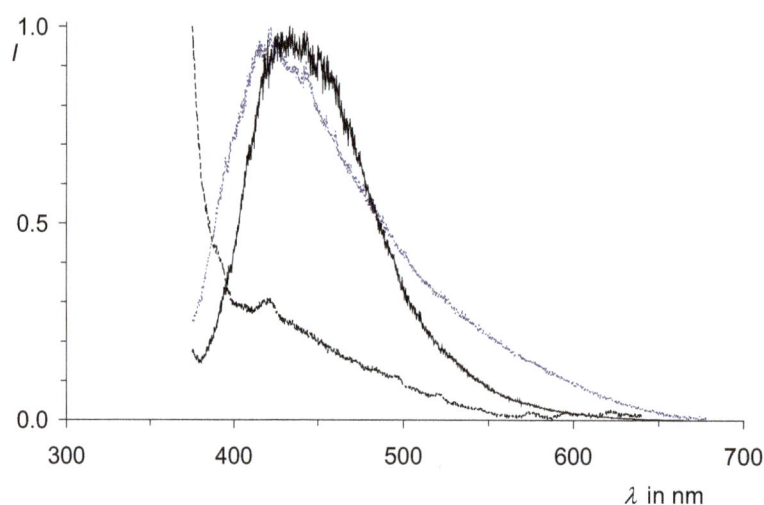

Figure 1. Fluorescence spectra of the auto fluorescence of Luran® (blue, dotted curve), Delrin® (red, dashed curve) and Ultramid® (black, solid curve) with optical excitation at 365 nm.

As an alternative, we investigated the fluorescence lifetimes of the auto fluorescence and found remarkable differences for various polymers; see **Table 1**, lines 1 to 3. Fluorescence decay essentially proceeds first order in time with the time constant τ. Minor, less important bi-exponential components (τ_{bi}) could be detected, however, a mono-exponential interpretation is sufficient for the identification by far. The decay curves can be easily splitted into two branches representing each of the single components of the fluorescence lifetime. Factors about two are between the decay times τ for Delrin®, Ultramid® and Luran® allowing an unambiguous identification of the polymers.

The decay curves of the auto fluorescence of the polymers are reported in **Figures 2(a)-(c)** and are clearly indicating their pronounced differences. These can be even more easily seen in the fitted function in **Figure 2(d)**. A simple logarithmic representation of the right branch of the decay curves is by far sufficient for the determination of the differences in lifetimes; right scales in **Figure 2**.

3.2. Fluorescent Labels

An additional labelling of the polymers by means of fluorescent dyes was taken into account for not only identifying the basic polymeric material but also special technological batches. We applied the perylene ester **1** (PTIE), the perylene carboxylic bisimide **2** (S-13) and the terrylene carboxylic bisimide **3** (S-13TBI) because of their light fastness and high fluorescence quantum yields. The fluorescence of these dyes proceeds in different spectral regions forming three channels for detection as can be seen from their fluorescence spectra in **Figure 3**. The spectra in various polymeric materials differ only slightly from the spectra in solution because solvatochromism of the dyes is weak. As a consequence, the three channels of fluorescence can be taken to be invariant with respect to the tested material. The labelling of polymers can proceed with a binary coding where the first or the second dye or both were applied and so on resulting in $2^n - 1$ possibilities for labelling with n as the number of applied fluorescent dyes: Thus, seven individual batches may be labelled for each polymeric material with the application on dyes **1** to **3**. The fluorescence spectra may be applied for the identification of the labelling with the individual dyes. The formation of the second derivative of the spectra improves [10] [11] the security of detection.

Furthermore, we found that the time constants for fluorescence decay vary both with the applied dye and the applied polymer; see **Table 1**. Such combinations can be taken as an additional pattern for the recognition of the

Table 1. Fluorescence lifetimes of genuine polymers, the fluorescence labels in chloroform solution and doped polymers.

Sample	τ in ns[a]	τ in ns[b]	λ ex[c]	λ em[d]
Delrin	0.74	5.78	365	573
Luran	3.53	8.42	365	573
Ultramid	1.96	7.83	365	573
PTIE (**1**) in CHCl₃	3.53		442	485
S-13 (**2**) in CHCl₃	4.06		490	573
S-13TBI **3** in CHCl₃	3.66		598	667
Delrin with PTIE (**1**)	3.90		442	485
Delrin with PTIE (**1**)	3.92		490	573
Delrin with S-13 (**2**)	3.74		490	573
Delrin with S-13TBI (**3**)	3.31		598	667
Delrin with S-13TBI (**3**)	3.34		490	667
Luran with PTIE (**1**)	4.08		490	573
Luran with S-13 (**2**)	4.56		490	573
Luran with S-13TBI (**3**)	3.53		598	667
Luran with S-13TBI (**3**)	3.96		490	667
Ultramid with PTIE (**1**)	2.44	7.53	442	485
Ultramid with PTIE (**1**)	2.26	5.75	490	573
Ultramid with S-13 (**2**)	1.83	5.37	490	573
Ultramid with S-13TBI (**3**)	2.34	6.98	598	667

[a]Fluorescence lifetime; [b]Additional biexponential component; [c]Wavelength of excitation in nm; [d]Wavelength of detection in nm.

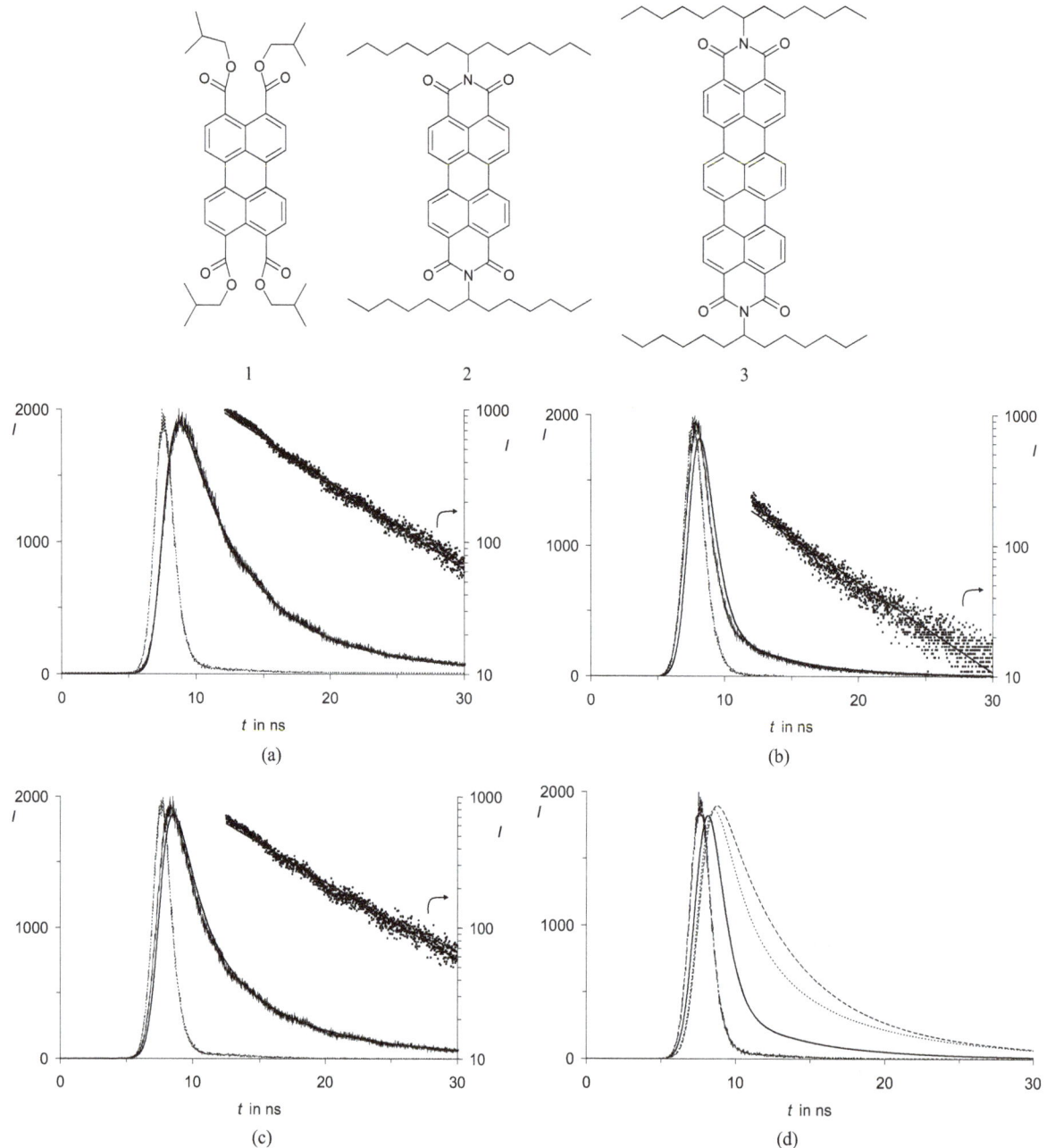

Figure 2. Fluorescence decay of polymers at 573 nm in linear (left) and logarithmic scales (right) and the characteristic of the light pulse of excitation at 365 nm as dotted lines. Mono-exponentially fitted functions of decay as solid lines. (a) Fluorescence decay of Luran®; (b) fluorescence decay of Delrin®; (c) fluorescence decay of Ultramid®; (d) comparison of the fitted functions for Delrin®: Solid line, Ultramid®: Dotted line and Luran®: Dashed line.

entire batch of a polymer for further improvement for the identification of polymers. Moreover, the determination of decay times needs no calibration concerning the fluorescence intensities (such a calibration may by applied with the auto fluorescence of polymers as internal standards), because the exponential decay remains similar independent from the starting intensity and some dead time before acquisition; this may be advantageous, even for very inhomogeneous flakes for recycling concerning size and shape.

We tested the reproducibility of the determined time constant of fluorescence decay and found standard deviation only for the second decimal; see **Table 2** for examples. As a consequence, the reproducibility is good enough by far for the unequivocally discrimination between the individual samples; on the other hand, even an absolute determination of the time constant is not necessary as long as the complete setup produces sufficiently reproductive values.

3.3. Time-Resolved Detection

The first order exponential decay curves need not be completely registered and fitted because there are well established mathematical procedures [12]-[14] for the determination of the time constant by means of the measurements of two points of the decay curve or even more appropriate by two integrated regions, preferment before and behind the half time ($t_{1/2}$). This is schematically shown with two Gaussian-shaped samplings in **Figure 4** where the integrating measurements improve the signal to noise ratio. One up to two ns time for integration time seem to be appropriate concerning a decay time of about 5 ns for the majority of fluorescent structures. The fluorescence is induced by the periodically pulsed light for excitation where one can expect a sufficiently complete fluorescence decay of 70 ns for the case of an unfavourable lifetime of 10 ns. As a consequence, an unproblematic frequency of about 15 MHz results for repetition. The two regions of integration may be selected by means of two phase sensitive detectors (PSD) and a phase shift between the two analyzing signals for sampling. These need not be applied for each pulse, but may be distributed, for example, between two consecutive pulses. A further improvement of the signal to noise ratio may be obtained by the accumulation of the signals of detection.

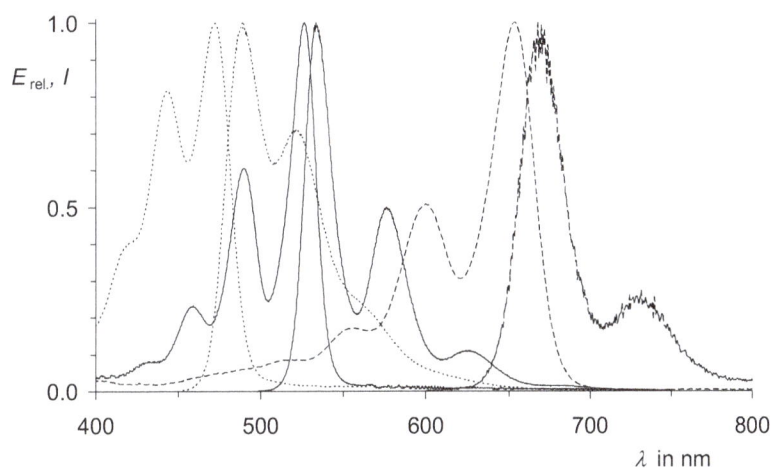

Figure 3. Absorption (left, scale E) and fluorescence spectra (right, scale I) of **1** (dotted lines), **2** (solid lines) and **3** (dashed lines) in chloroform.

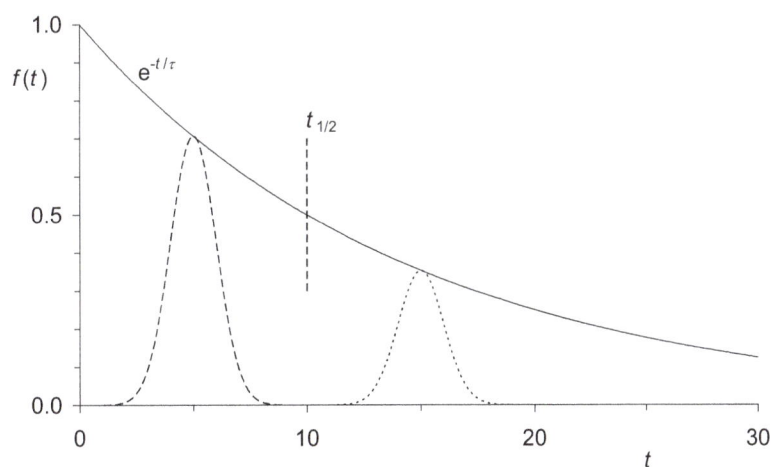

Figure 4. Schematic first order decay (solid line) with a time constant of $\tau = 14.4$ ns corresponding to a half life $t_{1/2}$ of 10 ns. Gaussian-shaped samplings at $t = 5$ ns (dashed curve) and at $t = 15$ ns (dotted curve).

Table 2. Test of reproducibility of the time constant of fluorescence decay including the applied method; measurements with individually prepared and re-oriented samples of labelled granulates.

Sample	τ in ns[a]	τ m, s[b]	λ ex[c]	λ em[d]
Delrin-PTIE (**1**)	3.92		442	485
Delrin-PTIE (**1**)	3.91		442	485
Delrin-PTIE (**1**)	3.95		442	485
Delrin-PTIE (**1**)	3.91		442	485
Delrin-PTIE (**1**)	3.90	3.92, 0.02	442	485
Delrin-S-13 (**2**)	3.78		490	573
Delrin-S-13 (**2**)	3.77		490	573
Delrin-S-13 (**2**)	3.79		490	573
Delrin-S-13 (**2**)	3.79		490	573
Delrin-S-13 (**2**)	3.73		490	573
Delrin-S-13 (**2**)	3.74	3.77, 0.03	490	573
Delrin-S-13TBI (**3**)	3.35		598	667
Delrin-S-13TBI (**3**)	3.45		598	667
Delrin-S-13TBI (**3**)	3.31	3.37, 0.07	598	667
Luran-S-13 (**2**)	4.55		490	573
Luran-S-13 (**2**)	4.56		490	573
Luran-S-13 (**2**)	4.43	4.51, 0.07	490	573
Ultramid-PTIE (**1**)	2.43		442	485
Ultramid-PTIE (**1**)	2.44	2.44	442	485

[a]Time constant of fluorescence decay; [b]Mean value, standard deviation s; [c]Wavelengths of excitation in nm; [d]Wavelengths of detection in nm.

One can roughly calculate the upper limit of the detection for sorting with industrial flakes for recycling of dimension of max. 10 mm. Minimal 20 mm space between the individual flakes seem to be realistic and a transport of maximal 500 m/s where 200 pulses for excitation at 15 MHz of repetition should be obtainable for a single flake; this should be more than sufficient for a good signal to noise ratio for an unequivocal sorting. An average mass of about 25 mg was found for standard industrial recycling flakes resulting in a sorting capacity of 1.5 tons of material per hour. This has to be taken as an upper technological limit for a permanent sorting of polymers covered by the described method. The bottleneck for such capacities seems to be more the mechanics than the methodology of detection. Both electronics and mechanics become much more simple for lower demand.

4. Conclusion

The unequivocal identification of technical polymers by means of their time constants of auto fluorescence decay is a promising method for their sorting for recycling. Time constants can be economically determined by phase-shifted integration of the fluorescence response of pulsed optical excitation. The auto fluorescence of polymers can be applied for the identification of the basic material where a doping with fluorescent dyes allows the further fine-classification of special batches. A binary coding of the doping with n fluorescent dyes results in $2^n - 1$ possibilities for the labelling of batches.

Acknowledgements

This work was supported by The BMBF, CIPSM cluster in Munich, and the Fonds der Chemischen Industrie. We thank Dr. Moritz Ehrl for technical assistance.

References

[1]   Nemeth, E., Schubert, G., Albrecht, V. and Simon, F. (2005) Procedure for Sorted, Electrostatic Separation of Mixed Plastics Wastes. *Aufbereitungs-Technik*, **46**, 35-46.

[2]   Nemeth, E., Simon, F., Albrecht, V. and Schubert, G. (2004) Procedure for Sorted, Electrostatic Separation of Mixed Plastics Wastes. Ger. Patent No. 102004024754 B3; *Chemical Abstracts*, **144**, 392348.

[3]   Gohs, U., Albrecht, V., Husemann, K., Reinsch, E., Schuenemann, R. and Simon, F. (2009) Procedure for the Electrostatic Separation of Mixed Plastics Waste. Ger. Offen. No. DE 102007055765 A1; *Chemical Abstracts*, **151**, 57663.

[4]   Corbet, E.C. Frey, J.G., Groce, R.I. and Hendra, P.J. (1994) An Investigation into the Applicability of Luminescent Tagging to Polymer Recovery. *Plastics, Rubber and Composite Processing and Application*, **21**, 5-11.

[5]   Rafi Ahmad, S. (2000) Marking of Products with Fluorescence Tracers in Binary Combinations for Automatic Identification and Sorting. *Assembly Automatation*, **20**, 58-65. http://dx.doi.org/10.1108/01445150010311617

[6]   Alam, M.K., Stanton, S.L. and Hebner, G.A. (1994) Near Infrard Spectroscopy and Neural Network for Resin Identification. *Spectroscopy*, **9**, 31-39.

[7]   Scott, D.M. (1995) A Two-Colour near Infra-Red Sensor for Sorting Recycled Plastic Waste. *Measurement Science and Technology*, **6**, 156-159. http://dx.doi.org/10.1088/0957-0233/6/2/004

[8]   General Electric Company; Hubbard, S., Potyrailo, R., Schottland, P. and Thomas, V. (2006) Tagging Materials for Polymers, Methods, and Articles Made Thereby. US Patent No. 2005/0095715 (Oct. 31, 2003); *Chem. Abstr.* 2005, **142**, 412287.

[9]   General Electric (inv. Hubbard, S.F., Potyrailo, R.A., Schottland, P. and Thomas, V.) (2003) Tagging Materials for Polymers, Methods, and Articles Made Thereby. PCT Int. Appl. (2003), WO 2003006965 A1 20030123; *Chemical Abstracts*, **138**, 107667.

[10]   Langhals, H., Schmid, T., Herman, M., Zwiener, M. and Hofer, A. (2013) Binary Fluorescence Labeling for the Recovery of Polymeric Materials for Recycling. *International Journal of Environmental Engineering*, **7**, 124-132.

[11]   Langhals, H., Schmid, T., Herman, M., Zwiener, M. and Hofer, A. (2012) Marking of Polymer Materials with Fluorescence Dyes for Their Clear Automatic Sorting. Ger. Offen. No. DE 102012012772.3; *Chemical Abstracts*, **160**, 63983.

[12]   Ballew, R.M. and Demas, J.N. (1989) An Error Analysis of the Rapid Lifetime Determination Method for the Evaluation of Single Exponential Decays. *Analytical Chemistry*, **61**, 30-33. http://dx.doi.org/10.1021/ac00176a007

[13]   Woods, R.J., Scypinski, S., Cline Love, L.J. and Ashworth, H.A. (1984) Transient Digitizer for the Determination of Microsecond Luminescence Lifetimes. *Analytical Chemistry*, **56**, 1395-1400. http://dx.doi.org/10.1021/ac00272a043

[14]   Meier, R.J., Fischer, L.H., Wolfbeis, O.S. and Schäferling, M. (2013) Referenced Luminescent Sensing and Imaging with Digital Color Cameras: A Comparative Study. *Sensors and Actuators B*, **177**, 500-506. http://dx.doi.org/10.1016/j.snb.2012.11.041

[15]   Alibert-Fouet, S., Seguy, I., Bobo, J.-F., Destruel, P. and Bock, H. (2007) Liquid-Crystalline and Electron-Deficient Coronene Oligocarboxylic Esters and Imides by Twofold Benzogenic Diels-Alder Reactions on Perylenes. *European Journal of Chemistry*, **13**, 1746-1753. http://dx.doi.org/10.1002/chem.200601416

[16]   Demmig, S. and Langhals, H. (1988) Readily Soluble Lightfast Perylene Dyes. *Chemische Berichte*, **121**, 225-230. http://dx.doi.org/10.1002/cber.19881210205

[17]   Langhals, H. and Poxleitner, S. (2008) Core-Extended Terrylene Bisimides—Novel, Strongly Red Fluorescent Broadband Absorbers. *European Journal of Organic Chemistry*, **2008**, 797-800. http://dx.doi.org/10.1002/ejoc.200701058

[18]   Langhals, H., Walter, A., Rosenbaum, E. and Johansson, L.B.-Å. (2011) A Versatile Standard for Bathochromic Fluorescence Based on Intramolecular FRET. *Physical Chemistry Chemical Physics*, **13**, 11055-11059. http://dx.doi.org/10.1039/c1cp20467j

17

Estimation of Waste Generation and Recycling Potential from Traditional Market: A Case Study in Hue City, Vietnam

Yasuhiro Matsui[1], Do Thi Thu Trang[1], Nguyen Phuc Thanh[2]

[1]Graduate School of Environmental and Life Science, Okayama University, Okayama, Japan
[2]Hitachi Zosen Corp, Osaka, Japan
Email: matsui@cc.okayama-u.ac.jp, dothithutrang209@gmail.com, nguyen_p@hitachizosen.co.jp

Abstract

This study was conducted to provide a detailed description of waste generation and characteristics from a traditional market in Hue city, located in central Vietnam. The authors conducted a waste generation survey and a waste composition survey for 309 stalls/vendors in five markets by 17 business categories for 10 consecutive days. The waste generation rates by stall/vendor and by floor area were assessed in three waste categories: general waste, recyclable, and food residues. The general waste that would be sent to a landfill site was classified into 10 physical categories and 77 sub-categories. For general waste, food waste accounted for the largest part, followed by plastic and grass. By multiplying the waste generation rate by stall/vendor by the total number of stall/vendors in 23 markets, the authors estimated the total amounts of general waste, recyclable, food residue and total waste by business category. The total waste generated from market was 17.0 tons/day, of which 4.6 tons (27.1%) were collected by pig farmers for feeding livestock and 0.6 tons (3.6%) were sold to the recycling market. The composting potential accounted for 55.2% of total waste generation from the traditional market in Hue. The recycling potential accounted for 5.1%. The total disposal amount sent to the landfill site would be reduced from 69.2% to 8.8% of the total. The 95% confidence interval (CI) of total waste amount from 23 markets was also estimated using Monte Carlo simulation based on the mean and standard error of the waste generation rate. The range of 95% CI was 14.9 - 18.9 tons/day.

Keywords

Traditional Market, Recycling Potential, Composting Potential

1. Introduction

In economically developing countries, the amount of municipal solid waste is approaching the capacity of existing waste facilities. It is impossible to continue waste disposal that is heavily dependent on landfill. Central and local governments must develop solid waste management (SWM) plans that include waste reduction and recycling.

Municipal solid waste is a growing problem in Vietnam, which is showing rapid economic growth year by year. Vietnam produces over 15 million tons of municipal solid waste each year from various sources. More than 80% (12.8 million tons/yr) derives from municipal sources: households, restaurants, markets, and businesses [1]. As the first step in designing integrated waste management systems, it is indispensable to ascertain detailed and reliable information related to waste generation, waste composition, and waste streams [2]. However, Vietnam lacks reliable and detailed data related to SWM. The latest report on SWM in Vietnam published in 2011 presented information related to overall waste generation and physical composition of MSW around Vietnam, but did not clarify details related to waste generation from different sources or details related to waste composition such as recycling or composting potential.

Regarding the municipal solid waste sources, traditional markets are known to be a considerable source of waste generation in economically developing countries. One earlier study [3] found that markets contributed about 20% of total waste generation in Indonesia. According to the Vietnam Retail Association, traditional markets still constitute the major channel of the retail sector. About 8550 traditional markets existed in Vietnam in 2011. However, few studies have specifically examined wastes from traditional markets. In Lao PDR, Byer *et al.* [4] surveyed one early morning market (EMM), and clarified the waste generation rate by six business categories: fruits and vegetables, packaged goods, meat/fish/eggs, food stalls, noodles and blood, rice, and charcoal. They assessed the physical composition, but reported no potential for recycling and composting. In Cambodia and Vietnam, two surveys [5] [6] have assessed the waste generation rate by the total waste amount divided by the total number of stalls in target market, but they did not address differences in waste generation rates among business categories.

This study was undertaken to present a detailed description of waste generation and characteristic of traditional market in Hue city, Vietnam. The authors chose five markets from three market classes, and allocated the targets by 17 business categories considering "*The System of Economic Branches of Vietnam*". The authors also estimated the total waste amount, the recycling and composting potentials from all markets in Hue as the basis of rational SWM planning including waste reduction and recycling.

2. Methodology

2.1. Research Area and Target Traditional Markets

Hue city, the latest imperial capital of Vietnam under the Nguyen dynasty, is located in the central region of Vietnam [7]. Hue city comprises 27 wards with area of 71.69 km^2 and a population of 350,345 people. There are two distinct seasons in Hue city: the dry season comes with the hot southwest wind for four months during April-August; the rainy season comes with high and unevenly distributed rainfall during September-March [8].

Regarding solid waste management, the amount of collected waste in Hue city is reported as approximately 210 tons/day. The general collection rate in the whole city was about 89%, and 90% - 95% in urban areas [9].

Hue city has 23 traditional markets with different scales. According to Government Decree No. 2/2003/NĐ-CP [10] on market development and management, 23 markets in Hue city are classified into three classes in terms of their scale, trade volume, and facility conditions. Among them, 3 markets belong to the first class, 6 markets belong to second class, and 14 markets belong to the third class. Markets sell widely diverse items such as food, vegetable, meat, clothes, and household equipment. There are businesses of two types in Hue, defined as follows.

➤ *Stall: a shop selling goods at a designated place with a contract for a certain period of time, normally located inside a market building*

➤ *Vendor: someone who is selling goods without contract for a certain period of time, normally located outside of market building*

Stalls and vendors in markets generate waste of many kinds such as rubbish from commodities, containers and packaging. Some of them separate recyclable items for recycling markets and food residues for animal feed in Hue. For this study, the authors chose Tay Loc market from first class, Vy Da and Xep markets from second class, and Phuoc Vinh and Thong markets from third class as the target considering location and scale. An outline of the five markets is shown in **Table 1**. Markets have widely diverse business categories. Prime Minister of the Government of Vietnam (VPM, 2007) issued "*The System of Economic Branches of Vietnam*" [11] and defined the business category in 642 branches. By referring the official definition, the authors defined 17 business categories for market as shown in **Table 2**. For target selection, the authors allocated target kiosks and vendors to cover 17 business categories. The total number of samples is shown in **Table 2**.

2.2. Outline of Survey

The authors conducted three surveys for all target samples: a waste generation survey by actual measurement, a waste composition survey, and a questionnaire survey. Surveys on Tay Loc, Xep and Phuoc Vinh market were conducted during 3 - 12 September. Surveys on Vy Da and Thong market were conducted during 9 - 17 September. The waste generation survey was administered to acquire data on the amount of waste generation for 10 consecutive days. Of them, the first three days were spent for practice; the authors used the data for the latter seven consecutive days. The target stalls and vendors were requested to keep their waste in three categories by their original customs: "*Recyclables*", "*Food residues*" and "*General waste*", defined as follows.

✓ *Recyclables*: items kept for recycling or sale to informal sectors or given to somewhere/someone by owners.

✓ *Food residues*: waste items kept for livestock (e.g. pigs) feeding; generally collected by livestock breeders.

✓ *General waste*: all remaining waste items excluding separated waste items described above. This type of waste is collected daily by an environmental company in Hue (HEPCO).

Table 1. Outline of five target markets.

Name	Tay Loc	Vy Da	Xep	Phuoc Vinh	Thong
Market class	1st	2nd	2nd	3rd	3rd
Number of cleanings	4	2	1	2	1
Number of stalls	682	280	365	179	112
Number of vendors	168	21	75	34	3 - 5
Number of target samples	91	66	63	58	30

Table 2. Definition of business category.

ID	Business category
1	Rice & powder
2	Meat and meat products
3	Chicken & Duck
4	Eggs
5	Fish & fish products
6	Vegetable
7	Betel and areca
8	Fruits
9	Coconut
10	Spice, Grocery, Cakes & candy
11	Food stalls
12	Beverages
13	Textiles, apparel, footwear
14	Fresh flowers, ornamental plants
15	Daily commodity, incense, porcelain
16	Service (hair cutting, foot repairing)
17	Recyclable, Second-hand clothes

A waste composition survey was also conducted during the survey period. To provide information related to the recycling and composting potentials, the authors analysed details of the waste composition of "General waste" for some representative targets. The waste was classified into 10 physical categories and 77 sub-categories. The classification categories were based on Materials (Plastic, Paper, Kitchen waste, Rubber & Leather, Grass, Textile, Metal, Glass, Ceramic, and Miscellaneous), Types (Container/Packaging, Product and Other), Recycling potential (recyclable and non-recyclable), and Composting potential (compostable and non-compostable). The recycling potential was defined based on the practical trading status of recycling market in Hue city. Recyclable items contained plastic, paper, glass, metal, and textiles that can be bought and sold at a recycling market. The composting potential was defined based on the acceptable items of some composting plants. Organic wastes are divisible into compostable and non-compostable wastes. Compostable items consisted of vegetables, food residue, grass, leaves, flowers, egg shells, fish bones, fruit, and fruit skins. Non-compostable items consisted of coconut shells, hard bones of animal, seashells, bamboos, large tree branches, and wood products. In this study, non-recyclable items were all items that could not be recycled or composted. Descriptions of waste classification categories are presented in **Table 3**. The authors also administered a questionnaire survey to assess attributes and the current status of businesses of target stalls and vendors.

2.3. Analytical Procedure

The authors calculated key statistics related to waste generation rates by business category. The authors also assessed the mean difference among market classes using analysis of variance (ANOVA). The waste composition by percentage (%) was calculated according to the physical category and by recycling and composting potentials.

By multiplying the waste generation rate by stall/vendors by total number of stall/vendors in Hue, the authors estimated the total waste generation amount from traditional market in Hue. The authors also calculated the 95% confidence interval of total waste generation amount from a Monte Carlo simulation (100,000 times) based on the mean and standard error of waste generation rate by business category. Monte Carlo simulations are used widely to assess error propagation for model parameters [12]. The authors inferred the sensitivity as a percentage of the contribution from each parameter to the variance of the final result [13].

3. Results and Discussion

3.1. Waste Generation Rate of Traditional Markets

3.1.1. Waste Generation Rate by Stall/Vendor

The waste generation rates of traditional market were calculated using business categories. **Table 4** presented the mean and standard deviation of waste generation rate (g/stall/day) of stall in three market classes. Regarding the waste generation rate by business category, "*Vegetable*", "*Fruits*", "*Beverage*" and "*Fresh flowers*" were higher in all wastes, whereas "*Rice*", "*Meat*", "*Service*" and "*Second-hand shop*" were identified as having lower generation rates. The result resembled that reported from a previous study by Byer *et al.* [4]. They reported that "Fruits and vegetables" generated the largest amount (6.49 kg/day) because of their high moisture contents.

Regarding fresh items that are easily be perishable after 1 - 2 days under normal conditions such as "*Meat & meat product*", "*Chicken & duck*", "*Fish & fish product*", "*Vegetable*", "*Fruit*", "*Fresh flowers*", the category of "*Meat & meat product*" generated the smallest amount with 289 (g/stall/day), whereas the other categories generated much higher waste amounts. In the category of "*Vegetable*" and "*Fruit*", even though the total waste amount was rather large, the amounts of food residues were quite small because the rotten or leftover vegetables and fruits were normally unsuitable for feeding animals. Conversely, the categories of "*Chicken & duck*", "*Fish & fish product*" generated a large amount of food residues because these categories had processing services on site. Therefore, normally the internal organs and unnecessary parts were separated as food residues.

Some dry-food items such as "*Rice/powder*" and "*Spice and grocery*" produced smaller amounts with 110 and 759 (g/stall/day), respectively. Regarding food services, "*Food stalls*" & "*Beverage*" respectively produced similar total waste amounts, with 3525 and 3323 (g/stall/day). However, "*Beverage*" generated small amounts of food residues, whereas a major amount of wastes in "*Food stalls*" were food residues. **Table 4** also shows that some business categories such as "*Meat*", "*Chicken & Duck*", "*Fish*", "*Vegetable*", and "*Fresh flowers*" did not separate recyclables. **Table 5** presents the mean and standard deviation of waste generation rate (g/vendor/day)

Table 3. Classification category of waste from market.

Category	Code	Details	Recycling potential	Category	Code	Details	Recycling potential
1. Plastic				**5. Grass and wood**			
Container & Packaging	101	PET bottle	Re	Container & Packaging	503	Containers & packaging	Co
	102	Other plastic bottle	Re		503*	Containers & packaging	NRe
	103	Tray	Re	Products and Others	504	Grass and wood products	Co
	103*	Tray	NRe		504*	Grass and wood products	NRe
	104	Tube	Re	**6. Textile**			
	104*	Tube	NRe		601	Clothes	Re
	105	Other shape	Re		602	Daily commodities	NRe
	105*	Other shape	NRe		603	Disposed commodities	NRe
	106	Shopping plastic bags	Re		604	Other product	Re
	107*	Other plastic packaging	NRe	**7. Metal**			
	108	Other C & P (e.g.: buffer)	Re		701	Containers	Re
	108*	Other C & P	NRe		702	Other containers and packaging	Re
Product	109	Plastic product	Re	Aluminum	702*	Other containers and packaging	NRe
	109*	Plastic product	NRe		703	Products and others	Re
Other plastics	110	Other plastics	Re		703*	Products and others	NRe
	110*	Other plastics	NRe		704	Containers	Re
2. Paper				Steel	704*	Containers	NRe
Container & Packaging	201	Carton	Re		705	Other containers and packaging	Re
	202	Containers	Re		706	Products and others	Re
	203	Cardboard	Re	Stainless	707	Products and others	Re
	204	Packaging	Re	Lead	707*	Products and others	NRe
	205	Other C & P	Re	Other metals	708	Other metals	Re
	206	Newspaper/poster	Re		708*	Other metals	NRe
	207	Books	Re	**8. Glass**			
	208	Notebooks	Re		801	Returnable bottle	Re
Product	209	Photocopy	Re	Container	802	Disposal bottle	Re
	210	Disposal paper products	NRe		803	Other containers	Re
	210*	Nappies/Diapers	NRe	Products and others	804	Thermometers, Fluorescent lamp	NRe
	211	Other paper product	Re		805	Products and others	NRe
	211*	Other paper product	NRe	**9. Ceramic**			
	212	Other Paper	Re		901	Containers	NRe
Other Paper	212*	Other Paper	NRe		902	Products and others	NRe
3. Kitchen waste				**10. Miscellaneous**			
Compostable	301	Kitchen waste	Co		1001	Combustibles	NRe
Non-compostable	301*	Coconut/Durian shells	NRe		1002	Liquids_edible	Re
	302	Hard bones of animal	NRe		1002*	Liquids_inedible	NRe
4. Rubber and leather					1003	Incombustibles (excluding ash)	NRe
	401	Rubber and leather	NRe		1004	Ash	NRe
5. Grass and wood					1005	Medical care syringe, needle, ...)	NRe
Garden waste	501	Garden waste	Co		1006	Batteries	NRe
	501*	Garden waste	NRe		1007	E-waste	NRe
	502	Flower	Co		1008	Others	NRe

[a]Re, Recyclable; Co, Compostable; NRe, Non-recyclable & non-compostable items, The recycling potential of each item was defined based on reports from two junk-shop owners. The compostable item and non-compostable item were defined based on the acceptable items in some composting plants.

Table 4. Waste generation rate by stall (g/stall/day).

Category	N	General waste	Recyclable	Food residues	Total waste
		Mean ± Standard deviation			
Rice, powder	6	89 ± 89	8 ± 20	13 ± 31	110 ± 95
Meat/meat product	13	268 ± 305	-	20 ± 40	289 ± 299
Chicken & duck	12	102 ± 165	-	2664 ± 3264	2766 ± 3194
Egg	4	445 ± 644	291 ± 314	-	737 ± 555
Fish & fish product	32	248 ± 269	-	2513 ± 2445	2761 ± 2486
Vegetable	40	4893 ± 5525	-	250 ± 727	5143 ± 5657
Betel & areca	2	1138 ± 597	140 ± 37	-	1278 ± 634
Fruit	31	3865 ± 2921	520 ± 853	184 ± 1023	4568 ± 3343
Spice & Grocery	37	392 ± 866	149 ± 301	218 ± 939	759 ± 1342
Food stall	21	762 ± 717	9 ± 42	2754 ± 3594	3525 ± 3801
Beverages	10	2829 ± 3023	170 ± 271	324 ± 1023	3323 ± 2955
Textiles & Apparel	16	195 ± 157	43 ± 166	-	238 ± 242
Fresh flowers	11	3209 ± 4009	-	-	3209 ± 4009
Daily commodity	18	271 ± 283	182 ± 514	-	453 ± 566
Service	14	190 ± 114	8 ± 27	-	198 ± 116
Second-hand shop	4	292 ± 491	124 ± 152	-	416 ± 409

Table 5. Waste generation rate by vendor (g/vendor/day).

Category	N	General waste	Recyclable	Food residues	Total waste
		Mean ± Standard deviation			
Rice, powder_vendor	1	268	-	-	268
Egg_vendor	2	412 ± 197	-	-	412 ± 197
Fish & fish product_vendor	2	-	-	1628 ± 462	1628 ± 462
Vegetable_vendor	11	3918 ± 3117	-	88 ± 212	4006 ± 3039
Betel and areca_vendor	2	4226 ± 4848	-	-	4226 ± 4848
Fruit_vendor	7	1811 ± 1911	71 ± 123	-	1882 ± 1867
Coconut_vendor	2	91801 ± 1040	-	-	91801 ± 1040
Spice & Grocery_vendor	3	291 ± 375	-	-	291 ± 375
Food stall_vendor	5	493 ± 418	-	876 ± 1206	1369 ± 868
Textiles & Apparel_vendor	1	23	-	-	23
Fresh flowers_vendor	2	1451 ± 84	-	-	1451 ± 84

of vendors in three market classes. The "*Coconut_vendor*" category was separated from "*Fruit_vendor*" because the waste generation rate of "*Coconut_vendor*" was extremely high, with 91,801 g/day, which was much higher than that of "*Fruit_vendor*" with 1882 g/vendor/day.

Comparing the waste generation rate of a stall with that of a vendor, some categories with higher waste generation rates such as "*Fish & fish product*", "*Vegetable_vendor*", "*Fruit*", "*Food stall*" and "*Fresh flowers*", the waste generation rate of a stall was higher than that of a vendor. That result is explainable that the space and business scale of vendors were normally smaller than those of stalls; the resultant waste generation rate was lower. The vendors did not separate recyclables in most business categories excluding "*Fruit_vendor*"; the food residue amount was small excluding "*Fish_vendor*" and "*Food_vendor*" as stalls, probably because they had insufficient space to keep such wastes.

3.1.2 Waste Generation Rate by Floor Area

In the traditional market, the stall areas differ among market classes and business categories. The authors calculated the waste generation rate by floor area for two markets for which the managers of markets provided the official data on floor area of each stall, Tay Loc as a first class market and Xep as a second class market. **Table 6** presents the means and standard deviations of waste generation rate of 17 business categories by floor area $(g/m^2/day)$ in two markets. The floor areas in most business categories in the first class market were larger than those of the second class market. For "*Rice & powder*", "*Fish & fish product*", "*Vegetable*", "*Spice & Grocery*", "*Beverage*", "*Textile & Apparel*", "*Fresh Flowers*", "*Daily commodity*" and "*Service*" categories, the waste generation rate of total waste by square meter was higher for the first class market than for the second class market. For "*Meat & meat product*", "*Fruit*" and "*Food stall*" categories, the waste generation rate of total waste by square meter was higher for the second class market, but the waste generation rate of total waste by stall was higher for the first class market.

In most of business categories, the stalls in first class market discharged more waste than those in the second class. In the first class market, no stalls existed for "*Egg*" and "*Betel & areca*" categories. These items were sold only at vendors in the first class market. Regarding the "*Chicken & duck*" category, the waste generation rate of total waste by square meter in the first class market was 178 $g/m^2/day$, which was much lower than that in the second class market at 2090 $g/m^2/day$. By the observation in both markets, the difference noted above can be explained by the habit of stall owner in keeping internal organs of "*Chicken & duck*" in each market. In the first class market, the owners generally separated internal organs and sold them to customers, whereas the owners in the second class market normally put internal organs into containers as food residues for livestock breeders and occasionally sold them to customers upon request.

Table 6. Waste generation rate by floor area.

Category	Tay Loc					Xep				
	Area	GW	Re	FR	Total	Area	GW	Re	FR	Total
	m²	g/m²/day				m²	g/m²/day			
Rice & powder	3.2	42 ± 24	8 ± 11	-	49 ± 34	1.69	1	1	-	2
Meat & meat product	2	250 ± 247	-	-	250 ± 247	1.69	272 ± 266	-	-	272 ± 266
Chicken & duck	3.6	-	-	178 ± 7	178 ± 7	2.25	40 ± 63	-	2050 ± 1670	2090 ± 1622
Egg	-	-	-	-	-	2.25	338 ± 408	167 ± 189	-	505 ± 218
Fish & fish product	3.6	126 ± 115	-	1209 ± 913	1335 ± 940	1.69	163 ± 164	-	739 ± 925	903 ± 861
Vegetable	3.6 - 10.8	2274 ± 1531	-	49 ± 63	2324 ± 1518	1.69 - 3.38	964 ± 575	-	78 ± 100	1042 ± 557
Betel & areca	-	-	-	-	-	2.25	693	74	-	767
Fruit	3.6	1248 ± 857	-	-	1248 ± 857	1.69 - 3.38	1362 ± 886	402 ± 625	211 ± 596	1975 ± 1288
Spice & Grocery	2 - 6.4	73 ± 47	80 ± 129	-	153 ± 159	2.25 - 4.5	23 ± 12	73 ± 90	-	96 ± 86
Food stall	3.6 - 7.2	199 ± 207	-	1085 ± 1030	1283 ± 949	2.25 - 4.5	353 ± 351	14 ± 35	1252 ± 2149	1619 ± 2464
Beverages	3.6	2095 ± 235	69 ± 11	-	2163 ± 246	2.25 - 4.5	397 ± 400	147 ± 69	-	544 ± 469
Textiles & Apparel	3.6 - 12	66 ± 38	-	-	66 ± 38	4.5	13 ± 9	-	-	13 ± 9
Fresh flower	3.6 - 7.2	1224 ± 1465	-	-	1224 ± 1465	2.25 - 7.29	444 ± 269	-	-	444 ± 269
Daily commodity	2 - 4	125 ± 126	29 ± 64	-	154 ± 182	1.69 - 14.58	83 ± 70	4 ± 7	-	88 ± 73
Service	2 - 3.6	53 ± 22	-	-	53 ± 22	7.29	23	14	-	37
Second-hand shop	3.2	91 ± 153	39 ± 48	-	130 ± 128	-	-	-	-	-

GW, General waste; Re, Recyclable; FR, Food residue.

3.2. Mean Difference in Waste Generation Rate by Market Class

Regarding comparison among three market classes, the authors assessed the mean difference in waste generation rates among the three market classes by 17 business categories using ANOVA. The authors found significant mean differences for "*Fish*" and "*Vegetable*". **Table 7** shows means and SDs of waste generation rates for "*Fish*" and "*Vegetable*" by three market classes and ANOVA results. The waste generation rate was the highest in the first class, and lower in the second class and the third class. The differences among three markets are explainable by the fact obtained from questionnaire survey that, in the first class market, the area was largest and the operation time was very long, whereas the area and operation time were smaller in the second and third class markets, which can be expected to influence the amounts of sales and waste generation.

3.3. Waste Composition of Market

3.3.1. Physical Composition

The physical composition of general waste by business category is shown in **Table 8**. The physical composition differs among business categories. The results indicated that food waste accounts for the largest part with sub-

Table 7. Waste generation rate by market class (g/stall/day).

Category	Market class	N	General waste	Food residues	Total waste
Fish & fish product	1	7	452 ± 412	4352 ± 3288	4804 ± 3384
	2	14	207 ± 224	1428 ± 1095	1635 ± 1018
	3	11	169 ± 137	2725 ± 2533	2894 ± 2493
ANOVA (F value)			**3.001**	**4.077**[*]	**4.74**[*]
Vegetable	1	9	11785 ± 7877	340 ± 564	12125 ± 8134
	2	14	2928 ± 2757	199 ± 358	3127 ± 2996
	3	17	2863 ± 1653	245 ± 1009	3108 ± 1504
ANOVA (F value)			**15.972**[***]	**0.098**	**15.351**[***]

[*]$p < 0.05$; [**]$p < 0.01$; [***]$p < 0.001$.

Table 8. Physical composition by business category (%).

ID	N	Plastic	Paper	Food	Rubber	Grass	Textile	Metal	Glass	Ceramic	Other
Rice, powder	6	6.8%	0.7%	52.6%	0.1%	21.5%	-	-	-	-	18.3%
Meat	15	48.2%	0.9%	44.8%	-	2.9%	-	-	-	-	3.3%
Chicken & duck	2	51.8%	9.9%	24.4%	11.3%	1.5%	-	1.0%	-	-	-
Egg	7	9.4%	2.1%	67.8%	-	20.5%	-	-	-	-	0.1%
Fish	30	32.6%	0.2%	45.1%	-	4.7%	-	-	-	-	17.4%
Vegetable	17	2.8%	0.3%	93.2%	-	1.3%	-	-	-	-	2.4%
Betel & areca	2	2.2%	0.1%	14.7%	-	83%	-	-	-	-	-
Fruit	14	1.6%	3.5%	81.7%	0.1%	12.7%	-	-	-	-	0.3%
Spice & Grocery	33	26.4%	14.7%	52.5%	1.6%	2.2%	0.1%	0.5%	-	-	2%
Food stalls	25	7.4%	8.3%	25%	0.2%	51.9%	2.9%	-	-	-	4.3%
Beverages	8	8.1%	0.1%	77.2%	-	8.6%	-	-	-	-	6.1%
Textiles & footwear	13	40%	15.1%	20.3%	21.2%	-	0.7%	2.1%	-	-	0.5%
Fresh flowers	6	2.1%	1.7%	21.7%	-	74.5%	-	-	-	-	-
Daily commodity Daily commodity	11	10.4%	12.7%	62.3%	0.2%	3.8%	2.3%	0.1%	-	8.0%	0.1%
Service	5	30.1%	2.4%	31.6%	0.1%	2.4%	11.6%	0.8%	-	-	21%
Second-hand shop	2	29.0%	69.5%	-	-	-	-	-	-	-	1.4%

sequent plastic or grass. The proportions of food waste were high, with 93.2%, 81.7% and 77.2%, respectively, in "Vegetable", "Fruit" and "Beverage". Previous reports have described that the major portion of solid waste in market is food waste and other organic matter. A survey conducted in Danang [6] reported that the organic part accounted for 81.5%. A survey in Thailand also presented that the organic from market was 85% [14].

The categories of "*Meat*", "*Fish*", "*Textile*" and "*Spice*" generated plastics with higher percentage, using mainly single-use plastic bags for packaging, whereas "*Betel & areca*", "*Fresh flower*" and "*Food stalls*" categories generated grass with higher percentages. These results directly reflected the fact that some parts of "*Betel and areca*" and "*Fresh flower*" such as stems, leaves, and un-needed parts are often removed according to requests from customers. "Food stalls" generated large amounts of grass because Vietnamese people have a habit of wrapping some foods in leaves such as banana or lotus leaves. They discard them after use. Glass was not found in all categories, and ceramic was only found in "*Daily commodity*" with 8%. Rubber was found in "*Chicken & duck*" because rubber bands are usually used to tie live chicken or duck legs. The "*Textiles & footwear*" category generated some rubber because rubber pieces were often used to repair shoes.

3.3.2. Recycling and Composting Potential of Waste from Traditional Markets

Table 9 presents waste composting and recycling potentials from general waste by business category according to the definition shown in **Table 3**. It is apparent that the composting potential was very high in some categories such as "*Betel & areca*", "*Vegetable*", "*Fruit*" and "*Fresh flowers*". The recycling potential was highest in the category of "*Second-hand shop*" with 98.6% followed by "*Textiles & apparel*" with 63.6%, "*Meat*" with 48.9% and "*Chicken & duck*" with 46.4%.

3.4. Estimation of Total Waste Generation from Traditional Market in Hue

3.4.1. Validation of General Waste at Three Markets

To validate waste generation rates in **Table 4** and **Table 5**, the authors estimated the 95% confidence intervals (CIs) of total amount of general waste for three target markets, Tay Loc in the first class, Xep in the second class and Phuoc Vinh in the third class, and compared the CIs with the measured waste amounts by actual measurement on site. By multiplying the waste generation rate by the stall/vendor by total number of stall/vendor in target markets, the authors calculated the total amounts of general wastes for the target markets. The 95% confidence interval for each market was estimated using Monte Carlo simulation (100,000 times) based on the mean and standard error of the waste generation rates shown in **Table 4** and **Table 5**. **Table 10** presents the 95% CIs

Table 9. Recycling and composting potential of general waste (%).

Category	N	Non-recyclable	Recyclable	Compostable
Rice/powder	6	18.1%	6.4%	75.5%
Meat	15	8.7%	48.9%	42.4%
Chicken & duck	2	26.1%	46.4%	27.5%
Egg	7	3.0%	10.5%	86.6%
Fish	30	18.0%	32.3%	49.7%
Vegetable	17	2.8%	2.9%	94.3%
Betel & areca	2	1%	1.3%	97.7%
Fruit	14	3.8%	4.1%	92.1%
Spice & Grocery	33	8.7%	37.8%	53.5%
Food stalls	25	14.4%	9.5%	76.1%
Beverages	8	6.2%	8.2%	85.7%
Textiles & footwear	13	11.9%	63.6%	24.4%
Fresh flower	6	2.7%	2.7%	94.5%
Daily commodity	11	14.5%	21.4%	64.1%
Service	5	34.6%	24.7%	40.7%
Second-hand shop	3	1.4%	98.6%	0.0%

Table 10. Interval estimation of three markets and actual amount.

Market	95% CI of general waste (kg)	Actual amount (kg)
Tay Loc market	1412 - 2163	1766
Xep market	286 - 502	473.8
Phuoc Vinh market	252 - 363	343.5

of general waste (kg/day) and the measured waste amounts in three markets. The results show that the measured waste amounts were in the 95% CI range for the three markets.

3.4.2. Total Waste Generation from Traditional Markets

Hue city has 23 traditional markets. Because of a lack of information exists for the floor area by each business categories in 23 markets, in this study, the author used only the number of stalls by business category as an indicator for estimation the total amount of waste from traditional market in Hue. By multiplying the waste generation rate by stall/vendor by the total number of stall/vendors in 23 markets, the authors calculated the total amount of general waste, recyclable, food residue, and total waste by business category. The general waste also included some amount of recyclable and compostable parts, as shown in **Table 8**. The authors also estimated the potentials of recycling and composting in general waste separately.

Table 11 presents details of waste generation from a traditional market in Hue: non-recyclable, recycling potential, composting potential contained in general waste: recyclable, food residues, and total waste by weight (kg/day) by business category. The total waste generation amount was 17.0 tons, of which 3.6% of waste was separated at the source as recyclable, 69.2% was general waste, and 27.1% was food residues. In general waste, the recycling potential accounted for 5.1%, composting potential accounted for 55.2%, and the remaining waste accounted for only 8.8%. This result indicates that waste generation in markets has high potential up to 82.3% of total for composting and livestock feeding.

The total disposal amount sent to the landfill site can be reduced from 69.2% to 8.8%. Regarding the "*Coconut*" category, the waste generation rate was the highest with 91.8 kg/stall/day and was mainly composed of coconut shell. Therefore, we did not bring it to the laboratory to examine the composition. The authors assumed that the total generated waste from the "*Coconut*" category was non-recyclable. Regarding composting potential, "*Vegetable*" accounted for the largest amount with 4422 kg/day, followed by "*Vegetable_vendor*", (1580 kg/day) and "*Fruit*" (1520 kg/day). These results suggest that these sources would contribute immensely to promoting waste reduction. Regarding about the recycling potential, the largest contribution category was "*Spice & grocery*" with 195 kg/day, followed by "*Vegetable*" (135 kg/day), "*Textiles & apparel*" (120 kg/day), "*Fish*" (86 kg/day), and "*Meat*" (83 kg/day). The results revealed that these categories should be considered heavily for promotion for recycling for market. The total estimated composting potential was 9.4 tons, which contributes mostly to the total general waste in the market. The large quantity of composting potential present in the MSW stream has a great impact on the production of high-quality compost and offers great potential for resource recovery. Such recovery is expected to play an important role in reducing total waste generation amount and in mitigating the negative effects on environmental quality and natural resource conservation.

3.4.3. Interval Estimation of Total Waste Generation from Traditional Market in Hue

The 95% confidence interval (CI) of total waste amount from 23 markets was also estimated using Monte Carlo simulation (100,000 times) based on the mean and standard error of waste generation rate shown in **Table 4** and **Table 5**. The results showed that the range of 95% CI was 14.9 - 18.9 tons/day.

The authors also examined the sensitivity as a percentage of the contribution from the waste generation rate of each business category to the variance of the total waste amount. **Figure 1** presents the result of sensitivity analysis. "*Vegetables of first class market*" was identified as the category with the largest contribution (34.8%) to the variance of the total waste amount, followed by "*Vegetable_vendor*" (13.5%) and "*Fish of first class*", (10.9%). To improve the reliability of total estimation, the sample size should be increased. Further investigation must be undertaken to clarify the factors affecting waste generation rate in these categories.

4. Conclusions

1) This study produced a detailed description of waste generation and composition by 17 business categories

Table 11. Waste amounts by business category from 23 markets in Hue (kg/day).

Category	N	General waste			Recyclable	Food Residues	Total
		Non-recyclable	Recycling potential	Composting potential			
Rice & powder	220	4	1	15	2	3	24
Meat	630	15	83	72	-	13	182
Chicken & duck	75	2	4	2	-	200	208
Egg	18	0	1	7	5	-	13
Fish	1045	48	86	133	-	2737	3004
Vegetable	951	132	135	4422	-	238	4928
Betel & Areca	50	1	1	56	7	-	64
Fruit	427	63	68	1520	222	79	1951
Coconut	5	459	-	-	-	-	459
Spice & Grocery	1315	45	195	276	196	286	998
Food stalls	265	29	19	154	2	730	934
Beverages	90	16	21	218	15	29	299
Textiles & Apparel	968	23	120	46	42	-	231
Fresh flowers	64	6	6	194	0	-	205
Equipment, watch	545	21	32	95	99	-	247
Service	181	12	9	14	2	-	36
Second-hand shop	61	0	18	0	8	-	25
Rice & powder_vendor	36	2	1	7	0	0	10
Egg_vendor	16	0	1	6	0	0	7
Fish & fish product_vendor	79	0	0	0	0	129	129
Vegetable_vendor	436	48	49	1611	0	38	1746
Betel and areca_vendor	20	1	1	83	0	0	85
Fruit_vendor	224	15	17	374	16	0	422
Coconut_vendor	6	551	0	0	0	0	551
Spice & Grocery_vendor	141	10	7	53	0	124	193
Food stall_vendor	31	0	0	0	0	0	1
Textiles & Apparel_vendor	16	1	1	22	0	0	23
Amount (kg)		**1502**	**872**	**9377**	**616**	**4605**	**16,972**
Amount (tons)		**1.5**	**0.9**	**9.4**	**0.6**	**4.6**	**17.0**
Percentage		**8.8%**	**5.1%**	**55.2%**	**3.6%**	**27.1%**	**100%**

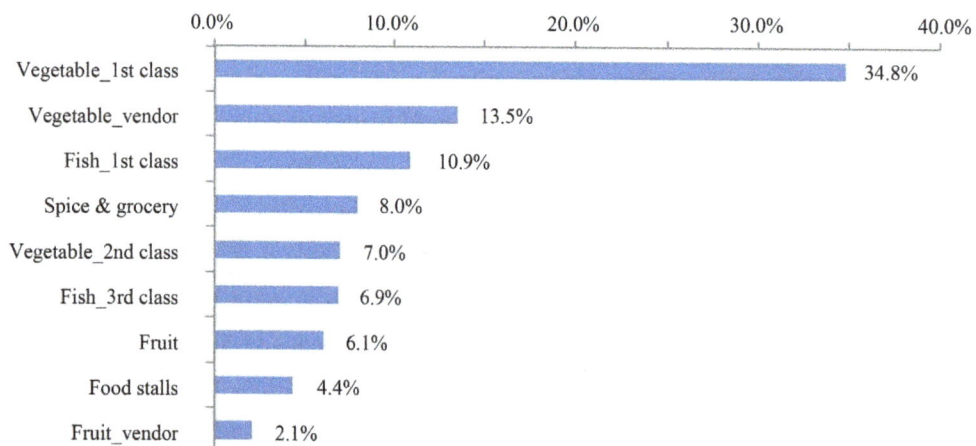

Figure 1. Sensitivity analysis of total waste generation.

in a traditional market in Hue city, Vietnam. In all, 309 stalls/vendors in five markets were surveyed for 10 consecutive days in the dry season.

2) The waste generation rate by stall/vendor was assessed by each business category by three waste categories: general waste, recyclable, food residues. The waste generation rates of "*Vegetable*", "*Fruits*", "*Beverage*" and "*Fresh Flowers*" were higher in all wastes, whereas those of "*Rice*", "*Meat*", "*Service*" and "*Second-hand shop*" were identified as having lower generation rates.

3) Significant mean differences of total waste amount were found in "*Fish*" and "*Vegetable*" categories. The waste generation rate was the highest in the first market class and lower in the second market class and third market class.

4) The waste generation rate by floor area was also calculated in two markets in first and second classes. In most business categories, the stalls in the first class market discharged more waste than those in the second class.

5) The general waste was classified into 10 physical categories and 77 sub-categories. Food waste accounted for the largest part, followed by plastic and grass. The results from sub-categories demonstrated the potential for composting and recycling.

6) As the validation of waste generation rate, the authors estimated the 95% confidence intervals (CIs) of total amount of general waste for three target markets and compared the CIs with the measured waste amounts by actual measurements conducted on site. The measured waste amounts were in the 95% CI range for the three markets.

7) The authors also estimated the total generated waste, the recycling and composting potential for 23 traditional markets in Hue. The total waste generated from market was 17.0 tons/day, of which 4.6 tons (27.1%) was collected by pig farmers for feeding livestock and 0.6 tons (3.6%) was sold to the recycling market. The composting potential accounted for 55.2% and the recycling potential accounted for 5.1% of total waste generation from traditional market in Hue. The total disposal amount sent to the landfill site would be reduced from 69.2% to 8.8% of the total.

By the Monte Carlo simulation, the confidence interval of total waste generation from traditional market in Hue was estimated. The 95% CI of total waste was 14.9 - 18.9 tons/day. By sensitivity analysis, "*Vegetable of first class market*" was identified as the category with the largest contribution to the variance of the total waste amount with subsequent "*Vegetable_vendor*" and "*Fish of first class*". To improve the reliability of total estimation, the sample size should be increased and further investigation is necessary to clarify the factors affecting waste generation rates in these categories.

Acknowledgements

The authors sincerely thank the staff and students from Hue University for their assistance in this study. Special thanks to Dr. Pham Khac Lieu and Mr. Tran Ngoc Tuan who supported our survey enthusiastically. The authors also express their appreciation to the statistical office in Hue as well as all of stall/vendor owners in five target markets for collaborating with us.

References

[1]   World Bank, MoNRE and CIDA (2004) Vietnam Environment Monitor. 65.

[2]   Forbes, R.M., Peter, R.W., Marian, F. and Peter, H. (2001). Integrated Solid Waste Management: A Life Cycle Inventory. Second Edition, Blackwell Science, Oxford.

[3]   Meidiana, C. and Gamse, T. (2010) Development of Waste Management Practices in Indonesia. *European Journal of Scientific Research*, **40**, 199-210.

[4]   Byer, P.H., Hoang, C.P., Nguyen, T.T.T., Chopra, S., Maclaren, V. and Haight, M. (2006) Household, Hotel and Market Waste Audits for Composting in Vietnam and Laos. *Waste Management & Research*, **24**, 465-472. http://dx.doi.org/10.1177/0734242X06068067

[5]   JICA and Kokusai Kogyu Co. Ltd. (2003) The Study on Solid Waste Management in the Municipality of Phnom Penh.

[6]   Otoma, S., Hoang, H., Hong, H., Miyazaki, I. and Diaz, R. (2013). A Survey on Municipal Solid Waste and Residents' Awareness in Da Nang City, Vietnam. *Journal of Material Cycles Waste Management*, **15**, 187-194. http://dx.doi.org/10.1007/s10163-012-0109-2

[7]   http://hueimperialcity.com/hue-introduction/

[8]   Hue Statistical Yearbook (2012) Statistical Yearbook, Hue city's Statistical Office, Hue, Vietnam.

[9]   HEPCO (2011) Report on Solid Waste Management of Hue City. Hue Urban Environment and Public Works State Company (HEPCO). (In Vietnamese)

[10] Government Decree No. 2/2003/NĐ-CP on Market Development and Management.

[11] Vietnam's Prime Minister (VPM) Decision 10-2007-QD-TTg of the Prime Minister: The System of Economic Branches of Vietnam, Dated 23 January 2007, Prime Minister of the Government of Vietnam. Obtained through the Internet: http://vbqppl.moj.gov.vn/

[12] Huijbregts, M.A.J., Gilijamse, W., Ragas, A.M.J. and Reijnders, L. (2003) Evaluating Uncertainty in Environmental Life-Cycle Assessment. A Case Study Comparing Two Insulation Options for a Dutch One-Family Dwelling. *Environmental Science & Technology*, **37**, 2600-2608. http://dx.doi.org/10.1021/es020971+

[13] Sonnemann, G.W., Schuhmacher, M. and Castells, F. (2003) Uncertainty Assessment by a Monte Carlo Simulation in a Life Cycle Inventory of Electricity Produced by Using a Waste Incinerator. *Journal of Cleaner Production*, **11**, 279-292. http://dx.doi.org/10.1016/S0959-6526(02)00028-8

[14] Ali, G., Nitivattananon, V., Abbas, S. and Sabir, M. (2012) Green Waste to Biogas: Renewable Energy Possibilities for Thailand's Green Markets. *Renewable and Sustainable Energy Reviews*, **16**, 5423-5429. http://dx.doi.org/10.1016/j.rser.2012.05.021

18

Performance of Green Aggregate Produced by Recycling Demolition Construction Wastes (Case Study of Tanta City)

Alaa El-Din M. Sharkawi[1], Slah El-Din M. Almofty[2], Eng. Shady M. Abbass[1]

[1]Department of Structural Engineering, Faculty of Engineering, Tanta University, Tanta, Egypt
[2]Department of Mining, Petroleum and Metallurgical Engineering, Faculty of Engineering, Cairo University, Giza, Egypt
Email: amsharka@hotmail.com, shady_abbass2020@yahoo.com

Abstract

Egypt has a high attitude in construction and demolition waste (CDW) amounts causing a negative impact on the environment. The use of such waste for infrastructures applications can be useful for each environment and in addition an economic benefit to it in the construction. This study explores the possibility of replacing natural coarse aggregate with recycled concrete construction and demolition waste aggregate for general purpose concrete (*i.e.* plain concrete and low strength structural concrete). Different samples of CDW were extracted from different demolition sites and landfill locations around Tanta city area for the experimental investigation. CDW was crushed with all its hard constituents (e.g. concrete, brick etc.) found in the landfill. Coarse size crushed CDW was used as a coarse aggregate for concrete. Main characteristics of CDW aggregate determined in addition to the main properties of concrete which was made using this aggregate were measured. The results showed that the CDW could be transformed into recycled concrete aggregate leading to reduction in the concrete compressive strength ranged from 37% to 62% depending on the type of the CDW constituents.

Keywords

Concrete, Demolition, Recycling Aggregate, Recycling, Compressive Strength

1. Introduction

Generally, construction makes a considerable environmental impact through extraction of its raw materials. The use of excessive energy in production processes and production of accumulated byproduct waste are sources for damage to environment and health in different phases of the life cycle of hazardous components. The disposal of CDW has become a major concern in recent years, especially in developing countries. Some building owners, waste haulers and demolition contractors are disposing of this waste illegally, in open landfills, in order to avoid transportation costs at waste disposal facilities. As major a hazardous, construction and demolition waste is defined as the solid waste generated by the construction, remodeling, renovation, repair, alteration or demolition of residential, commercial, government or institutional buildings, industrial, commercial facilities and infrastructures such as roads, bridges, dams, tunnels, railways and airports. Egypt is one of the main countries which has huge surge in construction and demolition waste (CDW) quantities harmfully affecting the environment. The approximate amount of construction and demolition waste (CDW) in Egypt is 4.0 million tons annually [1], while the only current method of managing such waste is through disposal in landfills causing large deposits of CDW. Despite of many researches for studying the feasibility of recycling construction demolition wastes in the production of construction materials, no such local integrated results are available for application. On the other hand, using the recycled CDW instead of the rapidly diminishing natural resources provides essence of sustainable development due to the critical shortage of these resources (e.g. aggregate materials).

This research aims to explore various possible structural applications of selected samples of the local construction demolition wastes in order to increase the waste recycling opportunity with greater added value. This exploratory study is an introduction for comprehensive program to provide practical opportunities for the CDW recycling in Egypt as a sustainable resource of environment-friendly applications as well as a reference for current studies for preparing code of practice for CDW quality standards. On the other hand, such researches provide trust on using such recycled materials in several infrastructural applications.

2. Previous Work

In Egypt, recycled CDW can be used as an alternative of the natural aggregate for concrete and infrastructure applications to provide environment friendly solution for their huge landfills and to reduce the abuse of natural aggregate resources. Providing such sustainable recycled resource of aggregate preserves the natural resources for longer life span as well as reduces their demand. The use of recycled CDW in construction had started since the end of World War II by using a demolished concrete pavement as recycled aggregate in stabilizing the base course for road construction [2]. The major materials found in CDW at Egypt are concrete, bricks, sand, mortar and tile residues in which concrete represents up to half of the total waste weight [3]. Weigh *et al.* [4] discussed the possibility to replace natural coarse aggregate (NA) with recycled concrete aggregate (RCA) in structural concrete. A total of 50 concrete mixes forming eight groups were designed to study the effect of recycled coarse aggregates quality/content, cement dosage, use of super plasticizer and silica fume. The results showed that the concrete rubble could be transformed into useful recycled aggregate and used in concrete production with properties suitable for most structural concrete applications in Egypt. A significant reduction in the properties of recycled aggregate concrete (RAC) made of 100% RCA was seen when compared to natural aggregate concrete (NAC), while the properties of RAC made of a blend of 75% NA and 25% RCA showed no significant change in concrete properties. The suitability of returned concrete aggregate for concrete applications had been investigated by many researchers. It was found that replacing virgin aggregates with returned concrete aggregate had resulted in 10% - 30% reductions in compressive strength, with the least impact being found in mixes that only include recycled coarse aggregates.

3. Research Objectives

The main objectives of this research are to explore the opportunities of using CDW for various infrastructure applications. Therefore the following points are investigated:

1) The main characteristics of commonly recycled locally available construction and demolition waste (CDW) as an aggregate.

2) The performance of concrete made of recycled CDW coarse aggregate versus that made using natural aggregate. Fine recycled aggregate was not considered for the production of recycled aggregate concrete (RAC) because its application in structural concrete is generally not recommended [5].

This research has a vital effect on the production of sustainable materials for infrastructure applications and protection of the local environment. In order to accomplish the aim of the exploratory study, an investigation was undertaken into the properties of CDW collected from one of the main landfills in Tanta city—Egypt as a targeted area for further extended research work.

Experimental Program—Selecting the Recycled CDW Samples

Four samples (*i.e.* A, B, C and D) of construction and demolition wastes (CDW) were selected-visually-from four different main landfill areas at Tanta city as shown in **Figure 1**. Each sample had 300 kg average weight to represent the most common types of such wastes. Samples had been cleaned from harmful material to concrete or fragile substances (*i.e.* paper, plastics, textile, wood and glasses). The hard constituents of each sample of construction and demolition waste were classified, visually, and **Table 1** presents the constituents ratio. Two approaches were considered in using the CDW as concrete aggregate; 1-using the classified constituents separately or 2-using the mixed constituents of each sample. Sample's constituents were crushed using hammer crusher. Recycled CDW coarse aggregate size (4.75 - 37.5 mm) was sieved to be prepared to simulate the grading curve of the dolomite natural aggregate to be used as concrete aggregate. Natural dolomite crushed stone was used, as control concrete aggregate, for comparison with the crushed CDW. Natural sand was the only fine aggregate used in all concrete samples. Aforementioned process was used to recycle both CDW sample and some of its constituents. **Figure 2** shows different phases for preparing recycled, crushing and sieved CDW.

Figure 1. Map showing the sites that contain landfills and forms of recycled CDW.

Table 1. Approximate constituents ratios (by volume) of the selected construction and demolition wastes (CDW) samples extracted from four different locations.

Sample ID	Red brick %	Cement tiles %	Concrete %	White brick %
A	30%	20%	40%	10%
B	30%	20%	50%	-
C	45%	30%	25%	-
D	40%	30%	30%	-

Figure 2. Different phases for preparing recycled, crushing and sieved CDW.

4. Testing Recycled Concrete Aggregate

Testing of the all aggregate types (*i.e.* recycled CDW and natural) was carried out as per Egyptian code of practice (ECP 203-2008, vol. 3) [6]. The following aggregate characteristics were measured during this research work:

1) Grading,
2) Specific gravity,
3) Bulk density,
4) Void ratio,
5) Water absorption,
6) Elongation index,
7) Angularity index,
8) Crushing value.

5. Concrete Mixes and Testing

CEM I 42.5 N was used in different concrete mixes and it was tested out as per the Egyptian Standard Specifications ESS 2421/2007. The used cement complied with the limits of ESS 4756-1/2009. Potable tap water was used for concrete mixing all through the study.

Seven concrete mixes—as shown in **Table 2**—had been designed to compare the properties of CDW with natural aggregate. Generally all mixes had constant cement content (300 kg/m^3), water/cement ratio (0.55) and coarse/fine aggregate by volume ratios (2/1) as regularly used mix for all purpose concrete. Saturated surface dry (SSD) aggregate condition and absolute volume mix design method was used for all mixes. **Table 2** shows different concrete mixes defining control mix and concrete mixes made of CWD as course aggregate. Control mix was made of dolomitic aggregate, while three mixes, RA, RB and RC, were made of mixed CDW extracted from sources A, B and C respectively. On the other hand, for mix R1-Con, crushed concrete, separated from recycled CDW of samples A, B, C and D were used as coarse aggregate. Similarly, for mix R2-Brick, crushed Red Brick, separated from recycled CDW of samples A, B, C and D, were used as coarse aggregate. For mix R3-Tiles, crushed Tiles, separated from recycled CDW of samples A, B, C and D were used as coarse aggregate. **Table 3** shows different concrete compressive strengths at 7 and 28 days of different mixes defining control mix, mix of concrete mixes from all sites, mix of cement tiles from all sites, mix of red brick from all sites, mix of all components in site (A), mix of all components in site (B) and mix of all components in site (C).

The following properties of concrete were selected to be measured:

- Bulk density of hardened concrete.
- Slump of fresh concrete.
- Compressive strength (age 7, 28 days).
- Indirect tensile strength (age 28 days).

6. Results and Discussions

6.1. Recycled Concrete Aggregate

Figure 3 shows the grading curves for different types of recycled CDW aggregate which were prepared to have grading curves close to the dolomite natural aggregate. Grading curves of all aggregate types are within the ECP 203 concrete aggregate limits, as shown in **Figure 3**. **Figure 4** shows the relative properties of all types of the recycled aggregate with respect to the natural aggregate.

(a)

(b)   (c)

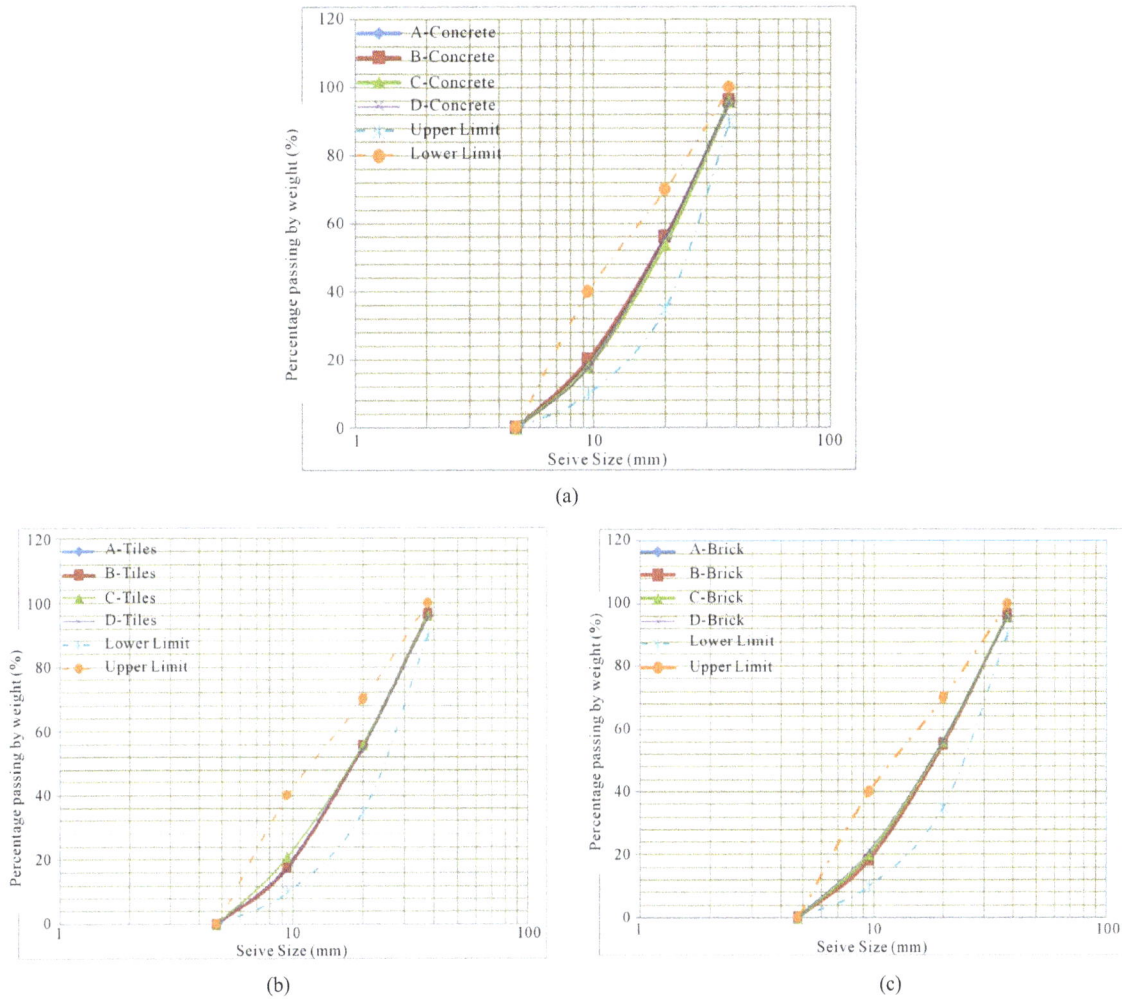

Figure 3. Grading curves for recycled aggregate with respect ECP limits. (a) Concrete in all sites; (b) Cement tiles in all sites; (c) Red brick in all sites.

Table 2. Shows different concrete mixes defining control mix (control) using dolomite and natural sand as a coarse.

Constituents/m³	Coarse aggregate source kg				Fine aggregate kg	Water kg	Cement kg
Concrete sample ID	Concrete	Cement tiles	Red brick	Dolomite			
Control	-	-	-	1284	642	165	300
RA	554	261	249	-	642	165	300
RB	528	382	202	-	642	165	300
RC	513	371	196	-	642	165	300

Constituents/m³	Coarse aggregate source kg					Fine aggregate kg	Water kg	Cement kg
Concrete sample ID	A	B	C	D	Total			
R1-Con	294	280	268	268	1110	642	165	300
R2-Brick	258	268	256	256	1038	642	165	300
R3-Tiles	268	280	280	268	1096	642	165	300

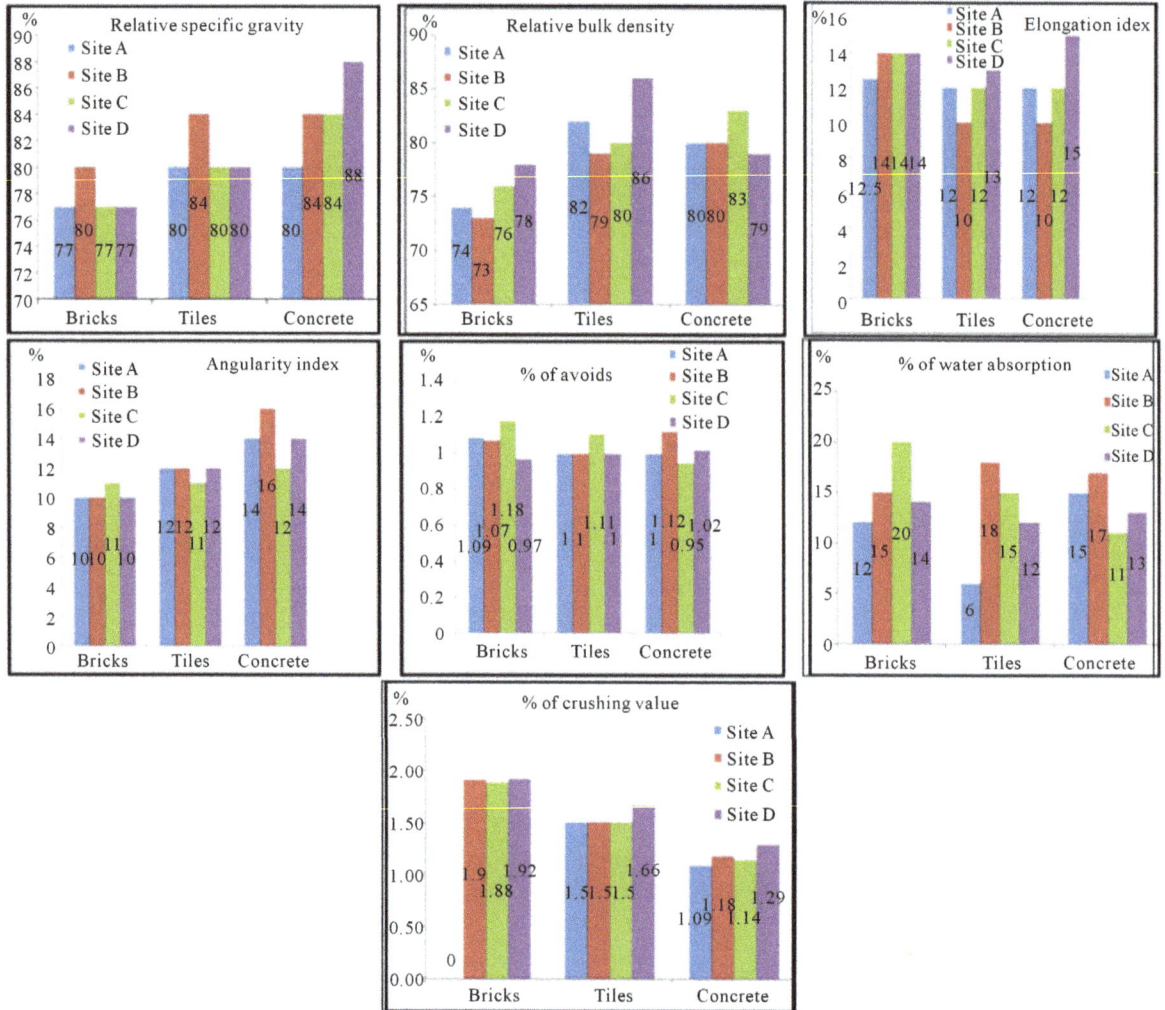

Figure 4. Shows the relative properties of all types of the recycled aggregate.

The relative specific gravity of recycled CDW was in the range of 77% - 88% of the natural course aggregate, while the bulk density of recycled CDW was in the range of 73% - 86% of the natural course aggregate. The water absorption of recycled CDW coarse aggregate was higher than the dolomitic aggregate by 6% to 20%. The crushing value of recycled CDW coarse aggregate was higher than the dolomitic aggregate by 1.09% to 1.92%. The angularity index of recycled CDW coarse aggregate was higher than the dolomitic aggregate by 10 to 16%. The elongation index of recycled CDW coarse aggregate was higher than the dolomitic aggregate by 10% to 15%. The percentage of avoids of recycled CDW coarse aggregate was higher than the dolomitic aggregate by 0.95 to 1.18%.

6.2. Concrete Mixes

Figures 5-7 show the properties of various concrete mixes which were made of different types of recycled CDW aggregate with respect to concrete made of natural aggregate properties. The maximum reduction of the compressive, indirect tensile, flexural and bond strengths were 37%, 16%, 26% and 64% respectively with using recycled CDW aggregate for making concrete with respect to natural concrete.

6.3. Slump Test Results

It is test done for measuring just coherence for the concrete mixture components and its result ranging between (2.8 - 3.1) cm.

6.4. Compressive Test Result

Table 3. Shows different concrete compressive strengths at 7 and 28 days.

N	Simple code	Age (day)	Compressive (kg/cm)
1	Control mix	7	230
2		28	355
3	Site (A)	7	183
4		28	224
5	Site (B)	7	112
6		28	184
7	Site (C)	7	93
8		28	134
9	Concrete Mix	7	132
10		28	211
11	Brick Mix	7	104
12		28	149
13	Tile Mix	7	117
14		28	152

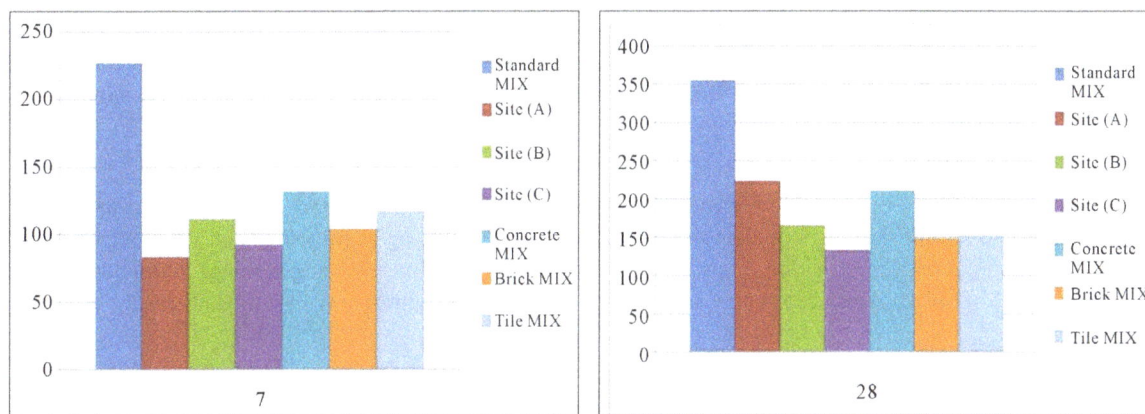

Figure 5. Shows the different compressive strengths at 7 and 28 days.

6.5. The Results of the Test of Flexural Strengths

Sample name	Strengths kg/cm^2
Mix control	48.36
Site (A)	35.76
Site (B)	-
Site (C)	26.64

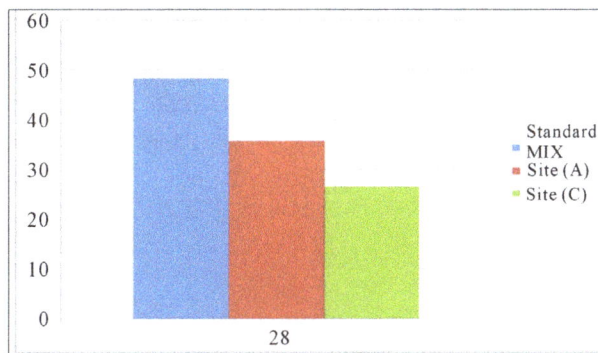

Figure 6. Shows the different flexural strengths at 28 days.

6.6. The Results of the Test of Indirect Tensile (Brazilian Tension)

Sample name	Strengths kg/cm^2
Mix control	21
Site (A)	17.6
Site (B)	13.5
Site (C)	12.8

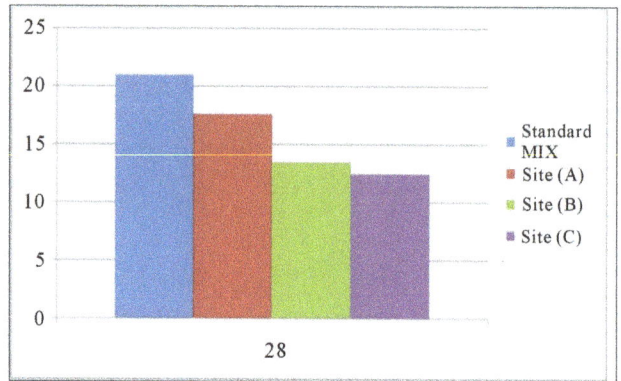

Figure 7. Shows the different indirect tensile strengths at 28 days.

7. Conclusions

Based on test results, the following conclusions are drawn:

1) Some properties of recycled concrete aggregate were lower than those required by Egyptian concrete code of practice such as crushing value, although it can be used for suitable low wearing/stress applications.

2) Construction and demolition waste (CDW) can be recycled to be used as concrete aggregate with properties suitable for infrastructure applications (e.g. pavement and sidewalk border). Few of the concrete mixes made of recycled CDW exceeded the lower strengths required for reinforced concrete (20 Mpa) where most of them exceeded the plain concrete strengths (17 Mpa).

3) In most of the concrete mixes made of recycled CDW aggregate, the regular ratios of natural aggregate concrete between both indirect tension and flexure strengths to compressive strength were fairly applied.

8. Further Study

Comprehensive study needs to be performed to investigate the following essential points to provide practical opportunity to recycle construction demolition wastes safely, providing economic added value which will cover the gap existing between providing promising research lab results for recycling residues and applying it in the main sources of the residues accordingly assure sustainable recycling process of such residues.

References

[1] Abou-Zeid, M.N., Shenouda, M.N., McCabe, S.L. and El-Tawil, F.A. (2004) Properties and Feasibility of Concrete made with Partial and Total Recycled Aggregates. *83rd Annual Meeting of the Transportation Research Board*, Washington DC, Submitted

[2] Afsa, A. (2011) Performance and Properties of Concrete Elements Made with Recycled Aggregate. M.Sc. Dissertation, Menoufia University, Egypt.

[3] Hansen, T.C. and Narud, H. (1983) Strength of Recycled Concrete Made from Crushed Concrete Coarse Aggregate. *Concrete International*, **5**, 79-83.

[4] Wagih, A., El-Karmoty, H., Ebid, M. and Okba, S.H. (2013) Recycled Construction and Demolition Concrete Waste as Aggregate for Structural Concrete. *HBRC Journal*, **9**, 193-200. http://dx.doi.org/10.1016/j.hbrcj.2013.08.007

[5] Hendriks, C. and Janssen, G. (2001) Application of Construction and Demolition Waste. *Heron*, **46**, 24-28.

[6] Ministry of Housing, Utilities & Urban Development (2008) Egyptian Code of Practice for Design and Construction of Reinforced Concrete Structures (ECP 203-2008, Vol. 3). Ministry of Housing, Utilities & Urban Development, Egypt.

19

Developing Sustainable High Strength Concrete Mixtures Using Local Materials and Recycled Concrete

Anthony Torres*, Alex Burkhart

Department of Engineering Technology, Texas State University, San Marcos, TX, USA
Email: *ast36@txstate.edu

Abstract

This study presents the development of high strength concrete (HSC) that has been made more sustainable by using both local materials from central Texas and recycled concrete aggregate (RCA), which has also been obtained locally. The developed mixtures were proportioned with local constituents to increase the sustainable impact of the material by reducing emissions due to shipping as well as to make HSC more affordable to a wider variety of applications. The specific constituents were: limestone, dolomite, manufactured sand (limestone), locally available Type I/II cement, silica fume, and recycled concrete aggregate, which was obtained from a local recycler which obtains their product from local demolition. Multiple variables were investigated, such as the aggregate type and size, concrete age (7, 14, and 28-days), the curing regimen, and the water-to-cement ratio (w/c) to optimize a HSC mixture that used local materials. This systematic development revealed that heat curing the specimens in a water bath at 50°C (122°F) after demolding and then dry curing at 200°C (392°F) two days before testing with a w/c of 0.28 at 28-days produced the highest compressive strengths. Once an optimum HSC mixture was identified a partial replacement of the coarse aggregate with RCA was completed at 10%, 20%, and 30%. The results showed a loss in compressive strength with an increase in RCA replacement percentages, with the highest strength being approximately 93.0 MPa (13,484 psi) at 28-days for the 10% RCA replacement. The lowest strength obtained from an RCA-HSC mixture was approximately 72.9 (MPa) (10,576 psi) at 7-days. The compressive strengths obtained from the HSC mixtures containing RCA developed in this study are comparable to HSC strengths presented in the literature. Developing this innovative material with local materials and RCA ultimately produces a novel sustainable construction material, reduces the costs, and produces mechanical performance similar to prepackaged, commercially, available construction building materials.

---

*Corresponding author.

Keywords

High Strength Concrete, Sustainability, Recycled Concrete Aggregate, Local Products, Construction Materials

1. Introduction

Green construction through sustainable building materials has been an important aspect in the concrete and construction field in the last decade. Using waste products in concrete production is beneficial environmentally and economically: environmentally by replacing a portion of the virgin components with waste materials and environmentally by clean disposal of waste materials. Combining these benefits with locally obtained materials ultimately increases these factors. As Texas is one of the largest producers of cement and aggregate in the nation, it is beneficial to develop a novel sustainable construction building material that utilizes locally available materials. Combining this factor with the reuse of concrete as Recycled Concrete Aggregate (RCA) can drastically increase the sustainable impact of concrete production. Reuse of concrete from demolished structures as aggregates was introduced into practice many years ago, and from the beginning it has been considered in two main environmental aspects: solving the increasing waste storage problem and protection of limited natural sources of aggregates [1]. Presently, most of the demolished concrete extracted from old structures is of relatively good quality [1]. This characteristic demonstrates the potential that recycled concrete aggregate could provide to high strength concrete.

In the past years, improvements have been occurring in concrete and construction material technology. Sustainable use of supplementary materials and revolutionary developments in chemical admixtures has facilitated improvements in the mechanical properties of concrete materials. Such mechanical properties that have drastically been impacted are the strength, density, and the modulus of concrete materials [2]. Through these developments higher strength concrete materials have emerged, known as High Strength Concrete (HSC). However, these chemical and material developments have consistently changed the definition of high strength concrete. The American Concrete Institute (ACI) has defined high-strength concrete as a concrete meeting a high strength that cannot always be achieved routinely, using conventional constituents and normal mixing, placing, and curing practice [2]. In the 1950s, concrete with a compressive strength of 34 MPa (5000 psi) was considered high strength [2]. Today, high-strength concrete is defined as concrete with a specified compressive strength of 55 MPa (8000 psi) or higher [2]. In many markets today, concrete having a specified compressive strength in excess of 69 MPa (10,000 psi) is routinely produced on a daily basis [2]. HSC itself is considered a sustainable construction building material due to the high requirement of silica fume and fly ash, both of which are waste products from other industries, which provides beneficial properties development of concrete [2]. Not only does HSC utilize a high percentage of fly ash and/or silica fume, the high strength requirement results in a higher specific strength of the material. The specific strength of a material is the ratio of the strength to its density. Since the density of HSC doesn't increase significantly to that of conventional concrete, but the strength does, the specific strength is much higher than that of conventional concrete. This aspect impacts sustainable construction through decreased transportation cost and emissions as a smaller structural member can be produced out of HSC, which will require less fuel and produce less carbon emissions during shipping.

In the present research, an attempt has been made to develop HSC mixtures with locally available materials and RCA. The material used in this study included Type I/II Portland cement, silica fume, dolomite, manufactured sand (limestone), and a High Range Water Reducing Agent (HRWRA). Factors such as the aggregate type and size, the curing regimen, the water-to-cement ratio (w/c), and age (7, 14, and 28-day) were investigated.

2. Literature Review

2.1. Background Information

HSC is traditionally composed of cement, coarse aggregate that is much smaller than conventional coarse aggregate $\leq$9.5-mm (0.375-in.) [2], fine aggregate, supplemental cementitious materials (SCM) such as silica fume, fly ash, granulated ground blast furnace slag (GGBFS), and quartz powder, fibers, and HRWRA. When used in optimum dosages, the HRWRA reduces the water-to-cement (or water-to-cementitious) ratio while improving the workability (viscosity) of the concrete. The addition of the SCMs enhances the mechanical properties of the cement paste by producing secondary hydrates, filling voids, and enhancing rheology [3]. Due to the burgeoning large structure industry, there are more and more requirements for higher strength concrete. HSC has been in development since the mid 1950's and many researchers are still investigating the optimum and most efficient manner to produce this material [2]. However, modern construction practices are moving towards sustainable construction through lower cost and sustainable materials, therefore, HSC improvements should also be focused on becoming more sustainable and more affordable.

The Environmental Protection Agency (EPA) has estimated that approximately 58% of landfill waste is from construction debris, which a significant portion is concrete and masonry rubble [5]. Recycling waste concrete in the production of new concrete has been investigated since the 1970's [4]-[12]. These studies report a wide variety of results concerning the impact of the RCA on the mechanical and durability properties of the new concrete. The studies have shown that the mechanical properties depend on the properties of the recycled concrete used as well as the amount used. For example, Ravindraraj [9] reported a 9% decrease in compressive strength with a 100% coarse aggregate replacement with RCA, whereas Yamato et al. [10] measured a 45% decrease. Reports have shown the same for the modulus of elasticity in which Gerardu and Hendriks [11] demonstrated a 15% decrease with a 100% replacement, while Frondistou-Yannas [12] reported a 40% decrease at high w/c (0.75) and no decrease at lower w/c (0.55). Due to the wide variation in the properties of concrete with RCA, more research is needed on the local materials used to produce the new and recycled concrete in order to better understand the combined mixture.

2.2. Sustainability

This study focuses on two aspects of sustainable construction building materials; local use of constituent materials and the use of recycled materials. A major concern in the production of HSC is the high cost from shipping such materials as quartz dust, steel or specialty aggregates, and fibers. Most of these constituents are often shipped long distances and internationally in many cases, which increases the cost of the material. It should also be noted that due to the chemical interaction requirements of the silica fume and cement drastically increases the cost of commercially available, prepackaged, HSC products [3]. The commercially available HSC from Ductal uses expensive materials such as ground quartz and fibers that are not traditionally available locally, which increases the cost of the final product. Therefore, the present work focuses on developing HSC mixtures using local materials so that HSC may be made more affordable to wide variety of construction applications. Using local products drastically decreases emissions associated with long shipping routes. The second focus of this study is producing HSC with the inclusion of RCA as partial replacement of virgin coarse aggregate. Replacing virgin materials used in the production of HSC with recycled materials drastically increases the sustainable impact of the material. By using discarded waste material in new construction, the strain for new, virgin, materials is slightly alleviated while also minimizing the demand for landfill space. Additionally, the Leadership in Energy and Environmental Design (LEED) provides a material credit for using building materials or products that have been extracted, harvested or recovered, as well as manufactured, within 500 miles of the project site for a minimum of 10% or 20%, based on cost, of the total material's value [4]. Therefore, using both local materials that make up 100% of the concrete product and using RCA will count double for this LEED requirement.

3. Experimental Program

3.1. Materials

The final HSC mix design consisted of dolomite coarse aggregate (1.18-mm [0.0469-in.]), manufactured sand, known as "man" sand, which is crushed from limestone with a size of 0.105-mm - 0.60-mm (0.0059-in. - 0.0232-in.), Type I/II cement, silica fume, recycled concrete aggregate with a size of 1.18-mm (0.0469-in.) to match the size of the virgin coarse aggregate, and HRWRA (Master Glenium 3030 from BASF Chemicals). All constituents were obtained from local providers within a 50-mile radius of San Marcos, TX. The aggregate was obtained from local quarries and sieved in the laboratory to achieve a specific size and gradation. The coarse aggregate size was minimized and held at an individual specific size based off the literature [2] [3] [5]-[8]. **Table 1** shows the grain size distribution for the fine sand. **Table 2** shows the chemical compositions of the Type

Table 1. Grain size distribution for the manufactured sand.

Sieve No.	Sieve size, mm (in.)	Percentage passing
16	1.18 (0.0469)	100
30	0.60 (0.0236)	42.3
50	0.30 (0.0118)	0.0

Table 2. Chemical composition of cement and silica fume.

Compound	Cement	Silica Fume
SiO_2	21.5%	95.8%
Al_2O_3	4.45%	0.18%
Fe_2O_3	3.15%	0.19%
CaO	64.10%	0.30%
K_2O	NA	0.29%
Na_2O	0.52% (Equiv.)	0.20%
MgO	1.90%	0.20%
SO_3	2.89%	0.11%

I/II Portland cement and silica fume. **Table 3** shows the physical properties of the coarse aggregate and the RCA. Information about the trial batches can be found in Section 3.2.

3.2. Concrete Mixtures

In order to develop a HSC mixture that used both local materials and RAC, three varying mixtures were developed to first produce a HSC baseline, once an optimum HSC mixture was reached a partial replacement of the coarse aggregate was replaced with RCA, completed in 10%, 20%, and 30% increments. The literature has shown that an upper limit of 30% be used in order to maintain the standard requirement of 5% absorption capacity of aggregates for structural concrete [13] [14]. The literature has also shown that replacement percentages of approximately 15% - 40% begin to diminish the strength of the concrete compared to the control mixture [6] [13] [14]. Therefore, three replacement percentages of 10%, 20%, and 30% were selected for this study. Section 4 describes the results obtained from the trail mixtures and the recommended final RCA-HSC mixture. All trial mixtures were developed based off the recommendations of the literature [1]-[3] [13] [14] and off of data obtained from each trial. The individual trial mixtures were categorized as follows:

Trial A: The mixtures in this category used Type I/II Portland cement, HRWRA, and silica fume. The aggregate for trial group A was Limestone coarse aggregate with a top size of 1.18-mm (0.0469-in.) and manufactured sand with an approximate size range of 0.105-mm - 0.60-mm (0.0059-in. - 0.0232-in.). Three water-to-cement ratios were investigated of 0.32, 0.30, and 0.28.

Trial B: The mixtures in this category used Type I/II Portland cement, HRWRA, and silica fume. The aggregate for trial group B contained Dolomite coarse aggregate with a top size of 2.36-mm (0.093-in.) and Dolomite fine aggregate with an approximate size of 1.18-mm (0.0469-in.). Three water-to-cement ratios were investigated of 0.32, 0.30, and 0.28.

Trial C: The mixtures in this category used Type I/II Portland cement, HRWRA, and silica fume. The aggregate for trial group C consisted of Dolomite coarse aggregate with a top size of 1.18-mm (0.0469-in.) and manufactured sand with an approximate size range of 0.105-mm - 0.60-mm (0.0059-in. - 0.0232-in.). Three water-to-cement ratios were investigated of 0.32, 0.30, and 0.28.

As shown in the three trial mixtures investigated, many of the mixture constituents were held constant in order to determine the impact of the local aggregates and aggregate size. Therefore, HRWRA, silica fume, and Type I/II cement were held constant and weighted based off of the literature. As the literature vastly differs on

Table 3. Physical properties of coarse aggregate and RCA.

Property	Standard	Unit	Limestone Coarse Agg.	Dolomite Coarse Agg.	RCA
Unit Weight	ASTM C29	kg/m^3 (lb/ft^3)	1442 (90.0)	1859 (116.0)	1411 (88.0)
Water absorption	ASTM C127	%	2.98	3.19	4.12
Bulk Specific Gravity$_{ssd}$[a]	ASTM C127	-	2.57	2.72	2.42
Bulk Specific Gravity$_{od}$[b]	ASTM C127	-	2.51	2.68	2.32

[a]ssd, saturated surface dry condition; [b]od, oven dried condition.

the water-to-cement ratio, three ratios were investigated per trial batch. Due to this change, the HRWRA may vary slightly to maintain a consistent flow-ability/workability of the concrete. As the results will demonstrate, trail C produced the highest results with a w/c of 0.28. Therefore, trial mixture C was used as the baseline HSC for coarse aggregate replacement of RCA. **Table 4** displays the mixture proportions for all the trial batches and the RCA-HSC mixtures.

3.3. Specimen Preparation

The aggregate used in this study (coarse and fine) were sieved to obtain the desired size needed as described previously. The aggregates were then thoroughly washed over a No. 200 sieve to remove any fine dust or debris. After washing, the aggregates were oven dried at 44°C (110°F) to achieve a 0% moisture content.

The constituents of each mixture were then mixed for approximately 20 minutes using a laboratory pan mixer. The dry constituents (aggregate, cement, silica fume) were mixed for the first 2 minutes and then 75% of the water was added. After thorough mixing, the HRWRA was added with the remaining 25% of the water. This preparation method was used based off of the literature and experience [1]-[3] [13]-[16].

3.4. Curing Regimens

In order to minimize as many variables as possible, three curing regimens were investigated on the first mixture developed (Trial A − w/c = 0.32). Trial mixture A was selected to determine the impact of the curing regimen, as it was the first mixture designed and batched. For the first regimen, concrete specimens were cured at room temperature (23°C [73°F]) for the first 24 hours. Once the specimens were demolded, they were moist cured at 23°C (73°F) and a relative humidity of 98% until the day of testing. This curing method is a traditional curing method for conventional concrete as outline in ASTM C192-15 [17]. For reporting purposes this curing method will be reported as Traditionally Cured (TC).

The second curing regimen consisted of curing the specimens at room temperature 23°C (73°F) for the first 24 hours. After the specimens were demolded, the specimens were heat cured in a water bath at a temperature of 50°C (122°F) until the time of testing. For reporting purposes this curing regimen will be reported as Heated Bath Cured (HBC).

The third curing regimen also cured the samples at room temperature 23°C (73°F) for the first 24 hours. After demolding, the specimens were heat cured in a water bath at 50°C (122°F) until 2 days prior to testing. At two days prior to testing, the specimens were removed from the water bath and dry cured at 200°C (392°F). For reporting purposes this curing method will be designated as Oven Dried Curing (ODC). The above curing regimens were developed based on the study by Shaheen (2006) *et al.* [18].

3.5. Compression Testing

Compressive strength specimens were molded using 50.8-mm (2-in.) cube molds. Cubes specimens were used to avoid problems with end preparation of cylindrical specimens [2]. After the specimens were properly cured they were individually tested according to BS 12390-3-2009 [19]. The British Standard was used as it provides greater detail to testing hardened concrete cubes in compression than ASTM C 39-15a [19] [20]. An average of three samples were tested per data point reported in the results section. Therefore, a minimum of nine specimens was produced per mixture per w/c in order to obtain a reportable value. Trail mixture A required additional samples as this was the mixture that was used to investigate the curing regimen.

4. Results and Discussion

4.1. Curing Regimen

The curing regimen results are discussed first as the curing variable was determined prior to curing the majority of the specimens. The first mixture listed in **Table 4** (Trial A – w/c = 0.32) was used as the baseline mixture to determine the impact of the curing regimen. Therefore enough specimens were prepared using the batch quantities for this mixture in order to determine the affect of the three curing regimens as outlined above. **Figure 1** shows the results of the three curing methods on Trial A – w/c = 0.32).

As seen in **Figure 1**, the curing regimen significantly influenced the compressive strength of the HSC. The specimens that were cured using the traditional curing method resulted in the lowest average strengths. This is expected as the literature suggests that water curing, such as "full immersion" curing is required for water-to-cement ratios below 0.40 [2]. This is due to the degree of hydration being significantly reduced due to less water in the mixture, therefore it is beneficial to fully immerse the specimens during curing to support additional hydration [2]. Specimens that were cured using the hot bath curing method demonstrated an increase in strength versus the traditional curing method. This curing regimen not only provided a full immersion curing method, but

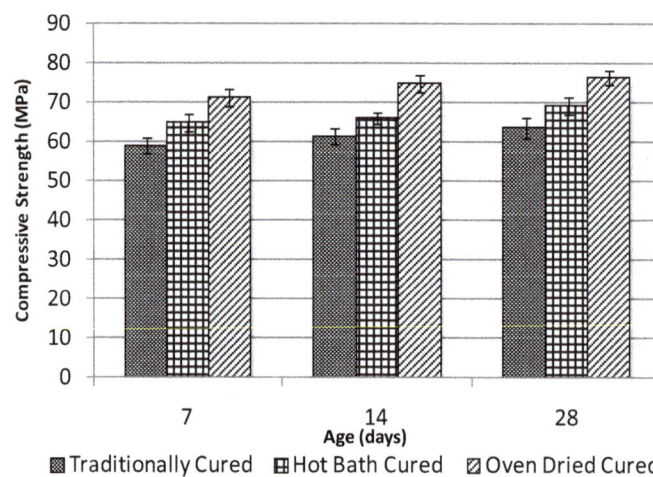

Figure 1. Impact of curing regimen on the compressive strength of Trial A – w/c = 0.32 at varying ages.

Table 4. Mixture proportions of the trail mixtures and RCA-HSC—kg/m³ (lb/yd³).

Mixture Name	w/c	Cement	Silica Fume	Coarse Aggregate	Fine Aggregate	RCA	Water	HRWRA
	0.32						166 (280)	16 (27)
Trial A	0.3	520 (876)	100 (69)	825 (1391)	685 (1155)	N/A	156 (263)	18 (30)
	0.28						145 (244)	20 (34)
	0.32						166 (280)	16 (27)
Trial B	0.3	520 (876)	100 (69)	825 (1391)	685 (1155)	N/A	156 (263)	18 (30)
	0.28						145 (244)	20 (34)
	0.32						166 (280)	16 (27)
Trial C	0.3	520 (876)	100 (69)	825 (1391)	685 (1155)	N/A	156 (263)	18 (30)
	0.28						145 (244)	20 (34)
RCA-HSC-10				742.5 (1252)		82.5 (139)		16 (27)
RCA-HSC-20	0.28	520 (876)	100 (69)	660 (1112)	685 (1155)	165 (278)	145 (244)	18 (30)
RCA-HSC-30				577.5 (973)		247.5 (417)		20 (34)

also an increase in curing temperature as the water was held at 50°C (122°F). This form of curing has being used by many authors in order to promote extra hydrates and develop higher early strengths [5]-[8] [12]. This curing regimen resulted in an average 7.3% increase in compressive strength over the traditional curing method. The last curing method, oven dried curing, which utilized a heated water bath and an oven for the last two days prior to testing produced the highest compressive strengths. The oven dried curing resulted in an average 18.8% increase over the traditional curing method and an average 10.5% increase over the hot bath cured. This increase of compressive strength is attributed to acceleration of the hydration reaction when the specimens were kept in the water bath and the formation of secondary calcium silicate hydrate (CSH) from the pozzolanic reaction of the silica fume when the specimens were kept in the oven at 200°C (392°F) for two days prior to testing. These results are consistent with similar results as observed from the literature [5]-[8] [12]. To determine if there is a statistical difference between the three curing regimens a sample t-test was performed. The test was performed with a 95% confidence level and the statistical significance (p value) considered at 0.05 level of confidence was used to analyze the data. All p values recorded were less than 0.05; therefore there is a difference between each curing method. Based off the observation of the curing regimen, all specimens prepared and discussed in the results section have been cured utilizing the oven dried curing regimen.

4.2. Impact of the Water-to-Cement Ratio

The next variable that was investigated in the development of a sustainable HSC mixture was the water-to-cement ratio (w/c). It has been well documented that a lower w/c results in a higher strength concrete and for HSC the w/c should be lower than 0.4 [2]. As previously stated, three w/c ratios were investigated for this study, which are 0.32, 0.30, and 0.28. **Figure 2** shows the impact of the w/c on the compressive strength of the specimens from Trial batches A-C.

As shown in **Figure 2**, the w/c has a significant impact on the strength of the three trial HSC mixtures. Although, the three w/c are only separated by two tenths, the impact is significant in such a controlled mixture as HSC. All three trial batches show that the highest strength occurs at a w/c of 0.28, which is the lowest tested for this study. This is also the case for the age of the HSC. These results are similar to w/c as presented in the literature [5]-[8] [12]. Due to the tight control of these mixtures, the average standard deviation for all three trials at each age was approximately 2.16. A t-test was also completed for the impact of the w/c ratio, which also revealed a difference between each grouping. Therefore, the w/c in fact has an impact on the developed mixtures, as is expected based off the literature and experience. **Figure 2** also compares the range of the three trial mixtures; such that Trial C demonstrates the highest recorded compressive strengths. Based off of these results a w/c of 0.28 was selected for the RCA-HSC. These results and results for each trial batch at this w/c are presented in subsequent sections.

4.3. Effect of RCA

The compressive strength was investigated for all trial batches and RCA-HSC specimens produced. To investigate the repeatability of the compressive strengths of these mixtures, testing was conducted on three specimens cast from each of the batches. As previously mentioned the compressive strengths were measured at 7, 14, and 28 days and 50.8-mm (2-in.) cubes were used. To facilitate a better coating of cement paste over the aggregate

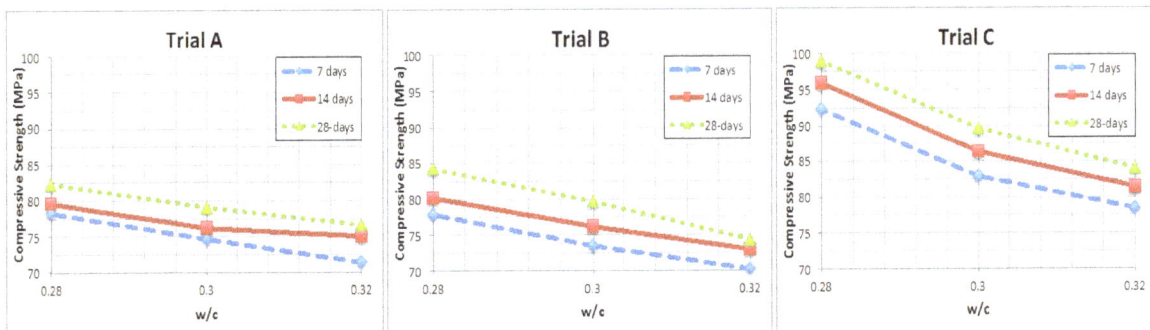

Figure 2. Impact of w/c on the compressive strength of HSC.

particles and increase the compressive strength, the dust was removed by washing the aggregates over a No. 200 sieve. The average compressive strengths of the trial mixtures can be seen in **Figure 3**.

Figure 3 demonstrates the impact of age versus the strength of the three trial HSC mixtures. Recall that all of these samples have a consistent w/c of 0.28 and are all cured using the oven dried curing regimen. As expected, the average compressive strength increases in strength as the concrete ages. Additionally, the average standard deviation for this data set was approximately 2.20 due to the tight control of the mixtures. Comparing the three trial mixtures reveals that Trial C produces the highest overall compressive strengths at every age. Trial C has an average percent increase of 25% versus the average compressive strengths of Trial A. Comparing Trials B and A demonstrates an average increase in compressive strength from B to A of only 5%. The increase in strength is due to the size and type of aggregate used. Trial A utilizes limestone coarse aggregate and manufactured sand that is produced from crushed limestone. Limestone is known to be a moderately strong aggregate, but is most commonly used in concrete construction due to it being readily available [21]. Dolomite aggregate was used in Trials B and C and it was selected as it's also readily available in central Texas, but it has a slightly higher density and strength compared to limestone [22]. However, Trial C uses both dolomite as a coarse aggregate and manufactured (limestone) sand as a fine aggregate. This was done due to the higher density of dolomite and the results obtained by Aquino *et al.* (2010) which shows that fine limestone, such as manufactured sand is known to improve the strength of concrete [21]. Therefore, Trial C combines the impact of both types of locally available aggregates. It is understood that a secondary cause of the minimal increase in strength from trial A to B was due to the size of the aggregates. The smallest aggregate used in Trial B was 1.18-mm (0.0469-in.), which was the largest aggregate size used in both Trials A and C. It should also be noted that all of the strengths at each age and from each trial were higher than 70 MPa (10,150 psi), which is above the range for consideration as HSC. The highest reported strength was from Trial C at a 28-day strength of 98.9 MPa (14,344 psi). Since Trial C had the highest performance, it was selected as the mixture to be used with the RCA replacement. Based off the literature, RCA replacement percentages for coarse aggregates should be less than or equal to 30% replacement by mass, therefore three replacement percentages were selected of 30%, 20%, and 10%. The compressive strength verses age for the RCA-HSC mixtures can be seen in **Figure 4**.

Figure 4 displays the compressive strength results for the developed sustainable high strength concrete using RCA and local materials. Trial C compressive strength data was included in this plot for comparison as this mixture was the baseline for all RCA-HSC mixtures. The results ultimately show a decrease in strength by including any percent replacement percentage investigated in this study. The decrease in strength increased with an increase in percentage of RCA used, *i.e.* the highest strength loss came from the 30% replacement of RCA coarse aggregate. The average strength loss was 7.9%, 14.4%, and 22.6% for the 10%, 20%, and 30% RCA inclusion respectively and an average standard deviation for each grouping of 2.63. This decrease in compressive strength is typical for most concrete mixtures containing RCA, as the quality of the RCA aggregate is significantly lower than virgin aggregates [5]-[8] [12]. The major issue when dealing with RCA is the fact that the

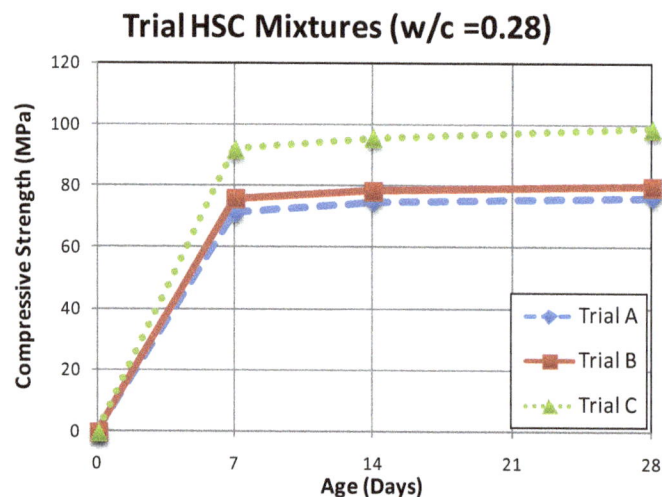

Figure 3. Average compressive strength of the trial HSC mixtures.

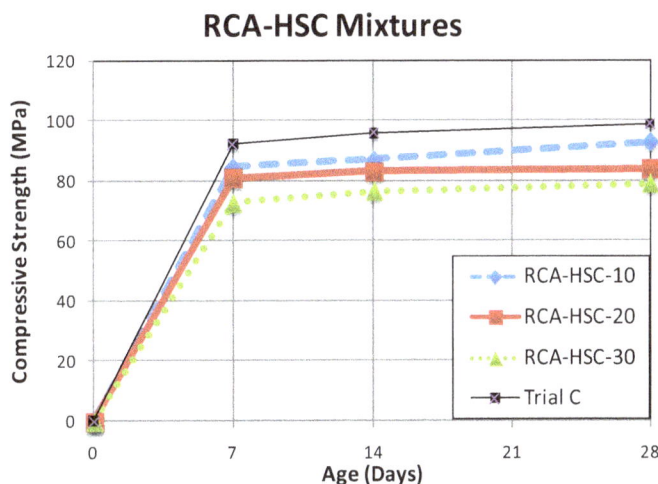

Figure 4. Average compressive strength of the RCA-HSC mixtures.

aggregate type and condition is unknown. Additionally, it is often difficult to remove debris or foreign matter that could be embedded in the RCA that could affect the performance of the new concrete. Although, there is a minor reduction in strength when using RCA as partial coarse aggregate replacement in HSC, it should be noted that the lowest recorded strength from RCA-HSC-30 was 79.1 MPa (11,476 psi) at 28-days. This strength is still high enough to be considered HSC and is also approximately the same strength at 28-days of the Trial B mixture that contained no RCA elements. A drop in strength was expected based off the literature; therefore the highest performing trial mixture was selected such that the novel sustainable construction building material would still have a high strength.

5. Conclusion

A novel sustainable construction building material has been developed that has high strength, uses recycled elements, and all the constituents were obtained locally. This study focused on developing a novel construction building material that can impact the sustainable construction building movement by not only developing a HSC, but the HSC has been made more sustainable by utilizing RCA and all local constituents. This study focused on maintaining consistent variables in order to assess the impact of the RCA on the HSC. This was developed by first assessing the impact of the curing regime on the strength of the concrete. Three regimens were investigated, which were cultivated based off the literature. Once an optimum curing regimen was reached, the w/c ratio was investigated. Again three different w/c ratios were investigated, which were also selected based off the literature. After the optimum curing regimen and w/c ratio was obtained, three different mixture designs were developed and compared. The three mixture designs consisted of two different locally available aggregate types and the size of the aggregate. The compressive strength measurements were compared at three testing ages of 7, 14, and 28 days. The highest performing HSC mixture was then used to investigate the impact of the RCA. RCA was used to partially replace the coarse aggregate in the top performing trial mixture at 10%, 20% and 30% replacement by mass. The results show that the RCA-HSC had a low end strength of 72.9 MPa (10,576 psi) at 7 days from the 30% RCA replacement and a high end strength of 93.0 MPa (13,484 psi) at 28 days from the 10% RCA replacement. Therefore, a novel sustainable HSC mixture has been produced that utilizes 100% local materials.

References

[1]   Andrzej, A and Alina, K (2002) Influence of Recycled Aggregates on Mechanical Properties of HS/HPC. *Cement and Concrete Composites*, **24**, 269-279. http://dx.doi.org/10.1016/S0958-9465(01)00012-9

[2]   ACI Committee 363 (2010) Report on High-Strength Concrete. ACI 363R-10, American Concrete Institute Committee 363, Farmington Hills, MI.

[3]   Dili, A.S. and Santhanam, M. (2004) Investigations on Reactive Powder Concrete: A Developing Ultra High Strength Technology. *The Indian Concrete Journal*, **74**, 33-38.

[4]   Azhar, S., Carlton, W.A., Olsen, D. and Ahmad, I. (2011) Building Information Modeling for Sustainable Design and LEED® Rating Analysis. *Automation in Construction*, **20**, 217-224. http://dx.doi.org/10.1016/j.autcon.2010.09.019.

[5]   Rahal, K. (2007) Mechanical Properties of Concrete with Recycled Coarse Aggregate. *Building and Environment*, **42**, 407-415. http://dx.doi.org/10.1016/j.buildenv.2005.07.033

[6]   Buck, A.D. (1977) Recycled Concrete as a Source of Aggregate. *ACI Journal*, **74**, 212-219.

[7]   Hansen, T.C. and Narud, H. (1983) Strength of Recycled Concrete Made from Crushed Concrete Coarse Aggregate. *Concrete International*, **1**, 79-83.

[8]   Forster, S.W. (1986) Recycled Concrete as Aggregate. *Concrete International*, **8**, 34-40.

[9]   Ravindraraj, R.S., Steward, M. and Greco, D. (2000) Variability of Recycled Concrete Aggregate and Its Effect on Concrete Properties—A Case Study in Australia. *International Workshop on Recycled Concrete*, JSPS 76 Committee on Construction Materials, Tokyo, September 2000, 27-42.

[10]   Yamato, T., Emoto, Y. and Soeda, M. (1998) Mechanical Properties, Drying Shrinkage and Resistance to Freezing and Thawing of Concrete Using Recycled Aggregate. *ACI Special Publication SP* 179-7, American Concrete Institute, Farmington Hills, MI, 105-121.

[11]   Gerardu, J.J.A. and Hendriks, D.F. (1985) Recycling of Road Pavement Materials in the Netherlands. Rijkswaterstaat Communications No. 38, The Hague.

[12]   Frondistou-Yannas, S. (1977) Waste Concrete as Aggregate for New Concrete. *ACI Journal*, **78**, 373-376.

[13]   Ajdukiewicz, A. and Kliszczewicz, A. (2002) Influence of Recycled Aggregates on Mechanical Properties of HS/HPC. *Cement and Concrete Composites*, **24**, 269-279. http://dx.doi.org/10.1016/S0958-9465(01)00012-9

[14]   Tu, T.-Y., Chen, Y.-Y. and Hwang, C.-L. (2006) Properties of HPC with Recycled Aggregates. *Cement and Concrete Research*, **36**, 943-950. http://dx.doi.org/10.1016/j.cemconres.2005.11.022

[15]   Rao, A., Jha, K.N. and Misra, S. (2007) Use of Aggregates from Recycled Construction and Demolition Waste in Concrete. *Resources, Conservation and Recycling*, **50**, 71-81. http://dx.doi.org/10.1016/j.resconrec.2006.05.010.

[16]   Evangelista, L. and de Brito, J. (2007) Mechanical Behaviour of Concrete Made with Fine Recycled Concrete Aggregates. *Cement and Concrete Composites*, **29**, 397-401. http://dx.doi.org/10.1016/j.cemconcomp.2006.12.004.

[17]   ASTM Standard C192-15 (2015) Standard Practice for Making and Curing Concrete Test Specimens in the Laboratory. *ASTM International*, West Conshohocken, PA. www.astm.org

[18]   Shaheen, E. and Shrive, N.J. (2006) Optimization of Mechanical Properties and Durability of Reactive Powder Concrete. *ACI Materials Journal*, **103**, 444-451.

[19]   BS EN 12390-3:2009 (2009) Testing Hardened Concrete Compressive Strength of Test Specimens B/517/1, 22.

[20]   ASTM Standard C 39-15a (2015) Standard Test Method for Compressive Strength of Cylindrical Concrete Specimens. ASTM International, West Conshohocken, PA. www.astm.org

[21]   Aquino, C., Inoue, M., Miura, H., Mizuta, M. and Okamoto, T. (2010) The Effects of Limestone Aggregate on Concrete Properties. *Construction and Building Materials*, **24**, 2363-2368. http://dx.doi.org/10.1016/j.conbuildmat.2010.05.008

[22]   Mikhailova, O., Yakovlev, G., Maeva, I. and Senkov, S. (2013) Effect of Dolomite Limestone Powder on the Compressive Strength of Concrete. *Procedia Engineering*, **57**, 775-780. http://dx.doi.org/10.1016/j.proeng.2013.04.098

Permissions

List of Contributors

C. G. Iñiguez
Departamento de Madera, Celulosa y Papel, Centro Universitario de Ciencias Exactas e Ingenierías, Universidad de Guadalajara, Guadalajara, México

C. J. J. Bernal, M. W. Ramírez and N. J. Villalvazo
Departamento de Ingeniería de Proyectos, Centro Universitario de Ciencias Exactas e Ingenierías, Universidad de Guadalajara, Guadalajara, México

Liangmou Yu
Faculty of Environmental Engineering, Kunming Metallurgy College, Kunming, China

Bo Shu and Shiwen Yao
Yunnan Copper Co., Ltd., Kunming, China

Heesup Choi and Myungkwan Lim
Department of Civil Engineering, Kitami Institute of Technology, Hokkaido, Japan
Graduated School of Engineering, Hankyong National University, Ansung, Korea

Hyeonggil Choi, Ryoma Kitagaki and Takafumi Noguchi
Department of Architecture, The University of Tokyo, Tokyo, Japan

Mats Zackrisson, Christina Jönsson and Elisabeth Olsson
Energy and Environment Group, Department of Materials, Swerea IVF AB, Mölndal, Sweden

Robert E. Baier
Industry/University Center for Biosurfaces, 110 Parker Hall, State University of New York at Buffalo, Buffalo, NY, USA

Francesco Di Maio and Peter Carlo Rem
Delft University of Technology, Delft, The Netherlands

Myungkwan Lim and Heesup Choi
Department of Archtecture Engineering, Hankyong National University, Gyeonggi-do, Korea

Amanda Louise Hill
Department of Development and Planning, Aalborg University, Aalborg, Denmark

Ole Leinikka Dall
Institute of Chemical Engineering, Biotechnology and Environmental Technology, University of Southern Denmark, Odense, Denmark

Frits Møller Andersen
DTU Management Engineering, Technical University of Denmark, Roskilde, Denmark

Rassel Raihan and Kenneth Reifsnider
University of Texas at Arlington, Arlington, USA

Fazle Rabbi and Vamsee Vadlamudi
University of South Carolina, Columbia, USA

Roberto Galindo, Isabel Padilla, Olga Rodríguez, Ruth Sánchez-Hernández and Aurora López-Delgado
National Centre for Metallurgical Research, CENIM-CSIC, Madrid, Spain

Sol López-Andrés
Department of Crystallography and Mineralogy, Faculty of Geology, UCM, Madrid, Spain

Heinz Langhals, Dominik Zgela and Thorben Schlücker
Department of Chemistry, LMU University of Munich, Munich, Germany

Bente Foereid
Bioforsk—Norwegian Institute for Agricultural and Environmental Research, Ås, Norway

Awadesh Kumar Mallik and Sandip Bysakh
CSIR-Central Glass & Ceramic Research Institute, Kolkata, India

Radhaballabh Bhar
Department of Instrumentation Science, Jadavpur University, Kolkata, India

Shlomo Z. Rotter and Joana Catarina Mendes
Instituto de Telecomunicações, Campus Universitário de Santiago, Aveiro, Portugal

Lorena Eugenia Sánchez Cadena and Q. Demetrio Quiroz
Departamento de Ingeniería Civil, Universidad de Guanajuato, DI, Av. Juárez 77, CP 36000, Guanajuato, Gto., México

Zeferino Gamiño Arroyo and Mario Alberto González Lara
Departamento de Ingeniería Química, Universidad de Guanajuato, DCNyE, Noria Alta s/n, CP 36050, Guanajuato, Gto., México

Nobuaki Yamaguchi, Hideaki Narusawa, Cho Han-Cheol and Takashi Miyazaki
Department of Conservative Dentistry, Division of Biomaterials & Engineering, Showa University School of Dentistry, Tokyo, Japan

Yoshiko Masuda and Yoshishige Yamada
Department of Conservative Dentistry, Division of Endodontology, Showa University School of Dentistry, Tokyo, Japan

Yukimichi Tamaki
Department of Dental Materials Science, Asahi University School of Dentistry, Gifu, Japan

Christian Dreyer, Dominik Söthje and Monika Bauer
Fraunhofer Research Institution for Polymeric Materials and Composites PYCO, Teltow, Germany

Mohammed El-Sayed El-Mahrouk
Horticulture Department, Faculty of Agriculture, Kafrelsheikh University, Kafr El Sheikh, Egypt

Yaser Hassan Dewir
Horticulture Department, Faculty of Agriculture, Kafrelsheikh University, Kafr El Sheikh, Egypt
Plant Production Department, College of Food and Agriculture Science, King Saud University, Riyadh, Saudi Arabia

Heinz Langhals, Dominik Zgela and Thorben Schlücker
Department of Chemistry, LMU University of Munich, Munich, Germany

Yasuhiro Matsui and Do Thi Thu Trang
Graduate School of Environmental and Life Science, Okayama University, Okayama, Japan

Nguyen Phuc Thanh
Hitachi Zosen Corp, Osaka, Japan

Alaa El-Din M. Sharkawi and Eng. Shady M. Abbass
Department of Structural Engineering, Faculty of Engineering, Tanta University, Tanta, Egypt

Slah El-Din M. Almofty
Department of Mining, Petroleum and Metallurgical Engineering, Faculty of Engineering, Cairo University, Giza, Egypt

Anthony Torres and Alex Burkhart
Department of Engineering Technology, Texas State University, San Marcos, TX, USA Abstract